ICME-13 Monographs

Series editor

Gabriele Kaiser, Faculty of Education, Didactics of Mathematics, Universität Hamburg, Hamburg, Germany

Each volume in the series presents state-of-the art research on a particular topic in mathematics education and reflects the international debate as broadly as possible, while also incorporating insights into lesser-known areas of the discussion. Each volume is based on the discussions and presentations during the ICME-13 Congress and includes the best papers from one of the ICME-13 Topical Study Groups or Discussion Groups.

More information about this series at http://www.springer.com/series/15585

Fulvia Furinghetti · Alexander Karp
Editors

Researching the History of Mathematics Education

An International Overview

Springer

Editors
Fulvia Furinghetti
DIMA—Dipartimento di Matematica
Genoa University
Genoa
Italy

Alexander Karp
Teachers College
Columbia University
New York, NY
USA

ISSN 2520-8322 ISSN 2520-8330 (electronic)
ICME-13 Monographs
ISBN 978-3-319-68293-8 ISBN 978-3-319-68294-5 (eBook)
https://doi.org/10.1007/978-3-319-68294-5

Library of Congress Control Number: 2017952935

Printed on acid-free paper

This Springer imprint is published by Springer Nature
The registered company is Springer International Publishing AG
The registered company address is: Gewerbestrasse 11, 6330 Cham, Switzerland

Contents

Editors and Contributors

About the Editors

Fulvia Furinghetti is professor emerita of Mathematics Education in the Department of Mathematics (University of Genoa). Her research concerns: beliefs, images of mathematics in society, proof, problem solving, use of history of mathematics in teaching, teacher professional development, history of mathematics education. She has organized the celebrations of the centenary of the journal *L'Enseignement Mathématique* and of ICMI and was one of the editors of the proceedings. She has developed the Web site on the history of the first hundred years of ICMI. She wrote the chapter "From mathematics and education to mathematics education" in the *Third International Handbook of Mathematics Education* and the chapter on the history of the international cooperation in mathematics education in the *Handbook on the history of mathematics education*. During 2000–2004, she chaired HPM (the International Study Group on the relations between History and Pedagogy of Mathematics affiliated to ICMI).

Alexander Karp is a professor of mathematics education at Teachers College, Columbia University. He received his Ph.D. in mathematics education from Herzen Pedagogical University in St. Petersburg, Russia, and also holds a degree from the same university in history and education. Currently, his scholarly interests span several areas, including the history of mathematics education, gifted education, mathematics teacher education, and mathematical problem solving. He served as the managing editor of the *International Journal for the History of Mathematics Education* and is the author of over one hundred publications, including over twenty books.

Contributors

Kristín Bjarnadóttir School of Education, University of Iceland, Reykjavík, Iceland

Elisabete Zardo Búrigo Federal University of Rio Grande do Sul, Porto Alegre, Brazil

Marion Cousin Lyons Institute of East Asian Studies (ENS Lyon), Lyon, France

Dirk De Bock Faculty of Economics and Business, KU Leuven, Leuven, Belgium

Athanasios Gagatsis University of Cyprus, Nicosia, Cyprus

Elena Kiliari University of Cyprus, Nicosia, Cyprus

Jenneke Krüger Freudenthal Institute, Utrecht University, Utrecht, The Netherlands

Vicente Meavilla-Seguí Departamento de Matemáticas, Universidad de Zaragoza, Saragossa, Spain

Antonio M. Oller-Marcén Centro Universitario de La Defensa de Zaragoza, Saragossa, Spain

Irene Papadaki University of Cyprus, Nicosia, Cyprus

Johan Prytz Department of Education, Uppsala University, Uppsala, Sweden

Sethykar SamAn Royal University of Phnom Penh, Phnom Penh, Cambodia

Gert Schubring Universität Bielefeld, Bielefeld, Germany; Universidade Federal do Rio de Janeiro, Rio de Janeiro, Brazil

Harm Jan Smid Delft University of Technology, Delft, The Netherlands

Annamaria Somaglia DIMA—Dipartimento di Matematica, Genoa University, Genoa, Italy

Geert Vanpaemel Faculty of Economics and Business, KU Leuven, Leuven, Belgium

Alexei Volkov National Tsing Hua University, Hsinchu, Taiwan

The Hamburg Score (In Lieu of an Introduction)

Alexander Karp and Fulvia Furinghetti

The title of this article is borrowed from a famous book written in 1928 by the Russian literary critic, art theorist, and prose writer Viktor Shklovsky. His book opens with a story about how wrestlers from different countries, in order to avoid getting confused by the outcomes of official competitions, which were constantly falsified in order to please the public and for financial reasons, would once a year gather in secret in Hamburg and there engage in "long, ugly, and difficult" wrestling matches behind closed doors, which however would allow them honestly to determine who among them was actually stronger or weaker and what the real ranking among them was, so that subsequently even outcomes whose details might be falsified should nonetheless correspond on the whole to the score obtained in Hamburg. Shklovsky further states that such a "Hamburg score" is also indispensable in the literature, which he invites his readers to discuss honestly and as it really is, discarding various possible extraneous considerations.

Although the subject of the present book is not the literature, but the history of mathematics education, and although the story cited by Shklovsky is apocryphal, his proverbial expression naturally comes to mind with the publication of a collection of materials drawn from the work of the Topic Study Group (TSG) on the History of Mathematics Teaching and Learning organized at the 13th International Congress of Mathematics Education (ICME 13) in Hamburg in 2016. Indeed, the work of the TSG in Hamburg presents a natural occasion to attempt to sum up the results obtained in our field thus far. The previously published Topical Survey (Karp and Furinghetti 2016), which appeared in preparation for the congress, was to a large extent also aimed at the same goal. Therefore, we can confine ourselves here to some concise conclusions.

The history of mathematics education as a scholarly field appeared effectively already in the nineteenth century, and an important stage in its development was constituted by the first American doctoral dissertations in mathematics education defended at the beginning of the twentieth century (Jackson 1906; Stamper 1906), which were devoted specifically to its history. Subsequently, however, attention to this field waned—quantitative, statistical approaches became dominant in some countries—and researchers came to believe that such "objective" studies were far more important and substantive, while history of mathematics education began to be perceived (at least, by some) as something that was left behind forever and unlikely useful to anyone who was looking ahead. This does not mean, of course, that these decades did not see any significant studies: Such studies were carried out, and they were carried out in different countries, but the attention that they received internationally was clearly insufficient.

Since the 1980s, the situation has been gradually changing. It has become evident everywhere that far from all questions can be clarified simply by collecting as much quantitative data as possible and by counting up various statistical characteristics. The natural response to this has been to ask how mathematics education develops, why arguments that would seem to have run their course flare up again and again, how textbooks and methods of instruction have changed and whether they have changed at all, whether old ideas might not be lurking beneath fashionable new names, and other questions that cannot be answered without history. Gradually, more and more books devoted to the history of mathematics education began to appear (such as Stanic and Kilpatrick 2003, for example). An important milestone was reached in 2004 with the appearance of a special TSG (Topic Study Group) devoted to the history of mathematics education in the program of ICME 10. This was followed by the appearance of a specialized *International Journal on the History of Mathematics Education* (which, unfortunately, after ten years ceased to exist), the organization of specialized conferences (Bjarnadóttir et al. 2009, 2012, 2015, 2017), the publication of a *Handbook on the History of Mathematics Education* (Karp and Schubring 2014).

It is important to note two processes that were taking place throughout these developments. The first consisted in the recognition of the history of mathematics education as a part of social history. In this regard, an important role was played by the studies of Schubring (1987, 1988), which methodically stressed the fact that merely describing various administrative changes or changes in the teaching of various topics was not enough. The described changes had to be seen in the context of the overall historical picture. Naturally, it would be naive to suppose that every minor methodological detail is necessarily determined by some general social trends, but it is nonetheless obvious that even the historians' favorite object of study—textbooks—does not exist and has never existed in an isolated world confined to schools: The numbers of books printed and their character were determined on the one hand by society's needs, and on the other hand by technological development—printing technology, for example. Textbooks were structured in new ways and supplemented with problems organized along new lines not only because their authors wished to come up with something unusual, but also and above all because of the appearance of new pedagogical demands—as when it became necessary, for example, to educate a far greater number of students than previously.

To repeat, the connection between education history and general history is not simple. Naturally, for example, certain details were sometimes borrowed from other authors without any special thought being given to why they originally appeared. Nonetheless, it must not be forgotten that individuals involved in mathematics education (including the authors of textbooks) live in society and thus come under its influence. Schubring (1987) suggested that the life of a textbook author could be regarded as a kind of self-contained unit, in which we might observe the combination of various influences experienced by mathematics education. Consequently, in recent decades, we have seen attempts at a deeper reading of what has transpired in school mathematics. This deeper view leads researchers to study not only the traditional objects of the history of mathematics education, such as textbooks, all conceivable curricula, or teachers' biographies, but also a far broader sphere of sources, including, for example, letters, diaries, or newspaper articles. A broader view of the history of mathematics education accords with the contemporary broader view of mathematics education itself, whose studied topics today include much that was never examined in the past (e.g., teachers' and students' beliefs).

Another process that has also been evident, and not only in the history of mathematics education, is internationalization. Although it cannot be said that a genuinely international history of mathematics education has already been written, it is clear that there is a growing interest in the study of what has taken place not only within the borders of a single country, but also in various large regions or even in the world as a whole. Centuries ago, textbooks, their authors, and methodological ideas were already crossing national borders. Naturally, a translated textbook was not by any means always identical to the original, but nonetheless the very existence of translations or students sent to study abroad, or teachers invited from abroad, forces researchers not to confine themselves to national frameworks. At times, what took place in one country cannot be understood without finding out about teaching in another.

Again, reductive simplifications are inadmissible. The same textbook could be used to teach the son of a Russian landowner, for whom, for example, a retired artilleryman might be hired as a private teacher; a cadet at an American military academy; or a French lyceum pupil. It would be naive to suppose that the mathematics education ended up being the same in all of these cases, but nonetheless the arising similarities between them must not be ignored, and the differences they exhibit even less so. In recent times, researchers have begun trying to write and think about international aspects of the historical development.

However, in discussing the development of the field and noting the obvious achievements of recent years, we cannot help pointing out persistent problems. Some of them are natural: Studying the history of mathematics education as a part of international social history places very high demands on the researcher. The researcher must, in addition to possessing the requisite methodological-mathematical preparation, also have command of historical methodology and ideally know several foreign languages. It is not surprising that not so many people enter this scholarly field. Moreover, only a very small proportion of those specializing in the teaching of mathematics study the history of mathematics education, even in its most abbreviated form: Usually, there simply is no room for it in the mathematics teacher education curricula, which are already overloaded as it is. Instead of history of mathematics education, certain introductory parts of lectures or workshops will occasionally contain some simplistic assertions to the effect that previously the norm was meaningless rote memorization, whereas now everyone has begun to aim for conceptual understanding and the development of reasoning skills. And even in this facile account reference is usually made only to the very recent past, and only in order to contrast it with the present day (and as for courses in the history of mathematics education specifically for history students, that is something that one can only dream about).

Oversimplification is in general likely the main danger for the historian of mathematics education. Oversimplification can happen in different ways: Sometimes the historian (we will allow ourselves to call such a person a pseudo-historian) wishes to construct a preconceived outline of the development of mathematics education in accordance with preconceived views. Focusing exclusively on the defects of education in the past (referred to as "traditional education"), in the manner described above, is only one possible example of oversimplification—in such cases, researchers do not even consider the traditions that actually exist in the countries which they are studying (Klein 2007 provided some examples). Sometimes, the history of mathematics education is structured and presented in a way that makes it possible to boast of national or other achievements—and in such cases, it is considered unnecessary to mention failures or problems; more generally, discussing the complexity of historical processes—and processes that involve millions of people cannot be simple—is something for which the authors of such studies have absolutely no need.

On the other hand, oversimplification can also manifest itself in renouncing any attempt to generalize what is observed or to discuss its causes. In this case, conversely, there may be a desire to avoid any subjectivity or ideologization: The researcher might claim to do no more than to establish facts, for example, in the past, textbooks allocated such and such a percentage of their overall volume to a certain topic, whereas now they allocate a different percentage. In such cases, researchers often express particular joy when they are able to count something up, which supposedly increases objectivity. Without even mentioning the fact that, for example, the obtained figures will have any meaning only if the analyzed selection was representative (which is not so easy to prove in historical studies), the obtained data can indeed be fully considered facts only when they are related to the remaining data (observations pulled out of context can have a very different meaning from the meaning that can be attributed to them by viewing them from a contemporary perspective). Above all, it is natural to expect that research will ultimately provide explanations and analysis of historical processes—their causes, their consequences, their influence, and so on (to be sure, any analysis must be based on the facts in their totality), but by abandoning analysis for the sake of simplicity and fight for objectivity, we will also abandon the point of our research.

The problems that we are discussing are typical of historical disciplines, in which arguments about methodological and theoretical issues have been going on literally for centuries. There is no reason to think that these arguments have no bearing on the history of mathematics education. They certainly do have a bearing, and indeed the history of mathematics education might even present certain extreme cases, such as those that are considered exceptions in "broader" social history.

Today, we can observe how in one or another history of antiquity, for example, with its centuries-long traditions, pseudo-theories of one kind or another suddenly arise, which boldly reject all former studies and results (usually due to ignorance), and which, despite the bewilderment of professional historians, find an audience prepared to acclaim the innovators, without entering too deeply into the gist of what they are advocating. Such exhortations to write a new history can be heard today also in the history of mathematics education.

And yet, not all of the problems connected with studying the history of mathematics education are common to all historical disciplines. What is specific to our field is its position between two disciplines, as we have already pointed out above—or indeed, even between three disciplines: mathematics, education, and history. And our field is certainly distinguished also by the fact that at present it contains an enormous number of topics that have not been investigated, including topics about which a considerable quantity of documents and various other evidentiary materials have survived and which await their scholars. The development of general studies, whose importance was discussed above, is hindered in no little measure by the lack of particular ones. The field needs new researchers, who will find much of interest in it.

The present collection, as we have said, documents the work of the TSG, which in 2016 also included so-called oral presentations. Thus, several dozen presentations were made all together. Not all authors chose to prepare articles on the basis of their presentations, and not all articles that were begun were completed. The collection contains 14 articles, whose subject matter in our view quite accurately reflects the basic directions of researchers' interest at this time.

The article by Gert Schubring, who was invited to give a special, expanded presentation at the conference, is devoted on the one hand to the methodology of research in the field, and on the other hand to the issue of the relationships between the international, the national, and the regional in the teaching of mathematics.

Probably the most popular research topics remain those connected with the study of reforms, above all the reforms of the 1950s–1970s, known in different countries by different names (New Math, Mathématique moderne, and others). This is not surprising: These reforms, which gave rise to heated arguments (whose echoes can be heard to this day), allow researchers to look at mathematics before the reforms; as well as at the new ideas introduced by them, which in many cases became an enduring part of school education, criticism notwithstanding; at the positions of various participants in the educational process; at the actual preparation of new textbooks; and at concrete methodological and mathematical details. The mutual influence of different countries is also instructive to consider. In different ways, and focusing on different countries, the following researchers devote their articles to topics in the history of the reforms: Kristin Bjarnadóttir, Elisabete Búrigo, Dirk De Bock, Geert Vanpaemel, and Johan Prytz.

Marion Cousin also writes about the reform of mathematics education, but a reform of a different kind, which occurred when Japan became connected to the values and achievements of Europe and the USA.

If not the reform of, then the formation of mathematics education is the subject of the articles by Jenneke Krüger and Harm Jan Smid. The first analyzes the development of the course in mathematics during the seventeenth century and the biography of one of its creators. The second examines the life of an important Dutch figure in mathematics education from the beginning of the nineteenth century.

The article by Elena Kiliari, Athanasios Gagatsis, and Irene Papadaki is also devoted to the initial stage of the development of mathematics education, but in a different part of Europe—it deals mainly with Greece and Cyprus, and with a surviving mathematical manuscript from the seventeenth century.

Studying the biographies and contributions of various figures in mathematics education, as has already been said, is one of the traditionally important directions of research—the article by Antonio Oller-Marcén and Vicente Meavilla-Seguí examines the lives and works of two Spanish authors from the end of the nineteenth century.

The article by Sethykar SamAn is, on the contrary, unusual in its subject matter: It addresses education in Cambodia during the colonial period; sources on the colonial period are inadequate in virtually all cases, but the case of Cambodia—which underwent a systematic destruction of its culture, including the destruction of written sources, during the reign of the Khmer Rouge—is unique. The attempt to reconstruct and describe what took place there deserves special attention.

Alexei Volkov's article is devoted to Chinese textbooks, but they are examined from an unusual angle: The focus of the study is visual representation in counting devices, so to some extent the article constitutes a study of the history of a kind of computational technology, while at the same time being a history of methodological techniques.

Finally, articles by Fulvia Furinghetti and Alexander Karp address topics that have been becoming increasingly popular in recent years. The first of them is devoted to one Italian mathematics education journal and its editors—the dissemination of ideas in mathematics education, including dissemination through the print media, attracts researchers from different countries. The second article constitutes an attempt to study Russian mathematics teachers during the years 1830–1880 as a single professional group: The creation of what might be described as group portraits, and more broadly, the study of various groups active in mathematics education, has also occupied and clearly will continue to occupy an important place in research.

As the reader can see, the topics of the articles included in the collection are quite varied. Also varied are their authors' methodologies and expository styles. One would like to hope that this collection, marking a certain milestone in the development of the field, will facilitate its further development.

In conclusion, we would like to express our gratitude to everyone who at various stages helped to work on the articles being published here. Their various drafts were repeatedly reviewed–this process began already during the preparation of presentations at the Congress, when practically all participants in the TSG also acted as reviewers. We express our deep gratitude to all of them. We would like to mention especially Heinrike Allmendinger, Elena Ausejo, Evelyne Barbin, Kristin Bjarnadóttir, Dirk De Bock, Nikos Kastanis, Jenneke Krüger, Johan Prytz, Gert Schubring, Man Keung Siu, Harm Jan Smid, and Alexei Volkov, who also reviewed the articles during later stages in their preparation. Finally, we must thank

Gabriele Kaiser, who initiated work on the series of which this volume is a part, for useful comments, suggestions, and recommendations.

References

Bjarnadóttir, Kristín; Furinghetti, Fulvia; & Schubring, Gert (Eds.) (2009). *Dig where you stand. Proceedings of the conference "On-going research in the history of mathematics education"*. Reykjavik: University of Iceland, School of Education.

Bjarnadóttir, Kristín; Furinghetti, Fulvia; Matos, José Manuel; & Schubring, Gert (Eds.) (2012). *Dig where you stand. 2. Proceedings of the second international conference on the history of mathematics education*. Lisbon, Caparica: UIED.

Bjarnadóttir, Kristín; Furinghetti, Fulvia; Prytz, Johan; & Schubring, Gert (Eds.) (2015). *"Dig where you stand" 3. Proceedings of the third international conference on the history of mathematics education*, Uppsala: Department of Education, Uppsala University.

Bjarnadóttir, Kristín; Furinghetti, Fulvia; Menghini, Marta; Prytz, Johan; & Schubring, Gert (Eds.) (2017). *"Dig where you stand" 4. Proceedings of the fourth International Conference on the History of Mathematics Education*. Rome: Edizioni Nuova Cultura.

Jackson, Lambert Lincoln (1906). *Educational significance of sixteenth century arithmetic from the point of view of the present time*. Doctoral dissertation. Columbia University.

Karp, Alexander, & Furinghetti, Fulvia (2016). *History of mathematics teaching and learning: Achievements, problems, prospects*. ICME-13 Topical Surveys edited by G. Kaiser. Switzerland: Springer Open.

Karp, Alexander, & Schubring, Gert (Eds.) (2014). *Handbook on the history of mathematics education*. New York: Springer.

Klein, David (2007). A quarter century of US 'math wars' and political partisanship. *British Society for the History of Mathematics Bulletin, 22*, 22–33.

Schubring, Gert (1987). On the methodology of analysing historical textbooks: Lacroix as textbook author. *For the Learning of Mathematics, 7*(3), 41–51.

Schubring, Gert (1988). *Theoretical categories for investigations in the social history of mathematics education and some characteristic patterns*. Bielefeld: Institut für Didaktik der Mathematik: Occasional papers # 109.

Stamper, Alva Walker (1906). *A history of the teaching of elementary geometry with reference to present day problems*. Doctoral dissertation. Columbia University.

Stanic, George M. A., & Kilpatrick, Jeremy (2003). A *history of school mathematics*. Reston, VA: National Council of Teachers of Mathematics.

Influences from the 1959 Royaumont Seminar. Proposals on Arithmetic and Algebra Teaching at Lower-Secondary Level in Iceland

Kristín Bjarnadóttir

Abstract This study investigates influences from the Royaumont Seminar on arithmetic and algebra textbooks at lower-secondary school level in Iceland as one of the Nordic countries. Icelandic and other Nordic New-Math textbooks are compared to both earlier and later textbooks, contrasting the introduction of set-theoretical concepts and notation together with related topics, structure of the number field, and the study of numbers. The conclusion is that set-theoretical concepts and notation dwindled soon after the New-Math was replaced, while the structure of the number field and in particular the study of numbers, obtained a permanent place in the syllabus. New topics, such as coordinate geometry, probability and statistics, received increased attention and others, such as divisibililty, were revived.

Keywords Royaumont seminar · New Math · Arithmetic · Algebra
Nordic countries · Textbooks

1 Introduction

In the post-World War II era, questions arose in many countries about school mathematics. An international reform movement in mathematics education arose with many sources of origin. The New-Math movement in the United States of America had been active for some time. The CIEAEM, the International Commission for the Study and Improvement of Mathematics Teaching was established in the early 1950s. Its main concern was a growing attention to the student and the process of teaching. There were also reform projects in Britain with initial emphasis on applied mathematics. Representatives of the different reform movements met at Royaumont, France, in November 1959 at a seminar for mathematicians, mathematics educators and mathematics teachers on reform of

K. Bjarnadóttir (✉)
School of Education, University of Iceland, Reykjavík, Iceland
e-mail: krisbj@hi.is

F. Furinghetti and A. Karp (eds.), *Researching the History of Mathematics Education*, ICME-13 Monographs,
https://doi.org/10.1007/978-3-319-68294-5_1

school mathematics. The meeting was organized by the Organisation for European Economic Co-operation, OEEC (OEEC 1961; De Bock and Vanpaemel 2015). OEEC was later superseded by the Organisation for Economic Co-operation and Development, OECD.

Waves from the Royaumont seminar spread around the world. Specifically, for the Nordic countries it was a seminal event. The Nordic participants agreed to organize Nordic cooperation on the reform of mathematics teaching. The Nordic countries soon joined in collaboration on curriculum and textbook development under the auspices of the Nordic Council, the official inter-parliamentary body in the Nordic Region (NKMM 1967; Gjone 1983).

The purpose of this paper is to analyse changes in the content of arithmetic and algebra textbooks at lower secondary level in Iceland, as a representative of the Nordic countries, in the follow-up to the Royaumont seminar. The analysis concerns broadly three topics: introduction of set theoretical concepts and notation and related topics, structure of the number field, and the study of numbers. The primary question is which of the various topics survived the backlash that followed first wave of enthusiasm, created by the reform movement, to become a permanent part of the curriculum.

2 The Royaumont Seminar

At the beginning of the report of the Royaumont seminar there is a rationale, called "The case for the reform":

> There is broad underlying fact that modern society is making increasing demands on all citizens for simple mathematical skills and appreciation of numerical significance … The demands for scientists and engineers – all of whom must have sound knowledge and understanding of mathematics—is growing. New applications of mathematics in industry and other branches of economic activity are leading to a demand for more mathematicians with new kinds of skills. All these demands are creating a need for a re-appraisal of the content and methods of school mathematics (OEEC 1961, p. 11).

The aim of the reform was thus to meet demands for more mathematicians with skills for new applications of mathematics, and to redefine school mathematics for this purpose. Among the goals of the seminar (OEEC 1961, p. 12) were to indicate procedures and means to obtain a more adequate supply of mathematically competent persons in science, industry and government.

Three educational goals were stated in the report of the seminar:

(a) Mathematics as liberal education. (Freedom of the mind.)
(b) Mathematics as a basis for living and work. (As the people's necessary tool.)
(c) Mathematics as propaedeutics. (As foundation for university study.) (OEEC 1961, p. 62).

Marshall Stone, president of ICMI and the chairman of the seminar, was concerned about mathematics teaching at primary school level. He said that illuminating psychological investigations, particularly those of Piaget, were fortunately pointing the way to hitherto unrecognized pedagogical possibilities (OEEC 1961, pp. 22–23). However, Stone's plea for research and reform of primary teaching was not heeded, and his programme of "mathematics for all" was not endorsed in the follow-up discussion nor mentioned in the final report of the meeting (Schubring2013).

The CIEAEM was well presented at the seminar. It emphasized the relevance of psychology in mathematics education, the key role of concrete materials and active pedagogy, and psychologist and biologist Jean Piaget's research of the relation between mental and mathematical structures (Furinghetti et al. 2008). Piaget had recently revealed his conclusions on the similarities of the structure of the mind and mathematical structures as introduced by the French Bourbaki group of mathematicians (Piaget 1955) which had awakened great interest among mathematicians. The Bourbaki group was a group of French mathematicians, who worked on a mathematical encyclopaedia, where the borders between the different mathematical topics were abolished. The Bourbaki group's central concept was "structure". When describing structures, the importance lay in the elements' relationships, as determined by axioms. The axiomatic method was essential for the mathematics and its language was that of the set theory.

Jean Dieudonné, a guest speaker at the seminar, spoke on new thinking in mathematics (OEEC 1961, pp. 31–45). He was a member of the CIEAEM, and of the Bourbaki group. Dieudonné criticised the teaching of Euclidean geometry and suggested that the whole system of Euclidean geometry could easily be replaced by an axiomatic system producing two-dimensional linear algebra.

Mathematics professor Gustave Choquet, chairman of the CIEAEM and guest speaker at the seminar, spoke on new thinking in mathematics education (OEEC 1961, pp. 62–68). Tendencies in modern mathematics to do away with boundaries between arithmetic, algebra, geometry and calculus could be realized through the study of structures, he said. The sets N and Z were endowed with numerous structures such as order, group or ring. The set Z constituted an excellent basis for study. Its "discrete" character made it tangible so that it might be used for introducing and studying such concepts as one-to-one correspondence, function, conversion and equivalence.

In secondary schools, Choquet said, one could explore the algebra of sets and its symbolism; functions; graphs; relation between algebra of sets and logic. Furthermore, the set of integers, developed by means of the Peano axioms; the ring of integers, Z; and the field of rationals, Q. From arithmetic could be taken divisibility; Euclidean algorism; prime numbers; irrationals; Diophantine equations; and geometric representation on the plane of $Z \times Z$. Examples of a group could be explored; ring of integers modulo n; and homomorphisms of Z over Z/nZ. Choquet mentioned series of squares with its successive differences as well as of cubes, probability, and some concepts of combinatorial analysis. No square roots were to be taught, but instead calculations in the binary, octal and duodecimal numeral systems. Teaching of algebra had mostly been concerned with how to do, neglecting the deductive aspects. The events and discoveries of the preceding century—matrices, sets, groups, rings, fields, vectors and vector spaces—had yet to penetrate secondary-school mathematics programmes. To what extent and the manner in which this may be done was outlined for the seminar by Willy Servais, Belgian secondary school teacher and the secretary of CIEAEM (Vanpaemel et al.2012).

Servais proposed reforms of algebra teaching (OEEC 1961, pp. 22–23, 62–63, 68–73). The notions of sets, functions, Cartesian product, relations and operations were prerequisite to all others, as sets form the basis of the construction of mathematics. The properties of the algebra of sets should be discovered rather than expounded. Sets also proved a good foundation for elementary notions of logic. The use of logic and logical symbols was essential in secondary school. There was, however, a danger that pupils would assume that symbols replace thinking while they should facilitate thinking.

According to recommendations of the seminar (OEEC 1961, pp. 105–125), new topics such as abstract algebra, vector spaces, theory of sets etc. were to enter the secondary-school programme. New applications of mathematics suggested new problem material, such as probability, statistical inference, finite mathematical structures, linear programming and numerical analysis. New mathematical ideas and new symbolism would lead to clarity, compactness and unity of mathematical exposition. Compartmentalisation was contrary to the recommendations and one of the aspects of modern mathematics was its unity. As the child processes through secondary school, linear algebra, vectors and coordinate geometry should grow into a unified body of knowledge.

3 Literature Review

This article builds on parts of the Ph.D. dissertation *Mathematical Education in Iceland in historical context—Socio-Economic demands and influences* (Bjarnadóttir 2007), an overview of the socio-economic effects of implementing New Math in Iceland in the latter half of the 1960 onwards. There, individual textbooks were discussed with respect to their importance and distribution but not analysed. This present article is a part of a larger research project of analysing textbooks from before, during and after the New Math era, and is as such complementary to the dissertation. In this project, mathematical topics through the school levels below the upper secondary level are analysed. The results are in preparation. Primary level textbooks with focus on a translated series by Agnete Bundgaard and Eeva Kyttä (1967–1968) and Bundgaard alone (1969–1972) were analysed with respect to arithmetic and introduction to algebra, according to Choquet's proposal on primary arithmetic (OEEC 1961, pp. 64–66). The results are presented in a sibling paper (Bjarnadóttir 2017). The primary level textbooks met an unaware society and turned out to bring new algorithms in addition to new concepts and notation. The approach and content were different on the two different school levels, primary and secondary, as were the receptions of teachers and the public. The two papers, however, do complement each other. The geometry content has partly been elaborated on in a third paper (Bjarnadóttir 2016).

An extensive research literature exists on the impact of the Royaumont seminar. Several notable works were published in the late 1970s and early 1980s when the international turbulence had settled. Bollerslev (1979) edited a collection of essays on the New Math in Denmark including Jens Høyrup's (1979) paper. Høyrup deemed the Bundgaard primary level textbooks as a most orthodox adjustment to the demands of the mathematics professors at Royaumont and in Denmark. Ole Skovsmose (1979), who also wrote about the Bundgaard textbooks, claimed that the sixties' mathematics was a clear and all-over brake with mathematics teaching that had stagnated around a definite set of methods and problem posing. The sixties mathematics had been too rigid but actually it could also realize pedagogical activities.

Barry Cooper (1985) studied the initial processes of the redefinition of English secondary school mathematics that occurred during the late 1950s and early 1960s. In his study, Cooper discussed the related issues of power and resources, and other socio-political issues. Exchanges between mathematicians, mathematics teachers and representatives from industry at several conferences created a basis for support from industry for reform of school mathematics. Initially, the emphasis was on applied mathematics. Other agents campaigned for the introduction of post-1800 algebraic ideas into the syllabi, while they also campaigned for pedagogical changes, legitimising both elements of that mission in terms of improving the child's "understanding" of mathematics. At the end, those who supplied the resources were unaware of the nature of redefinition which finally was implemented, that is the set-theoretical approach instead of applied mathematics.

Gunnar Gjone (1983) analysed the implementation of New Math in Norway. Initially, it was closely related to Nordic cooperation in NKMM. After some experiments with NKMM's products and other material, it continued into a process of reconsideration and subsequent debate. The New Math was first formally introduced nationwide when the most abstract concepts had retreated into the background. Individual textbooks were not analysed.

An interesting comment on the New Math in general is found in Richard Skemp's (1978) paper. He expressed concerns that superficial instrumental understanding was often promoted at the cost of deeper relational understanding. He suspected that much of what was taught under the description "modern mathematics" alias New Math was taught and learnt just as instrumentally as under syllabi which had been replaced. In England as elsewhere, the New Math was introduced under the cover that it promoted deeper understanding, the relational understanding in Skemp's terms.

Of more recent studies, papers by Schubring (2013), De Bock and Vanpaemel (2015), Prytz and Karlberg (2016) and Kilpatrick (2012) warrant a special mention. Schubring found that there were significant differences between the intended setup of the Royaumont Seminar and the actual published report. There were conflicts between two directions for the conception of the seminar—between the pedagogical-empirical and the subject matter one. This is corroborated by De Bock and Vanpaemel (2015). They think though that mathematicians were aware of the biased message promulgated by the Royaumont Seminar. In most reports and reviews it was recognized that the Royaumont view was actually derived from the Bourbaki approach of mathematics. Prytz and Karlberg (2016) investigated earlier claims that the New Math reform in Sweden had negative effects on students' learning, but their conclusion was that the claim is in most parts incorrect. Kilpatick (2012) concluded that although the New Math is often deemed a failed endeavor, it changed not only school mathematics but also the way people and countries viewed school mathematics.

4 The Nordic Committee for Modernizing Mathematics Teaching

Nordic participants at Royaumont agreed to organize Nordic cooperation on the reform of mathematics teaching under the auspices of the Nordic Council, the official interparliamentary body in the Nordic Region, formed in 1952. The Council's elected members were and are from Denmark. Finland, Iceland, Norway and Sweden as well as from the Faroe Islands, Greenland and Åland. Shared linguistic heritage is one of the key threads that bind the nations together, even if the Icelandic language diverged early away from the other North Germanic languages, and Finnish is of different origin. The cooperation has been particularly prominent in culture and education.

The Nordic Council set up the Nordic Committee for Modernizing Mathematics Teaching, NKMM, chaired by Lennart Sandgren, one of the Swedish representatives at the Royaumont seminar. The committee, which was active from 1960 until 1967, appointed teams of textbook writers. According to the final report, the committee had the Royaumont report as a starting point. As a confirmation, it quoted verbatim the seminar's aforementioned educational aims together with its questions about what mathematics to teach, to whom and how to teach it (NKMM 1967, pp. 12–13). Joint Nordic manuscripts were produced in succession until early 1966 and were translated into the various languages. The experimental texts were tested in school classes during 1961–1966. The material for grades 7–9 was tested out in 450 classes (Gjone 1983, II. pp. 78–80; NKMM 1967; Prytz and Karlberg 2016).

In the academic year 1962–63, a series of algebra texts, produced by the NKMM, was ready for grades 7–9 in the four countries. In Sweden, a somewhat shorter version in a simpler language than the Danish one was subsequently produced (NKMM 1967, p. 111). The Swedish version, originally a total of 556 pages (p. 119) seems to have been shortened further when translated into Icelandic, a total of 274 pages (NKMM 1970; Björnsdóttir 1972). The NKMM-material is located in an archive in Stockholm, while no detailed analysis of it is presently known (Prytz, personal communication).

Denmark was one of the countries which went furthest in introducing the New Math into university and secondary school programs (Karp2015). According to the programme of the Royaumont seminar, but not mentioned in its report, the Danish mathematician Svend Bundgaard spoke on its second last day on the mathematical competencies of teachers and the consequent problem of training and retraining (Schubring, personal communication). His sister, Agnete Bundgaard, was a member of the writing team of NKMM, producing the aforementioned Bundgaard series for primary school level. Another Danish participant was Ole Rindung whose New Math upper secondary level textbooks were in use Iceland during the 1960s and 1970s. Finland did not participate in the Royaumont seminar, while the Finnish Eeva Kyttä was A. Bundgaard's co-writer of the primary level material for grades 1–2. The Norwegian participants did not have noticeable influence in the writings of NKMM although the Norwegian Kay Piene, who died untimely, was influential through his writing in mathematical journals.

The proceedings of an international working session on new teaching methods for school mathematics, held in Athens in November 1963 (OECD 1964), reveal that an Icelandic participant, Björn Bjarnason, later one of the main proponents of mathematics reform of the college-bound level, was a participant in the session. All the Nordic participants were those most involved in the NKMM activities, and Sweden and Denmark presented their "modern" curricula. Thus there were certainly continued interactions between the Nordic Committee and Iceland with the rest of international reform community.

5 Iceland's Educational Policy

Iceland was a part of the Danish Realm until 1944. In the post-WWII era, Icelandic society of 130 thousand inhabitants, after having broken with Denmark, was geographically and culturally isolated, thousands of miles out in the Atlantic Ocean, speaking and writing own language. Moreover, it lacked foreign currency to keep up international interaction. In the early 1960s, the country was still culturally dependent on Denmark, using untranslated Danish textbooks for upper secondary education. Textbooks for compulsory school levels were published by a state publishing house, established during the Great Depression, in order to provide students with free learning material. The enterprise had been kept in frugal conditions under sustained inflation. Up to 50-year-old textbooks were reprinted. Efforts in revising mathematics textbooks consisted merely of updating prices and social conditions described in exercises. Most university or higher vocational training was sought in Denmark, and it was also the source of most cultural currents.

Iceland began to open up in the 1960s. At that time, educators, mathematicians, physicists and psychologists had learnt about the New Math reform movement, promoted by the OEEC, later by the OECD. A move was made towards implementing similar reforms in Iceland. Reform experiments, made by individuals, were initially exclusively aimed at college-bound students, as had also been the original plans in the U.S. and England and at the Royaumont Seminar in 1959. The news was channelled through personal contacts to Danish educators. Guðmundur Arnlaugsson, who stayed in Denmark for education and work during 1933–36 and 1939–45, knew personally Svend Bundgaard who e.g. informed him about primary level material produced by the NKMM and written by his sister, Agnete Bundgaard (Gíslason, personal communication).

At the same time, the Icelandic authorities were introduced to OECD theories about education contributing substantially to economic and social progress and stability, and education being as important a sector of society and of the national economy as the traditional sectors (Efnahagsstofnunin 1965). At this time, reforms seemed feasible. The economic benefits had been clarified by the policy of the OECD, and experts were ready to propose solutions in the form of new educational material, along similar lines as had proved promising in other OECD countries that Iceland identified with, both the USA and the Nordic countries.

Iceland neither sent a representative to Royaumont nor participated in the NKMM cooperation but all the Danish representatives in the seminar and the Nordic committee made an impact in Iceland through their writings. Arnlaugsson (1966) wrote his textbook *Tölur og mengi* [*Numbers and sets*] for lower secondary school level, and simultaneously presented the NKMM-material for primary level to Icelandic authorities in 1966. The New Math was introduced for experimental reasons at both school levels from 1966, and the A. Bundgaard primary level material was implemented in the majority of Reykjavík primary schools in 1967 (Bjarnadóttir 2017). In the early 1970s, when students having studied that material entered the lower secondary level at the age of 13, they were met by translated NKMM-material (Bergendal et al. 1970; NKMM 1970; Björnsdóttir 1972), for that level. The first New Math textbooks published in Iceland were thus a part of the joint Nordic initiative and the same textbooks were also published in other Nordic languages, Danish in the case of the primary level, and Swedish in the case of lower secondary level. The foreign New Math material, however, created turbulence and very soon, a sequence of new domestic textbooks were created for both levels.

When the Icelandic parliament agreed to allocate funding to curriculum changes in 1966, it certainly was not aware of the nature of the changes ahead concerning mathematics. Decisions on content were left to specialists who, due to their high prestige, could promote their version of change, which in turn was highly influenced by the theories promoted at the Royaumont Seminar, diffused through Danish and Swedish contacts and textbooks.

Syllabus changes that followed the introduction of the New Math in lower secondary schools were based on a draft of a national curriculum document, written by Bjarnason who had participated in the OECD meeting in Athens, preparing the introduction of the New Math to the country-wide entrance examination into upper secondary level (Landsprófsnefnd 1968). The aim was to base school mathematics on the basic concepts of the set theory, which simultaneously were simple and general, and to increase emphasis on the meaning and nature of numbers and of number computations. National curriculum documents for compulsory schools, including lower secondary level, were only published in drafts until 1989 and in a detailed edition in 1999 (Menntamálaráðuneytið 1989, 1999). They therefore did not reflect the reform waves of the 1960s and 1970s.

6 The Development of the New Math in Retrospect

Niss (1996) defined three fundamental reasons for the society to provide mathematics education: Firstly, contributing to the technological and socio-economic development of society; secondly, contributing to society's political, ideological and cultural maintenance and development, and thirdly, providing individuals with prerequisites which may help them to cope with life. These fundamental reasons harmonize with the Royaumont seminar's educational goals.

The proposals, made by Choquet and in the final recommendations of the seminar, aim at meeting needs of the society for scientifically-educated intellectuals. Proposed changes were legitimized by reference to the nations' need for scientific and technological manpower. This aspect was welcomed in Iceland in the 1960s. The common hope was economic gain in line with OECD's paradigm of (mathematical) education as an economic resource (Efnahagsstofnunin 1965).

Those points of view seem to have won greater interest at the Royaumont seminar than a desire for providing individuals with tools to cope with their life on their own premises. The agenda of the CIEAEM, such as attention to teaching methodology and the key role of concrete materials and active pedagogy, was not very visible in the recommendations of the seminar (Schubring 2013).

Clear expectations of emphasis on applied mathematics were behind the initiation of the New Math. This was e.g. the case in Britain, where support from industry for reform of school mathematics was based on expectations of creating more mathematicians with skills for new applications of mathematics. This is echoed in the rationale for the seminar, quoted in section 2: "New applications of mathematics in industry and other branches of economic activity are leading to a demand for more mathematicians with new kinds of skills." (OEEC 1961, p. 11). Somehow, those with power and influence were led to believe that the new concepts would bring increased clarity in thinking. Practical and pedagogical aspects gave way, while theoretical aspects dominated the reform wave. The Royaumont report, which did not entirely reflect the original intention or the discussion at the seminar (Schubring 2013), did lead the work and products of the NKMM committee, and was arguably the dominating influence in the Nordic countries, in particular in Iceland.

The Royaumont seminar was held in late 1959. Enthusiastic mathematicians and educators used the succeeding years to arrange meetings and prepare writing projects. But as the 1960s drew to an end, the focus on purpose and goals of education, mathematics education in particular, began to shift. Preparation for technological developments of society received less attention and possibly there were disappointments that the version of New Math introduced in most countries, the university version, did not bring the expected improvements. Enthusiasm began to fade. Students were rising up in many countries with different aspects on their agenda. Personal development, helping individuals to cope with their own life, received increased attention.

The teaching and learning process also brought difficulties according to Skemp (1978). Even if the set theoretical concepts and symbolic language could lead to clarity, compactness and unity of the different mathematical topics, such as arithmetic, algebra and geometry, they had to be memorized and students be trained to distinguish them from each other. This can also be spotted in Iceland. In many cases Icelandic school teachers missed the point of the reform while dealing with new concepts, and saw only yet another method in addition to the old ones (Arnlaugsson 1967). Ultimately, university people had most to say about the content and Arnlaugsson was one of them, but they were also among the first to react negatively when they realized the implementation problems, as exemplified by his reaction already in 1967.

7 Research Question and Research Method

In the turbulent environment of new approach to school mathematics and frequent changes of teaching material, many questions arise. The questions that still are being asked by the public are (1) if the New Math was a complete disaster, ruining mathematical learning of a whole generation, and (2) if it ultimately led to any positive changes. Prytz and Karlberg (2016) have answered the first question for Swedish students. This present study is intended to address question (2) by posing one more question:

> What permanent changes in content of arithmetic and algebra textbooks did the implementation of New Math bring at lower secondary school level in Iceland?

The question is of interest to clarify if the process of implementing the New Math was a part of an on-going development, bringing Icelandic mathematics education into the international forum of exchanging ideas for the improvement of teaching and learning mathematics. It is also of international interest to study how a relatively isolated and sparsely populated society with own language met the challenges of adopting and integrating revolutionary new educational ideas.

The research method consisted of choosing topics, mentioned in the Royaumont report as proposals for reforms of lower secondary school level arithmetic and algebra. This is mainly confined to the summary of remarks by Gustave Choquet and a discussion of professor Servais' remarks (OEEC 1961, pp. 67–74). Choquet and Servais did not separate their proposals for lower and upper secondary school levels so the choice was confined to what was presented in the Icelandic textbooks. Given the list of topics, typical mathematics textbooks in use at lower secondary level during the period 1950–2015 were examined with respect to these topics. This process was iterated a couple of times by the author of this article, and subsequently amended and updated as was considered necessary during the preparation of the paper.

The results are presented in three tables. The focus is on three textbooks or textbook series, presented within the framework of the New Math. These were compared to two earlier textbook series and three more recent in order to trace changes during the sixty-five years' period. As the changes were originally channelled to Iceland from the Nordic cooperation, NKMM, the analysis also concerns the NKMM.

8 Analysis of Textbooks

By law (no. 22/1946) the lower secondary level was divided in 1946 into two streams, one for prospective vocational training, and the other for college-bound education. Both had a mathematics syllabus, represented by textbooks, here named ÓD/JSG (Daníelsson 1938; Gissurarson and Guðmundsson 1949–1950), with a review of fractions, proportions and percentages in the form of Rule of Three, and equations, as well as computations of area and volume. In addition, the college-bound stream, for those who wished to attempt an entrance examination into higher education, had a special syllabus of introduction to algebra by a textbook, referred to here as ÓD (Daníelsson 1951). The two streams were united in 1976.

The process of implementing New Math may be divided into three steps. There were also three reaction steps:

1. The textbook *Numbers and Sets* [*Tölur og mengi*], here named GA, by Arnlaugsson (1966), intended for college-bound students, marked the first step of implementation. It was written as a complementary text to the ÓD algebra textbook. The topics of GA were on one hand numbers and number relations, such as the nature of numbers: divisibility and number notation to other bases than ten. On the other hand, there were sets and set algebra, and some logic. In his forewords, the author stated that the basic concepts of logic and set theory would facilitate understanding, even for young children. who easily internalized the new set theoretical concepts and enjoyed them. Furthermore, the concepts seemed to contribute to clarity and exactness in thinking and arithmetic. This seems to echo an interpretation of Piaget‘s theories that similarities between mental structures and mathematical structures in that the essential notions which characterize modern mathematics were much closer to the structures of "natural" thought than were the concepts used in traditional mathematics (Piaget 1973). This textbook was ground-breaking for the New Math in Iceland, written under Danish and American influences. Arnlaugsson even led a 17-episode television series on its content at the national television station (Bjarnadóttir 2015).
2. The second step was translating the Nordic NKMM-material into Icelandic. From 1970, a textbook series for lower secondary level, here named NK (Bergendal et al. 1970; NKMM 1970; Björnsdóttir 1972), translated from Swedish, was in use for a few years.

3. The third step was domestic material, here named HL, by H. Lárusson (1972–1976), somewhat tailored after the NK-series but more comprehensive. The series dominated the market until the late 1980s.
4. A series by Kristjánsdóttir et al. (1988–1991), here named KMB, less oriented towards set theory than earlier series, marked the first reactions. The series was not completed. By the time of its publication, the New Math had been presented to the majority of pupils entering lower secondary level, and therefore less introduction of basic set theoretical concepts was needed, while it kept the emphasis on the study of numbers and divisibility.
5. In the late 1980s, a Swedish series, here named BBB (Björk, Björksten, Brolin, Ernestam, and Ljungström 1987–1989), was translated and used into the 2010s. It neither mentioned set-theoretical concepts nor notation.
6. In the early 2000s, a domestic series, PG, by Pálsdóttir and Gunnarsdóttir (2005–2008) was published.

The comparison of the content of the Icelandic lower secondary level textbooks to Choquet's and Servais' proposals, and the recommendations of the Royaumont seminar about arithmetic and algebra at secondary level, is split into three parts. The three areas of study are:

- set theoretical concepts and notation, their connection to geometric presentation of algebra in the coordinate system, and applications
- structure of the number field
- the study of numbers.

The focus is on the three New-Math texts, GA, NK and HL. They are compared to two older Icelandic textbook series, ÓD/JSG and ÓD, and three more recent ones, the KMB, BBB and PG-series.

9 Results

In Tables 1, 2 and 3, the numbers indicate the age level when the topic in question was introduced for the age level 13–15. Normally, the topics were readdressed regularly after they had been introduced. Some topics may already have been mentioned at primary level but are also listed here if they were devoted theoretical attention.

One can read between the lines that mathematicians and later textbook authors were enchanted by Piaget's idea that the structure of the mind aligned with the structure of modern mathematics, and at Royaumont, Choquet proposed numerous structures to study at schools within the framework of sets and sets algebra. The New Math textbook series, GA, NK and HL all presented them systematically.

Table 1 Set-theoretical concepts, the coordinate system and applications in Icelandic textbooks in use during 1950–2015

Textbooks	ÓD/JSG	ÓD	**GA**	**NK**	**HL**	KMB	BBB	PG
In use during	1950–1976	1927–1976	**1966–1976**	**1970–1975**	**1972–1990**	1979–1999	1987–2015	2005–2015
Age level	13–15	15	**15**	**13–15**	**13–15**	13–15	13–15	13–15
Sets			**15**	**13**	**13**			13
Subsets			**15**	**13**	**13**			
Union			**15**	**13**	**15**			13
Intersection			**15**	**13**	**13**			13
Set difference			**15**					
Modern symbolism			**15**	**15**	**13**			14
Algebraic properties of set operations			**15**					
Cartesian product			**15**	**15**	**14**			
Mappings				**15**	**14**			
Functions					**14**			
1-to-1 correspondence			**15**	**15**	**13**			
Coordinate system			**15**	**15**	**14**	13	14	13
Coordinate geometry				**15**	**14**	13	14	13
Vectors					**14**			
Graphically represented linear equations				**15**	**15**	13	14	13
Logic			**15**					(14)
Logical symbols			**15**	**(14)**	**(14)**			
Statistics, introduction					**15**	13	13	13
Probability, introduction					**15**	14	15	13

Age within parentheses indicates that the topic was only marginally presented and not elaborated further on in what followed

Table 2 Structure of the number field in Icelandic textbooks in use during 1950–2015

Textbooks	ÓD/JSG	ÓD	**GA**	**NK**	**HL**	KMB	BBB	PG
In use during	1950–1976	1927–1976	**1966–1976**	**1970–1975**	**1972–1990**	1979–1999	1987–2015	2005–2015
Age level	13–15	15	**15**	**13–15**	**13–15**	13–15	13–15	13–15
The commutative laws		15		**14**	**13**	13		13
The associative laws		15		**14**	**13**	13		13
The distributive laws		15		**14**	**13**	13		13
Negative numbers	14	15		**14**	**13**	14	14	<13
Additive identity		15		**14**	**13**			13
Additive inverse		15		**(14)**	**13**	14		14
Multiplicative identity		(15)		**14**	**13**	13		13
Multiplicative inverse		(15)		**14**	**14**	14		15
0 in multiplication		15		**14**	**13**			13
0 in division		15		**15**	**15**	(14)		13

Age within parentheses indicates that the topic was only marginally presented and not elaborated further on in what followed

Table 3 Study of numbers in Icelandic textbooks in use during 1950–2015

Textbooks	ÓD/JSG	ÓD	**GA**	**NK**	**HL**	KMB	BBB	PG
In use during	1950–1976	1927–1976	**1966–1976**	**1970–1975**	**1972–1990**	1979–1999	1987–2015	2005–2015
Age level	13–15	15	**15**	**13–15**	**13–15**	13–15	13–15	13–15
Even and odd numbers			**15**		**13**	13		13
Other classes of numbers			**15**		**13**			
Prime numbers	(14)		**15**	**14**	**13**	13		13
Factoring	14	15	**15**	**14**	**13**	13		13
Prime factoring	(14)		**15**	**14**	**14**	13		13
Squares, cubes			**15**			13	13	13
Divisibility			**15**	**14**	**13**	13	(13)	13
Series of squares, etc.			**15**			13		
Bases other than ten			**15**	**14**	**13**			
Modular arithmetic						13		
Scientific notation	(14)		**15**	**14**	**14**	14	14	14
Approximation				**14**	**14**	13	13	13
Square roots	14				**15**	14	15	15

Age within parentheses indicates that the topic was only marginally presented and not elaborated further on in what followed

9.1 Introduction of Set-Theoretical Concepts and Logic, and Their Connection to Topics in Applied Mathematics

To this list of set-theoretical concepts and logic, the coordinate system, coordinate geometry and graphically represented linear equations in two variables, have been added with respect to their relations to the Cartesian product, mapping and functions. Similarly, introductions to statistics and probability are listed here with respect to set-theoretical presentation of probability and its relations to logic.

The textbook GA, written for the college-bound stream for the 15-year age group as complementary to ÓD, introduced all relevant set-theoretical concepts with modern symbolism, and the algebraic properties of sets, i.e. introduction to set algebra, in addition to logical symbols and introduction to logic, all of which had neither appeared in ÓD/JSG nor in ÓD. At Royaumont, Servais emphasized that the properties of the algebra of sets should be discovered rather than expounded and Arnlaugsson followed his advice, presenting them in exercises.

The Nordic NK-series and the HL-series introduced set-theoretical concepts and notation except set difference; Cartesian product, represented geometrically on the plane of $Z \times Z$; mappings and one-to-one correspondences during age 13–15; but did not explicitly address algebraic properties of sets.

The NK-series used some logical symbols but did not venture further into logic. The set-theoretical concepts and notations were not mentioned in later texts after HL, such as KMB or in the Swedish BBB, until some were again introduced in PG as aids to counting and probability.

9.2 Structure of the Number Field

The new GA did not touch algebra of numbers since it complemented ÓD which presented explicitly all axioms of Q except multiplicative identity and inverse.

The axioms of the number field were systematically constructed in NK and HL and successively shown to apply to natural numbers, whole numbers and rational numbers.

The axioms were referred to in KMB and PG as aids to computations but rather for explaining computation and not for structural purposes. They were not at all mentioned in the translated Swedish BBB.

9.3 The Study of Numbers

The introduction of the New Math reinvigorated the study of numbers that had appeared in early 1900s textbooks but was only marginally touched upon in ÓD/JSG. As for fractions, the main emphasis had earlier been on a partition of the unit, and for example factors in denominators were only mentioned in the context of

Bjarnadóttir, Kristín (2016). Was Euclid in Iceland when he was supposed to go? In K. Krainer & N. Vondrová (Eds.), *Proceedings of CERME9 – the Ninth Congress of the European Society for Research in Mathematics Education* (pp. 1790–1796). Prague. Available at https://hal.archives-ouvertes.fr/CERME9-TWG12/hal-01288381v1.

Bjarnadóttir, Kristín (2017). Recommendations of the Royaumont Seminar on primary school arithmetic. Influences in the Nordic countries. In K. Bjarnadóttir, F. Furinghetti, M. Menghini, J. Prytz, & G. Schubring (Eds.), *"Dig where you stand" 4. Proceedings of the fourth International Conference on the History of Mathematics Education.* Rome: Edizioni Nuova Cultura.

Bollerslev, Peter (Ed.) (1979). *Den ny matematik – en essaysamling. [The new mathematics – an essay collection*]. Copenhagen: Gyldendalske Boghandel.

Cooper, Barry (1985). *Renegotiating secondary school mathematics. A study of curriculum change and stability.* London: The Falmer Press.

De Bock, Dirk, & Vanpaemel, Geert (2015). Modern mathematics at the 1959 OEEC seminar at Royaumont. In K. Bjarnadóttir, F. Furinghetti, J. Prytz, & G. Schubring (Eds.) *"Dig where you stand" 3. Proceedings of the third international conference on the history of mathematics education* (pp. 151–168). Uppsala: Uppsala University.

Efnahagsstofnunin (1965). *Nokkur efnisatriði erinda og umræðna frá fundum um menntaáætlanagerð 2. og 3. júní, 1965. [Several items from presentations and discussions in a meeting on educational planning June 2nd and 3rd 1965.*] Reykjavík: Efnahagsstofnunin.

Furinghetti, Fulvia, Menghini, Marta, Arzarello, Ferdinando, & Giacardi, Livia (2008). ICMI renaissance: the emergence of new issues in mathematics education. In M. Menghini, F. Furinghetti, L. Giacardi, & F. Arzarello (Eds.), *The first century of the International Commission on Mathematics Instruction (1908–2008)* (pp. 131–147). Rome: Instituto della Enciclopedia Italiana.

Gjone, Gunnar (1983). *"Moderne matematikk" i skolen. Internasjonale reformbestrebelser og nasjonalt læreplanarbeid ["Modern mathematics" in school. Reform attempts and national curriculum work*], I–VIII. Oslo.

Høyrup, Jens (1979). Historien om den nye matematik i Danmark – en skitse. [The history of the new mathematics in Denmark – a sketch]. In P. Bollerslev (Ed.), *Den ny matematik i Danmark*, pp. 49–65. Copenhagen: Gyldendal.

Karp, Alexander (2015). Interview with Mogens Niss. *The International Journal for the History of Mathematics Education, 9*(1), 55–76.

Kilpatrick, Jeremy (2012). The new math as an international phenomenon. *ZDM Mathematics Education 44*, 563–571.
DOI 10.1007/s11858-012-0393-2.

Landsprófsnefnd [National Examination Board] (1968). *Drög að námsskrá í landsprófsdeildum miðskóla [A draft to a curriculum for a national examination for middle schools].* Reykjavík: Ministry of education.

Law no. 22/1946 on a school system and official duty to provide education.

Menntamálaráðuneytið (1989). *Aðalnámskrá grunnskóla. [A national curriculum for compulsory school level*]. Reykjavík: Ministry of education.

Menntamálaráðuneytið (1999). *Aðalnámskrá grunnskóla. Stærðfræði. [A national curriculum for compulsory school level. Mathematics*]. Reykjavík: Ministry of education.

Niss, Mogens (1996). Goals of mathematics teaching. In A.J. Bishop, K. Clements, C. Keitel, J. Kilpatrick, & C. Laborde (Eds.), *International handbook of mathematics teaching* (Vol. 1. pp. 11–47). Dordrecht: Kluwer Academic Publishers.

NKMM (1967). *Nordisk skolmatematik [Nordic school-mathematics*]. Stockholm: Nordisk råd.

OECD (1964). *Mathematics To-Day. A guide for teachers.* Paris: OECD.

OEEC (1961). *New thinking in school mathematics.* Paris: OEEC.

Piaget, Jean (1955). Les structures mathématiques et les structures opératoires de l'intelligence. In J. Piaget, E.W. Beth, J. Dieudonné, A. Lichnerowicz, G. Choquet, & C. Gattegno (Eds.), *L'enseignement des mathématiques* (pp. 11–33). Neuchatel and Paris: Delachaux and Niestlé.

Piaget, Jean (1973). Comments on mathematical education. In A.G. Howson, (Ed.), *Developments in Mathematical Education* (pp. 79–87). Cambridge: Cambridge Univ. Press.

Prytz, Johan, & Karlberg, Martin (2016). Nordic school mathematics revisited. On the introduction and functionality of New Math. *NOMAD*, *21*(1), 71–93.

Schubring, Gert (2013). The road not taken—The failure of experimental pedagogy at the Royaumont Seminar 1959. *Journal für Mathematik-Didaktik, 35*(1), 159–171.

Skemp, Richard (1978). Relational and instrumental understanding. *Arithmetic Teacher*, *26*(3), 37–50.

Skovsmose, Ole (1979). 60□er-matematiken – Idé og virkelighed. [The sixties' mathematics - Ideas and reality]. In Bollerslev, P. (Ed.), *Den ny Matematik i Danmark*, pp. 49–65. Copenhagen: Gyldendal.

Vanpaemel, Geert, De Bock, Dirk & Verschaffel, Lieven (2012). Defining modern mathematics: Willy Servais (1913–1979) and mathematical curriculum reform in Belgium. In K. Bjarnadóttir, F. Furinghetti, J.M. Matos, & G. Schubring (Eds.), *"Dig where you stand" 2. Proceedings of the second "International Conference on the History of Mathematics Education"* (pp. 485–505). Caparica: UIED.

Textbooks

Arnlaugsson, Guðmundur (1966). *Tölur og mengi.* [*Numbers and sets*]. GA. Reykjavík: Ríkisútgáfa námsbóka [State Textbook Publishing House] (RN).

Bergendal, Gunnar, Hemer, Ove, & Sander, Nils (1970). *Rúmfræði* [*Geometry*]. NK. Reykjavík: RN.

Björk, Lars-Erik, Björksten, Christina, Brolin, Hans, Ernestam, Arne, & Ljungström, Lars Fredrik (1987–1989). *Almenn stærðfræði fyrir grunnskóla* [*General mathematics for compulsory schools*]. (Swedish title unknown). Reykjavík: National Centre for Educational Material (NCEM).

Björnsdóttir, Helga (translator) (1972). *Stærðfræði handa gagnfræðaskólum* [*Mathematics for secondary schools*]. NK. Author unknown. Reykjavík: RN.

Bundgaard, Agnete (1969–1972). *Stærðfræði. Reikningur.* [*Mathematics. Arithmetic*]. *3a-6*. Reykjavík: RN.

Bundgaard, Agnete & Kyttä, Eeva (1967–68). *Stærðfræði. Reikningur.* [*Mathematics. Arithmetic*]. *1-2b*. Reykjavík: RN.

Daníelsson, Ólafur (1938). *Reikningsbók* [*Arithmetic*]. 5th ed. Reykjavík: Ísafoldarprentsmiðja.

Daníelsson, Ólafur (1951). *Kennslubók í algebru* [*A textbook in algebra*]. 3rd ed. ÓD. Reykjavík: Ísafoldarprentsmiðja.

Gíslason, Sigurbjörn Á. (1911–1914). *Reikningsbók* [*Arithmetic*] *IV*. Reykjavík: Sigfús Eymundsson.

Gissurarson, Jón Á., & Guðmundsson, Steinthór (1949–1950). *Reikningsbók handa framhaldsskólum.* [*Arithmetic for lower secondary schools*] *II A*. JSG. Reykjavík: Ísafoldarprentsmiðja.

Kristjánsdóttir, Anna, Magnúsdóttir, Ásgerður, Birgisdóttir, Björg, Halldórsdóttir, Hildigunnur, Ármannsson, Ingólfur & Bjarnadóttir, Kristín (1988–1991). *Stærðfræði handa grunnskólum* [*Mathematics for compulsory schools*] *7–9. Talnaspegill, Hornalína, Sjónarhorn, Skuggsjá, Baugabrot*. KMB. Reykjavík: NCEM.

Lárusson, Hördur (1972–1976). *Stærðfræði fyrir grunnskóla* [*Mathematics for compulsory schools*]. HL. Reykjavík: RN.

NKMM (1970). *Algebra unglingaskóla* [*Algebra for lower secondary schools*]. NK. Translators: Hördur Lárusson and Anna Kristjánsdóttir. Reykjavík: RN.

Pálsdóttir, Gudbjörg, & Gunnarsdóttir, Gudný H. (2005–2008). *Átta-tíu stærðfræði* [*Eight-ten mathematics*] *1–6*. PG. Reykjavík: NCEM.

Author Biography

Kristín Bjarnadóttir is professor emerita at the University of Iceland. She completed a BA degree in physics and mathematics at the University of Iceland in 1968. She pursued her graduate education to a master's degree in mathematics at the University of Oregon in 1983 and a Ph.D. degree in mathematics education at Roskilde University Center, Denmark in 2006. She taught mathematics and physics at lower and upper secondary schools in 1965–2003 and mathematics education at the University of Iceland, School of Education in 2003–2013. Her research interests concern secondary level mathematics education and the history of mathematics education.

Real Numbers in School: 1960s Experiments in France and Brazil

Elisabete Zardo Búrigo

Abstract The study of real numbers was introduced in the curriculum of lower-level secondary schools in a number of countries in the 1960s, under the influence of the modern mathematics movement and the guiding principle of bridging the gap between secondary school and higher education mathematics. Despite this common drive, the dynamics and the outcomes of this innovation were not the same everywhere. The cases of France and Brazil, in particular, deserve to be contrasted because, despite the similarities between the programs up to the early 1960s, the ways in which the two countries experimented with and implemented the teaching of real numbers were different. A brief comparison between the two reforms suggests that the international dimension of the movement itself had more effect on lending legitimacy to the realization of local experiments than on the format they took and their consequences.

Keywords History of mathematics education · Real numbers
School curriculum · Modern mathematics movement · Secondary education

1 Introduction

In many countries during the 1960s and 1970s, the secondary school curricula underwent substantial changes under the impact of the so-called modern mathematics movement. Research of the implemented changes reveals very unique processes in those countries or even in different regions. The local and diverse character of these reforms contrasts with the wide dissemination of the curriculum proposals produced at the time of the movement, such as those carried out by the OECD (Organization for Economic Cooperation and Development) and UNESCO (United Nations Educational, Scientific and Cultural Organization) (Kilpatrick

E.Z. Búrigo (✉)
Federal University of Rio Grande do Sul, Porto Alegre, Brazil
e-mail: elisabete.burigo@ufrgs.br

F. Furinghetti and A. Karp (eds.), *Researching the History of Mathematics Education*, ICME-13 Monographs,
https://doi.org/10.1007/978-3-319-68294-5_2

2012; Vanpaemel et al. 2012). Cross-sectional studies on the history of arithmetic, algebra, and geometry teaching indicate, on the other hand, some common traits to the movement's effects on the programs, textbooks, and teacher training processes (see Part V in Karp and Schubring 2014).

The introduction of real numbers as a topic of the curricula of the first level of secondary school (11–15 years) is one of these features concerning various countries (Fehr et al. 1971; Servais and Varga 1971). This paper addresses this innovation as it occurred in France and Brazil.

The comparison is part of the broader context of a comparative research, begun in 2011, between mathematics education reforms that have taken place in both countries in the 1960s and 1970s. From the point of view of the Brazilian case, such comparison is relevant, since, until the 1950s, French secondary education was the main external reference for the design of Brazilian mathematics curricula, in terms of educational programs and books. The approach taken here, however, is not that of studying the influences or connections between the two countries. Considering their different outcomes, we argue that they are two interesting cases of being contrasted, based on the assumption that they were components of the same international movement (Kilpatrick 2012; Gispert 2010).

There were some similarities between the programs of both countries at the beginning of the 1960s, and by the end of the decade, both underwent major changes under the impact of modern mathematics (D'Ambrosio 1991; D'Enfert and Gispert 2011; Trabal 1998). Introducing lower-level secondary students to real numbers was a component of such changes. The processes of curricular innovation in the two countries, however, were quite different, as will be shown below. Additionally, in Brazil, the teaching of real numbers remained a topic of this school stage program after the decline of the movement, while in France it was discarded after a few years. To what extent can one consider these changes to be effects of a cross-national movement? Why were changes temporary in one case and long-lasting in the other? How did the international and the local dynamics intertwine to produce reforms, both similar and distinct?

Curriculum changes are complex and produced in several instances, as shown by Goodson (1988), Gimeno Sacristán (1998) and other scholars in the field. Thus, it is not intended, from a brief assessment of the two cases, to definitively answer the issues raised above. On the other hand, as Chervel (1988) argues, studying reforms can be very enlightening, as multiple actors come into play and thus express their views. In this article, the comparative approach is mainly a starting point from which, relying on previous research and the documentation produced in that period, we seek to glean new insights and contribute to studies regarding the impact of the modern mathematics movement.

2 The Expansion of Secondary Schooling in France and Brazil

The innovations addressed in this paper occurred in the context of change within groups of teachers and students, as well as in the purposes assigned to post-elementary education in France and Brazil.

In France, until the 1930s, primary and secondary school constituted two parallel and disjointed systems, both accepting children from six years of age, but with different purposes, terms, rules, populace, and sets of teachers (Prost 2004). The primary school system, which covered not only elementary, but also some post-elementary classes, welcomed the people's children, and had as its declared purpose, preparing students for "life"; the secondary school system, which taught a humanistic culture and aimed at higher education, received the children of the elites. Then, by the 1950s, several measures had contributed to more openness in secondary schooling: it became progressively free, was partitioned into two cycles, and expanded into new branches.

According to Prost (2004), however, the Berthoin and Fouchet reforms, implemented respectively in 1959 and 1963, were the decisive steps leading to the erosion of that early compartmentalization. First, compulsory schooling was expanded to the age of 16. Throughout the 1960s, elementary education, which lasted five years, was unified, and the first cycle of secondary education became the task of the *collèges,* which were autonomous establishments, separate from the traditional *lycées*. The enrollment rate at this stage rose from approximately 50% of the population aged 12–15 in the academic year 1960–1961, to 90% of that same population in the school year 1970–1971 (ibid., p. 294).

By the end of the decade, therefore, the first cycle of secondary education included nearly the entire school-age population in France. Remnants of a bipartite system, however, were still present: until 1969, the first cycle of secondary education comprised four distinct branches; three of these originated from the primary school system and were less prestigious, taught by teachers with less training, and led to a shorter school career. The process of unification of the first cycle of secondary education was not completed until 1976, with the declaration of the so-called Haby reform (Prost 2004).

In Brazil, secondary education was sparse until the 1920s, when it became a prerequisite for access to higher education. It was reorganized in 1931 and then revamped again in 1942 into two cycles that lasted four and three years respectively. Until the 1960s compulsory schooling was limited to primary education, offered to children aged seven years and older, and lasted four or five years. A small portion of the students was able to overcome the barriers to enter into secondary education. Only 20% of children entering the first grade of primary school in 1957 reached the fourth grade in 1960 (Cunha 1980). Students who had completed the primary level could apply to take a very selective entrance exam to enter the first cycle of secondary school, called *ginásio* (Nunes 2000). In 1960, less than 10% of the Brazilian population aged 12–18 years attended secondary courses (Gouveia and Havighurst 1969).

This picture changed somewhat throughout the 1960s. In 1965, Circular 973 eliminated the examination requirement for students whose primary education certificate was recognized by competent authorities. In 1971, the enactment of Law 5692, merging the old primary level and *ginásio*, definitively abolished the exam (Bastos and Ermel 2014). Enrollments in *ginásio* increased from around 825,000 in 1961 to 3,053,000 in 1970; that is, it expanded more than 3.5 times over the decade (CNE 1961; IBGE 1972). In 1970, the average schooling of the population aged 15–19 reached four years (Pinto et al. 2000).

Access to secondary education in the late 1960s was still much more restricted in Brazil than in France. Among other factors, this discrepancy can be attributed to the differing stages of economic development, the educational policies implemented in both countries, and the high degree of social inequality and income concentration in Brazil (Cunha 1980; Pinto et al. 2000). Nonetheless, in both countries in the 1960s, the first cycle of secondary education, following five years of elementary schooling and lasting four years, was expanded and democratized. In both cases and in differing ways, primary school teachers were called to teach at the secondary level. Diverse school cultures mingled in this transition from an elitist system to a school for all. The reforms implemented in the 1960s occurred during this changeover scenario, and teachers from different backgrounds were challenged to teach new mathematics to incoming groups of students (Búrigo 2009; Barbazo and Pombourcq 2010; D'Enfert and Gispert 2011; Gispert 2014).

3 Real Numbers in the Early 1960s Programs

Before the 1960s in both France and Brazil, the curricula of the first level of secondary school (*collège* and *ginásio*, respectively), which taught 11–15-year olds, did not address real numbers.

The French program established in 1958 and 1960, slightly altered in the following years, was organized into the classic sections of arithmetic, algebra, and geometry. In the second grade of *collége*, the *cinquième*, students were required to learn about fractions and decimal numbers and how to work with them (MEN 1960); in the last grade, named *troisième*, they were to deal with square roots and their decimal approximations. The algebra section of the third grade, the *quatrième*, included the "algebraic measurement of a segment oriented on an axis" and the "localization of a point on an axis" (MEN 1958, p. 2513). The program included the manipulation of some irrational numbers, such as roots or number π, and the notion of real numbers was implicit in such concepts as vector length or abscissa of a point on the line. However, it was not expected that there would be a discussion on the nature of the numbers corresponding to these geometric notions and the properties of operations with them. Belhoste (1997) goes further, noting that real numbers were not even a topic of secondary school curricula up to 1960: "And if the algebra and analysis courses in the *seconde* class de facto presuppose the existence of real numbers, the question is not dealt with in the second cycle except by alluding to it in the last class of the mathematics branch.[1]" (ibid., pp. 383–4).

In Brazil until the 1930s, secondary education followed this same division of arithmetic, algebra, and geometry. The main references were French textbooks and their translations. Euclides Roxo, an advocate for modernization, wrote the mathematics programs of 1931 under the inspiration of Felix Klein's ideas. According to these programs, the concept of a function should be the axial notion of mathematics teaching, articulating the different branches of the discipline (Valente 2004).

The reform of 1931 triggered conservative reactions from influential figures, among them the military schools' teachers and some professors who were educated in the ancient tradition, and Catholic educators who advocated a classical, humanistic education and the reduction of the focus on scientific disciplines. The program, established by the reform of 1942, restored the division between arithmetic, algebra, and geometry, and postponed the study of functions until the second cycle of secondary (Soares et al. 2004).

The Ministerial Ordinances 966 and 1045 of 1951 changed these programs, eliminating topics and prioritizing practical, rather than the theoretical approaches to content. According to Marques (2005), this simplification took into account the expansion of secondary education and foresaw the possibility of the development of the programs at varied levels, according to the regions, and the public served.

In these new programs, drawn up by the teachers of traditional *Colégio Pedro II* in the city of Rio de Janeiro, integers, fractions, and decimals were the first year's subjects; square and cubic roots pertained to the second year of *ginásio*. Irrational numbers were mentioned, but operations were limited to those with radicals. The program did not advise discussing real numbers or working with them. The arithmetic section was therefore quite similar to French, regarding the presentation and manipulation of different types of numbers (MEC 1952).

The study of algebra and geometry, on the other hand, did not involve the localization of points on the line or in the plane according to coordinates. Irrational numbers were to relate to the distinction between commensurable and incommensurable lengths, but this evocation of Greek thought did not lead to any discussion of a numerical set representing these quantities. One could say that the Brazilian program corresponded to an older, more simplified version of the French programs, in which analytical geometry was consigned to a more advanced educational level. The notion of real numbers was implicit in several topics of the second cycle of secondary education, especially in the study of functions, but it did not appear as a subject (MEC 1952).

---

[1]All translations of texts originally published in French or in Portuguese are the responsibility of the author.

4 Real Numbers from the Perspective of the Modern Mathematics Movement

A large number of experiments, curricular proposals, texts, and events, constituted what is known as the modern mathematics movement, its international character stemming from an intense exchange between actors from different countries. In the 1950s, the CIEAEM (Commission for the Study and Improvement of Mathematics Teaching) played a key role bringing together mathematicians, teachers engaged in classroom experiments in different European countries, and such scholars of cognitive psychology as Jean Piaget (Gispert 2010; Félix 1986). During the same period, efforts were being made in the United States to reform secondary school mathematics (Kilpatrick 2012). In 1959, the Conference of Royaumont (Paris), chaired by the ICMI (International Commission of Mathematical Instruction of the International Mathematical Union) president Marshall Stone, and organized by OECE (now OECD), brought together those different groups. From this point on, the sharing attained a new dimension, through the initiatives of OECD and other international and governmental agencies, such as the OAS (Organization of American States) and UNESCO, that sponsored publications of the ICMI (Furinghetti 2014).

In the midst of a wide range of proposals developed and conveyed at that time, it is possible to identify the convergence of general ideas. One of the key intents was to bring school mathematics closer to the academic mathematics of the twentieth century (Kilpatrick 2012). This concern implied including new themes, such as probability or vector geometry, but also the adoption of language and procedures closer to those adopted in higher education. Overcoming the traditional separation between algebra and geometry, the program was to be organized around unifying concepts such as sets, relations, and functions. And at school, there was to be room for mathematical activity on the part of the students, through problem-solving, the discovery of models, or the construction of new entities (Fehr 1964; Fehr et al. 1971).

Bringing real numbers forward would become a requirement of broader guidelines of the movement. On one hand, the subject was highlighted by the earlier study of functions and vector geometry. On the other, it topped all approaches of different types of numbers by the bias of algebraic structures. Some proposals for studying real numbers in lower secondary education circulated in the 1960s, before reforms in France and Brazil.

The *Synopses for modern secondary school mathematics* (OECD 1961), drafted by a meeting of experts in Dubrovnik in 1960, was a reference book at the time; it appeared in the aftermath of the famous seminary of Royaumont, promoted by OECD in 1959. In the section devoted to rational numbers for the lower secondary, it recommended constructing sequences of nested intervals with decimal endpoints, so that the student could realize that the procedure is infinite and could be applied to the coordinate of any point on the axis. This is a guideline that could be considered preparatory to the introduction of real numbers, but the intention is not explicit in the text. In the algebra chapter for the upper secondary, real numbers were mentioned without any guidance on how to teach them.

The OECD session on new teaching methods held three years later in Athens, Greece, reached the conclusion that it would be unrealistic to propose a standard program for all countries. Instead, the session's report advised studying the already-scrutinized or official programs (Fehr 1964). Two of the proposals presented at the seminar dealt explicitly with real numbers: one was introduced by Begle (1964), leader of the North American SMSG (School Mathematics Study Group), and the other was presented by the Belgian, Papy (1964).

In the Belgian experiments, students aged 13–14 (*cinquième* class) explored the idea of locating a point on a straight line by successive sub-graduations, reaching to the match between point and unlimited expansion, and the resulting notion of the x-coordinate. Since, from modern mathematics standpoint, the understanding of numbers was not to depend on their decimal representation, this exploration started from binary grading, the figures zero or one corresponding to the notions of "before" or "after" a marked point. In this approach, the study of real numbers did not have to be preceded by that of rational numbers, which were defined simply as quotients between integers (Papy 1964, 1967).

In a brief presentation on the teaching of algebra, Begle (1964) argues that students should "be aware of the algebraic structure their set is fitted with" (ibid., p. 30). He gives examples that would allow the presenting of real numbers as an extension of rationals, just as one can treat rationals as integer quotients, and integers as differences between natural numbers. He notes that in most cases within rational numbers, given any particular one, it is not possible to obtain its square root. Instead, in the set of real numbers, one can obtain the square root of any positive number. The description is somewhat misleading since it suggests that real numbers are roots, or roots of roots of rational numbers.

Indeed, we infer from the SMSG's textbooks that it was not expected that teachers would develop a discussion on the nature of real numbers. The book named "Mathematics for high school – first course in Algebra" mentions the existence of numbers that have no decimal representation: "Perhaps you believe that $\sqrt{2}$ is 1.4. Test for yourself whether this is true by squaring 1.4. […] In the same way, test whether $\sqrt{2}$ is 1.41" (SMSG 1959, p. 121). The matter of the correspondence between numbers and points of the line is soon closed with the statement that "There are many more points on the real number line which have coordinates which are not decimal fractions" (ibid., p. 121). The following chapters, dealing with operations with real numbers, are only concerned with the discussion of how to deal with positives and negatives.

In 1965, ICMI organized a congress in Dakar on mathematics teaching in conjunction with the teaching of other sciences (Furinghetti 2014). Unlike the cautious Athens seminar, the Congress of Dakar drafted a consensus program to be adopted in all countries "for the next ten to twenty years" (Fehr 1967, p. 68). Real numbers was a topic of this program. Fehr (1967) proposed studying sets of numbers as algebraic structures departing from the field of rational numbers. Infinite representations were to be introduced by the binary or decimal graduation of the line. Then the set of real numbers would be established by an axiom of continuity, the real line and the set of real numbers being "established simultaneously" (ibid., p. 49).

5 Real Numbers in French Reform and Experiments

Since the early 1950s, French secondary teachers and mathematicians have been engaged in international debates on the modernization of mathematics teaching, such as those promoted by the CIEAEM, UNESCO or the CIEP (International Centre for Educational Studies) de Sèvres. The APMEP (Association of Public Education Mathematics Teachers) brought together militant teachers who claimed, at the same time, the appreciation of the discipline, the renewal of its contents and the teaching methods. However, the resistance of some sectors, including that of the General Inspectors, who until then had been responsible for the drafting of the programs, postponed the *collège* reform (D'Enfert 2010). The change was finally launched in 1966 when the CEEM (Commission for the Study of Mathematics Teaching) was created by a ministerial decision.

The Commission became known as the Lichnerowicz Commission, for it was chaired by mathematician André Lichnerowicz, a participant in the CIEAEM meetings in the 1950s and the president of the ICMI from 1963 to 1966. In the beginning, the CEEM brought together around twenty members, among them teaching inspectors, university professors, and secondary school teachers; in 1971 it had grown to about 40 members.

The main task of the CEEM was to draft new curricula for secondary education and later for elementary education. However, since the beginning of the work, there had been agreement that any reform needed to be accompanied by efforts to improve initial and continuing teacher training. Under the pressure of APMEP, and following an ongoing international trend, it was also decided to test the programs beforehand. The subcommittee responsible for the experimentation was coordinated by André Revuz, a professor at the Faculty of Sciences in Poitiers, former president of the APMEP from 1960 to 1962. The experimentation process, of which the IPN (National Pedagogical Institute) was in charge, was developed two years ahead of the implementation and comprised the test of new teaching methods (CEEM 1967, p. 249). Initially, 80 teachers gathered in five research centers in the towns of Bordeaux, Lyon, Marseille, Paris and Poitiers. The following year, another ten centers were created, bringing together about 300 teachers. The experimenters participated in local, regional, and national seminars throughout the school year (Revuz 1972).

The program proposals were drafted by subcommittees and considered at plenary meetings of the commission. The debate on the programs for the first cycle of secondary education, the *collège,* began as early as April 1967. The commission members agreed that the study of real numbers should be conducted in the *quatrième*, the third grade of secondary schooling, intended for students aged

13–14 (CEEM 1967). The first draft program for *quatrième* was published in June 1968. It proposed an intuitive approach to real numbers, starting from decimals and their representation on the number line (CEEM 1968).

The experiments involving curricula for *quatrième* began in September 1969, since the new programs were to be implemented in the 1971–1972 school year. In a lecture to the experimenters in March 1970, Dehame, a member of the APMEP committee, justified the introduction of real numbers in preparation for analysis. He argued that "it doesn't have to do with a theory of the real numbers, but with a series of exercises designed to develop in students a correct intuition of R" and with "an intuitive contact with the infinite, under its different aspects" (Dehame 1970, p. 455). For the intuition of the number line, it would be necessary to "disentangle the numbers from the concrete support used to introduce them" (ibid., p. 464). The motivation for the extension of the set of decimals noted as $\mathbb{D}$ was to be algebraic, and involve the study of equations unresolvable in this set.

Reports of the trials conducted in the various research centers, were published by APMEP and IPN in 1971. These reports show a variety of paths followed and a positive assessment of their viability. The approaches had the shared purpose of "leading students to feel the need to create new numbers" (Brailly 1971, p. 27) and a concern that students should be confident in their ability to calculate with approximations. In all cases, the starting point was the set of decimals and a search for solutions to equations, such as $3x = 1$ or $x^2 = 2$ having no decimal solutions.

From then on, the approaches were diversified. At the Lyon experiments, for instance, the students were challenged to "make these new numbers exist" and insert them between the decimals so that the whole set, gathering the decimals and these new numbers, would stay well-ordered. Successive upper and lower approximations were obtained with the aid of calculators. After this familiarization with the notions of sequence, convergence, and real numbers, the axiom of $\mathbb{R}$ structure as a group (concerning addition and multiplication, in this case excluding zero) was presented (Anguenot 1971). In the Bordeaux experiments, after students calculated the approximations, the set of real numbers was defined as an extension of the set of decimals with the same properties, as well as the existence of the inverse multiplicative for non-zero numbers, and of a common element in any sequence of nested intervals with decimal endpoints (Becker 1971).

In addition to the emphasis on approximation calculations, and with approximations themselves, experiments also involved the use of graphical representations, such as those in Figs. 1 and 2, which illustrate the study of the number 1.202002000200002000002…

A geometric approach to real numbers was put forward by Morin from Montpellier, having as motivation "to give all points of the line an x-coordinate." (Morin 1971, p. 306). This approach, however, was not given support, since, as we shall discuss below, the Commission's program stipulated that the straight line itself should be defined from the set of real numbers.

At the same time that a broad agreement was reached on the study of real numbers, the deductive approach to geometry raised controversy within and outside the Commission, and experimenters raised complaints (D'Enfert and Gispert 2011;

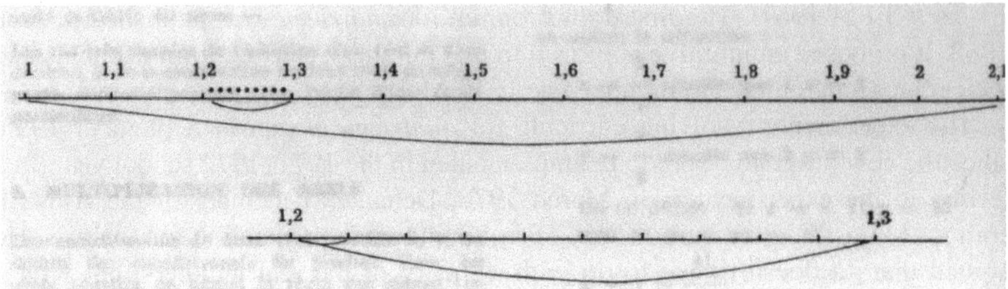

Fig. 1 Scheme at a given scale (Dehame 1971, p. 18)

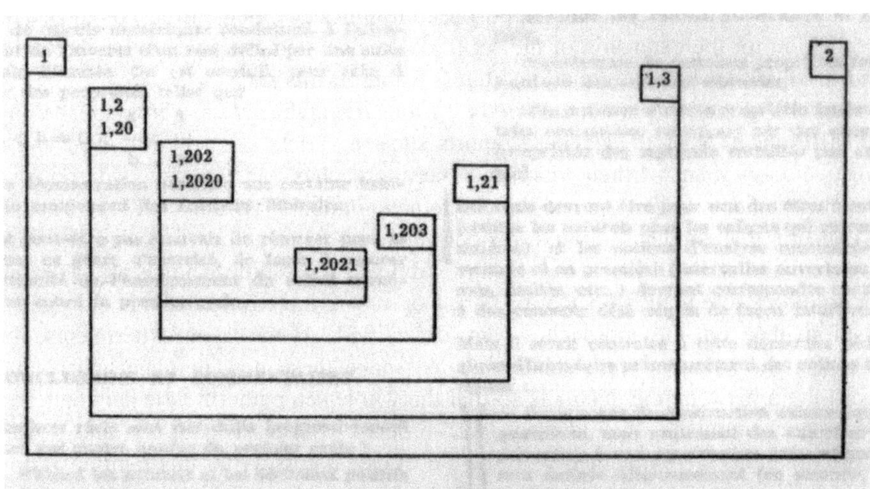

Fig. 2 Scheme without scale (Dehame 1971, p. 19)

APMEP 1972). The program projects for the *quatrième* underwent a long and tumultuous revision process until its approval for application in the school year 1971–1972. In November 1970, the Commission adopted a proposal incorporating the experimentation approach to real numbers, which was to come from inserting decimal numbers (which can be written as $a \cdot 10^p$, with a and p being integers) and from building unlimited decimal sequences (CEEM 1971a). The distinction between rational and irrational numbers was to be addressed later, in *troisième*.

The controversy over geometry persisted, however, and the program proposed by CEEM received only four votes in favor and ten abstentions at the CEGT (Council for General and Technical Education) meeting in December. By ministerial demand, the program was revised again in February 1971. Miss Stouls, a secondary teacher and a participant of the trials, and M. Pisot, a professor at the Faculty of Sciences in Paris, proposed an inversion in the algebra program, so that the study of rational numbers should precede that of real numbers. The upholding of the previous proposal, which received 34 votes out of 39 voters, is another indication of the broad agreement on innovation in the teaching of real numbers (CEEM 1971b). According to Revuz (1972), it was "'the experimenters' painstaking work that pointed to the possibility" (ibid., p. 6) of innovatively approaching real numbers, based on approximate calculations with decimals.

The new approach was welcomed by APMEP (1971), as it overcame an approach that, until then, was limited to presenting the numbers in the same order in which they had historically emerged—fractions, algebraic, and then some transcendent numbers—"extending the notion of number without ever showing how it was possible to extend operations to these numbers" (ibid., p. 261). Along the same lines, Anguenot (1971) from Lyon, argued that it was inconsistent to speak of numbers as $\sqrt{2}$ or π without introducing the concept of real numbers. He claimed, "It is expedient to do so with the highest intellectual honesty: that what is admitted is clearly expressed, that great ideas are correctly suggested even if their theoretical development is not perfect" (ibid., p. 22). In the articles published by the APMEP and the IPN, few put into question these pretensions of rigor: "The motivation is a little fragile: why not go directly to the complex numbers, so that the equation $x^2 + 1 = 0$ be solvable? […] The student with some imagination may put this question…" (Chayé and Gauthier 1972, p. 60).

New programs were published at the end of July, and guidance comments followed, but not until early December after the beginning of the school year (MEN 1971a, b). The comments replicated the approach adopted in the experiments, including looking for a multiplicative inverse to a given decimal, obtaining embedded intervals, constructing unlimited decimals, and defining $\mathbb{R}$ by checking certain properties. The obtaining of square roots was to be postponed until the *troisième*. The comments also stipulated that the construction of $\mathbb{R}$ become a prerequisite for geometry, since the effort to distinguish the Euclidean line from the physical line led to its definition as "A set E provided with a family F of bijections on $\mathbb{R}$" which are distinguished from each other only by a constant (MEN 1971b, p. 2886).

At the end of the school year 1971–1972, the controversy surrounding the introduction of axiomatic geometry had widened. The study of real numbers was also questioned. According to Circular 72,220 of May 30, 1972, "a number of groups extended more than suited on the decimals or nestings on the approximations of real numbers" (MEN 1972, p. 1562). Thus teachers were pressured to reduce the time students spent on the exploration of these approximations. Finally, Circular 73,087 of February 19, 1973, greatly reduced the ambition of the programs, presenting as "necessary acquisitions" only the "approximate calculations on the real numbers" and recommending a return to physical space as a reference: "If nesting by decimals […] is presented in the abstract, the students are not interested enough, and then it will be necessary to borrow concrete examples from the measurement activities" (MEN 1973, p. 629). In the same year, the CEEM dissolved.

The study of real numbers in *collège* was put aside in 1978, in the context of the above-mentioned Haby Reform which established a single path to the first level of secondary school in France (Chevallard 1985). The new programs' guidelines were established by the Circular 78,392 of November 16, 1978: "As regards $\mathbb{R}$, it is sufficient that students know that it is an expansion of $\mathbb{D}$ or $\mathbb{Q}$ over which the addition, the multiplication, and the order relation are extended preserving the same properties, and throughout which any positive element admits a square root" (MEN 1978, p. 80).[2] Eventually, the *collège* programs, established in 1985, made no reference to real numbers, and radicals were the only irrational numbers mentioned, just as in the programs preceding the movement of modern mathematics.

Referring to the 1990s, Bronner (2000) points out the existence of an institutional vacuum,[3] with regard to the negotiation of the transition from decimal and rational numbers to real numbers. The reform of modern mathematics, however, seems to have print traces in the conceptions and practices of teachers. Based on interviews with teachers of the *troisième* class about the teaching of radicals, Bronner describes their attitudes towards the topic "real numbers" as quite diverse, ranging from no discussion about the appearance of "new numbers" to that of dealing with radicals as non-decimal numbers, thus reducing the aforementioned vacuum.

6 The Introduction of Real Numbers in *Ginásio*

In 1961, the new National Education Law (*Lei de Diretrizes e Bases da Educação Nacional*) granted autonomy to the states in developing their own curricula, terminating the national standardization in Brazil. Such withdrawal established in practice the deregulation of the programs, since the regional curricula would not appear before the comprehensive reform of education set by the Law 5692 of 1971 which extended compulsory schooling to an eight-year cycle. The movement of modern mathematics emerged in Brazil in this context, on the initiative of groups of teachers who worked together to carry out experiments and disseminate proposals for the renewal of programs and teaching methods. Although they received some government support, these groups enjoyed broad autonomy. International agencies such as IBECC (Brazilian Institute for Education, Science and Culture), a UNESCO local agency, and OAS, encouraged exchanges with North American universities, but there were also exchanges with teachers and authors, like Lucienne Félix and Georges Papy, from European countries (D'Ambrosio 1991; Búrigo 2009, 2015).

Among these groups, the Study Group on Mathematics Teaching (*GEEM—Grupo de Estudos em Ensino de Matemática*), created in São Paulo in 1961, was the one that achieved the greatest audience in the country. GEEM was made up of university professors and secondary and primary school teachers involved in modernizing the teaching of mathematics. Like others at that time, the group

[2]The author thanks Renaud d'Enfert for his comments and for sharing scanned versions of the French ordinances and CEEM proceedings.

[3]*Vide institutionnel* (Bronner 2000, p. 3).

members aspired to an improvement in teaching, in a context of a shortage of licensed teachers and competition for access to secondary education, in which the mathematics exam played a decisive role. GEEM's discourse promised to overcome this elitist and ineffective teaching, promoting interest, curiosity, and learning (Búrigo 2009).

In 1962, the IV Brazilian Congress of Mathematics Education, held in Belém do Pará, approved a GEEM's proposal concerning "Minimum Topics in a Modern Mathematics Curriculum for *Ginásio* and *Colégio*" (GEEM 1965). An adapted version of the proposal was publicized as "Suggestions for a program script for the chair of Mathematics" by the São Paulo Board of Education in 1965.

The proposal of GEEM included real numbers and linear functions among topics for *ginásio*. For the study of real numbers, the orientation was to "emphasize the permanence of properties already introduced with rational numbers," "solve first degree equations and systems with real coefficients," and address the graphical representation of real numbers on the straight line (GEEM 1965, p. 95). As for the functions, it was proposed students study the linear function $y = ax + b$ and quadratic functions.

Although these programs were not mandatory, collections of so-called "modern" textbooks referencing them were produced from 1964 on. The dissemination of these books was initially an enterprise of the publishers who financed the authors' lectures on modern mathematics throughout the country (Valente 2008). As of 1966, some of these collections were supported by COLTED (Technical Books and Textbooks Committee), a governmental book-distribution program funded by an agreement with the USAID (United States Agency for International Development). In that same period, many experiments were carried out in different places, under the inspiration of modern mathematics. In some public schools in the cities of Salvador and Rio de Janeiro, the experiments were inspired by the proposals of Georges Papy (Freire and Dias 2010; Soares et al. 2004; Dassie et al. 2015). But in most schools, the modernizing proposal came through these books, which teachers felt compelled to follow. It is in this context that one can consider there had been wide experimentation of these new programs in *ginásio* in Brazil.

The best-selling collection in the country was "Mathematics—Modern Course" by Osvaldo Sangiorgi, chair of GEEM (Valente 2008). The book dedicated to the third grade of *ginásio* (for students aged 13–14) begins with a chapter on real numbers, preceding the study of radicals. The author asserts the existence of irrational numbers by presenting a few non-repeating decimals, such as 0.52522522252222…, "where there is always an 'additional' 2 after you write a 5." He explains that an infinite number of endless decimals can be generated and resorts to the metaphor of a "special watch" whose hand is activated an infinite number of times and always falls back on a digit from zero to nine. The set of real numbers is then defined as the union of the sets of rational and irrational numbers. After several exercises involving the distinction between rational and irrational numbers, the bi-univocal correspondence between the real numbers set $\mathbb{R}$ and the straight line, called "real line," is mentioned. At the end of the chapter, the properties of addition and multiplication featuring $\mathbb{R}$ as a field structure are presented (Sangiorgi 1968).

A collection that also achieved best-selling status was the "Mathematics in renewed school" by Pierro Netto (1970). In this book, the author mentions some square roots as examples of irrational numbers and displays a line graph, marked with the points that have these numbers as coordinates. Next, he mentions the following: that the union set of rational numbers with the irrational numbers is the set of real numbers ℝ; that there is a one-to-one correspondence between the set of real numbers ℝ and the straight line; that the properties of operations with rational numbers also apply to real numbers.

In the collection "Mathematics for *Ginásio*," produced under the direction of Lydia Lamparelli (Lamparelli et al. 1971), measuring the hypotenuse of the isosceles triangle with a unit long leg is the starting point of discussing the square root of 2 and its approximation with decimal numbers. The authors demonstrate that the square root of 2 cannot have a repeating decimal expansion and that therefore, it is not a rational number; on the other hand, they note that it "can be approximated by lack or excess by rational numbers." This approach culminates with a short discussion about a bijection between the set of real numbers and the "real number line." The collection, coordinated by Lydia Lamparelli, did not circulate as far as Sangiorgi's and Scipione's but was adopted in an extensive teacher training program, PREMEM (Program of Teaching Expansion and Improvement), launched in 1970.

In these three collections, one can identify traces of the SMSG approach, such as the reference to the number line and the presentation of the set of real numbers as an extension of the rationals through examples. In Pierro Netto's book, this discussion is rather simplified and evokes Begle (1964) when it presents only square roots as examples of irrational numbers. Sangiorgi's approach seems to gather elements from various backgrounds. There is an attempt to discuss real numbers as unlimited decimals, which evokes Papy (1964), who participated in the V Brazilian Congress of Teaching Mathematics in 1966, at the invitation of GEEM. The above-mentioned ten-figure clock metaphor indicates an effort to present a generic description of real numbers, appealing at the same time to intuition. But this description is centered on the distinction between periodic and non-periodic sequences, without calling into question the need for acceptance of these numbers or what they represent. Both approaches indicate the authors' concerns with presenting a simplified view of the new set that could be intelligible to teachers who, for the most part, had no university education in mathematics. In each approach, including that of Lamparelli et al. (1971), the bi-univocal correspondence between the line and the set of real numbers is stated, but without an in-depth discussion of their shared properties.

Lydia Lamparelli was also one of the co-authors of the drafting of the mathematics section of the "Proposed Curricular Guidelines for School Common Subjects" (São Paulo 1973). This was the first official program published in São Paulo after the elimination of the national program in 1961, and it carried national significance because São Paulo housed the largest school network in the country as well as the most important publishers of textbooks. According to the document, the approach of real numbers should begin with infinite decimal representations, distinguishing rational from irrational, according to the occurrence (or not) of the periodicity of this representation. The document also proposed the objectives of "applying the Pythagorean theorem to represent real numbers on the line," "establishing a bijection between line and the set $\mathbb{R}$," "relating the order of $\mathbb{R}$ with the number line," "highlighting the completeness of $\mathbb{R}$," "recognizing in the set of real numbers the structure of an ordered field," and "acquiring greater ability to calculate with real numbers using their properties" (São Paulo 1973, pp. 207–208).

The references to the field structure were abandoned in later programs. But the "Curriculum Proposal for the Teaching of Mathematics" (São Paulo 1992) restates the approach proposed by Lamparelli (Lamparelli et al. 1971) of starting from the isosceles right triangle with a unit long leg in order to establish the need to expand the set of rational numbers. The document also mentions the objectives of "intuitively working the completeness of the number line," without specifying how to do it, and "creating the conditions for the student to have an overview of the numerical sets and the inclusion relationship they maintain."

The latter objective was emphasized by textbooks, engendering a tradition that persists to this day. The 1998 National Curriculum Standards, which are not mandatory but influence the production of textbooks, proposed the study of "situations whose solutions can not be given by rational numbers," the "identification of an irrational number as having an infinite, non-periodic decimal representation, and the location of some of them on the number line, with ruler and compass" (MEC 1998, p. 87).

7 Final Considerations

While the movement of modern mathematics resulted from the convergence of many initiatives, reactions to the movement also had multiple origins (Kilpatrick 2012). Prior to Kline's (1973) noted failure diagnosis, at the Second International Congress on Mathematical Education in 1972, the mathematician René Thom prophesied that "the movement now appears to have reached its zenith" (Thom 1973, p. 194).

By the early 1970s, some of the central ideas of the movement, such as the emphasis on algebraic structures, had lost support. But more importantly, participants in the international debate on mathematics education recognized that the process of curriculum building was much more complex than was thought a decade earlier (Howson 1973). In the words of René Thom, the promise that "thanks to the new syllabuses and new methods, the most average pupil would be able to complete his secondary studies in mathematics" was no longer sustainable (Thom 1973, p. 209). From then on, the discussion about program renewal gave way to a variety of research and efforts related to the improvement of teaching.

Modern mathematics can thus be seen as a temporary interval, during which multiple actors have bet on the international exchange as a way of building better and more effective programs. These programs, however, have assumed unique features in each context.

In France, the set of real numbers was introduced in the program as an expansion of the set of decimals, from equations and extensive calculations of approximations of their solutions. This approach was different from those proposed earlier at international events, based on the graduation and the postulate of continuity of the line or the expansion of the set of rational numbers. The French version for the real numbers approach can be understood as the product of a broad consensus between mathematicians and secondary teachers to reconcile active methods and concerns with rigor and generalization. The dissolution of this alliance, the pressures for compliance with the program, and the reductions of the modernizing ambitions are some of the possible explanations for the abandonment of a curricular proposal whose experimentation had been considered successful.

In Brazil, in a context of deregulation, the most widespread versions of modern mathematics, and those that produced the greatest impact in schools, were those published in textbooks. So-called modern textbooks presented different versions for the introduction of real numbers. The one that persisted from the 1980s, was the description, without discussion, of the set of real numbers as the coordinates of the points of the line, continuity being an implicit assumption. Compatibility with entrance exams at the university, which are composed of multiple-choice questions, and the acceptance by users of books—teachers and secondary school students—are some of the explanations for the prevalence of this simplified presentation of real numbers.

The modern mathematics movement lent legitimacy to the curricular innovation projects carried out in several countries, but its accomplishment and outcome were the work of the local actors. Apart from the widespread evaluation of a "failure of modern mathematics," the legacies of the reforms are unique, especially as far as the memories of the experiments are concerned. In relation to the introduction of real numbers in the first cycle of secondary education, the final effects of the changes are similar, for, in France and Brazil, the discussion about the nature of real numbers was deferred to more advanced levels of schooling. This apparent convergence, however, did not result from an international debate about the programs, but rather from various adaptations to the demands of preparation for higher education and the pressures for the democratization of secondary education.

Since the late 1980s, international assessments have pushed systems toward new convergences, as they compare the performance of students from different countries under the same criteria. How do these international pressures and local actions intersect and what effects does this combination produce? This question, like those regarding the modern mathematics movement, deserves further research by mathematical educators.

References

Anguenot, Bernard (1971). Le point de vue de Lyon [The Lyon point of view]. *Recherches Pédagogiques, 48*, 22–23.

APMEP (Association des Professeurs de Mathématiques de l'Enseignement Public). (1971). Les décimaux et les réels [The decimals and the real numbers]. *Bulletin de l'APMEP*, 279, 261.

APMEP (Association des Professeurs de Mathématiques de l'Enseignement Public). (1972). Les données sur un problème: le programme de la classe de quatrième [Facts about a problem: the program of the *quatrième* class]. *Bulletin de l'APMEP*, 284, 645–679.

Barbazo, Eric, & Pombourcq, Pascale (2010). *Cent ans d'APMEP* [One hundred years of APMEP]. Paris: APMEP.

Bastos, Maria Helena Camara, & Ermel, Tatiane de Freitas (2014). Ritos de passagem, classificação e mérito: os exames de admissão ao Ginásio (1930 a 1961) [Rites of passage, classification and merit: entrance exams to *ginásio*]. In N. Dallabrida & R. F. de Souza (Eds.), *Entre o ginásio de elite e o colégio popular: estudos sobre o ensino secundário no Brasil* (1931–1961) (pp. 115–159). Uberlândia: Edufu.

Becker, C. (1971). Le point de vue de Bordeaux [The Bordeaux point of view]. *Recherches Pédagogiques, 48*, 25–27.

Begle, Edward Griffith (1964). L'algèbre avant l'entrée à l'université [Algebra before entering university]. In H. F. Fehr (Ed.), *Mathématiques modernes – guide pour enseignants* (pp. 30–33). Paris: OCDE.

Belhoste, Bruno (1997). L'histoire de l'enseignement mathématique au collège et au lycée [The history of mathematical education in secondary education]. In P. Legrand (Ed.), *Les Maths en collège et en lycée* (pp. 368–387). Paris: Hachette.

Brailly, Anne (1971). Conclusion. *Recherches Pédagogiques, 48*, 27.

Bronner, Alain (2000). Analyse des contraintes et degrés de liberté de l'enseignant de mathématiques en formation initiale des professeurs [Analysis of the constraints and degrees of freedom of the mathematics teacher in initial teacher training]. *Tréma, 18*, 1–9.

Búrigo, Elisabete Zardo (2009). Modern Mathematics in Brazil: the promise of democratic and effective teaching. *The International Journal for the History of Mathematics Education, 4*(1), 29–42.

Búrigo, Elisabete Zardo (2015). Lucienne Félix and Osvaldo Sangiorgi: Interchanges between two Bourbakist militants in the 1960s. In K. Bjarnadóttir, F. Furinghetti, J. Prytz, & G. Schubring (Eds.), *"Dig where you stand"* 3. *Proceedings of the third conference on the history of mathematics education* (pp. 93–107). Uppsala: Uppsala University.

Chayé, Jacques, & Gauthier, René Louis (1972). La rigueur et l'enseignement des mathématiques dans le premier cycle [Rigor and the teaching of mathematics in the first cycle of secondary education]. In INRDP. *Mathématiques en 3e: bilan d'une expérimentation dans le 1er cycle* (pp. 60–66). Paris: INRDP.

Chervel, André (1988). L'histoire des disciplines scolaires. Réflexions sur un domaine de recherche [The history of school disciplines. Reflections on a research field]. *Histoire de l'Éducation, 38*, 59–119.

Chevallard, Yves (1985). Le passage de l'arithmétique à l'algébrique dans l'enseignement des mathématiques au collège [The transition from arithmetic to algebra in mathematics teaching in *collège*]. *Petit x, 5*, 51–94.

CEEM (Commission d'étude pour l'enseignement des mathématiques) (1967). Rapport préliminaire de la comission ministérielle [Preliminary report of the ministerial commission]. *Bulletin de l'APMEP*, 258, 246–271.

CEEM (Commission d'étude pour l'enseignement des mathématiques) (1968). Un projet de programme pour le premier cycle secondaire [A draft program for lower secondary education]. *Bulletin de l'APMEP, 263-4*, 361–370.

CEEM (Commission d'étude pour l'enseignement des mathématiques) (1971a). Programmes pour les classes de Quatrième et de Troisième. *Bulletin de l'APMEP*, 278, 203–208.

CEEM (Commission d'étude pour l'enseignement des mathématiques) (1971b). *Compte rendu analytique de la réunion plénière du 1er février* 1971 [Summary record of the plenary meeting of 1 February 1971]. Unpublished document.

CNE (Conselho Nacional de Estatística). *Anuário estatístico do Brasil – 1961* [Statistical Yearbook of Brazil - 1961]. Rio de Janeiro: CNE.

Cunha, Luiz Antonio (1980). *Educação e desenvolvimento social no Brasil* [Education and social development in Brazil]. Rio de Janeiro: Francisco Alves.

D'Ambrosio, Beatriz Silva (1991). The modern mathematics reform movement in Brazil and its consequences for Brazilian mathematics education. *Educational Studies in Mathematics, 22*(1), 69–85.

Dassie, Bruno Alves, Costa, Letícia Maria Ferreira da, & Carvalho, João Bosco Pitombeira Fernandes de (2015). A trajetória de Dom Ireneu Penna e suas escolhas como educador matemático [The trajectory of Dom Ireneu Penna and his choices as a mathematical educator]. *Zetetiké, 23*(44), 395–410.

Dehame, Edouard (1970). Les réels en Quatrième [The real numbers in *quatrième* class]. *Bulletin de l'APMEP, 275-6*, 455–465.

Dehame, Edouard (1971). Le point de vue de Poitiers [The Poitiers point of view]. *Recherches Pédagogiques, 48*, 14–21.

D'Enfert, Renaud (2010). Mathématiques modernes et méthodes actives: les ambitions réformatrices des professeurs de mathématiques du secondaire sous la Quatrième République [Modern Mathematics and active methods: the reforming ambitions of secondary mathematics teachers under the Fourth Republic]. In R. d'Enfert & P. Kahn (Eds.), *En attendant la réforme: disciplines scolaires et politiques éducatives sous la Quatrième République* (pp. 115–129). Grenoble: Presses Universitaires de Grenoble.

D'Enfert, Renaud, & Gispert, Hélène (2011). Une réforme à l'épreuve des réalités: le cas des "mathématiques modernes" au tournant des années 1960–1970 [A reform put to the test of realities: the case of "modern mathematics" at the turn of the years 1960–1970]. *Histoire de l'éducation*, 131, 27–49.

Fehr, Howard Franklin (1964). Sommaire et conclusions [Summary and conclusions]. In H. F. Fehr (Ed.), *Mathématiques modernes – guide pour enseignants*. Rapport d'une Session d'Étude Internationale sur les nouvelles méthodes d'enseignement des mathématiques, Athènes, 17–23 Novembre 1963 (pp. 236–249). Paris: OECD.

Fehr, Howard Franklin (1967). Mathematics instruction. In UNESCO, *New trends in mathematics teaching, vol. 1. Prepared by the International Commission on Mathematical Instruction* (pp. 32–84). Paris: UNESCO.

Fehr, Howard Franklin, Camp, John, & Kellog, Howard (1971). *La revolucion en las matematicas escolares (segunda fase)* [The revolution in school mathematics (second phase)]. Washington: OAS.

Félix, Lucienne (1986). *Aperçu historique sur la Commission internationale pour l'étude et l'amélioration de l'enseignement mathématique* (CIEAEM) [Historical background on the International Commission for the Study and Improvement of Mathematical Education]. Bordeaux: IREM de Bordeaux.

Freire, Inês Angélica, & Dias, André Luís Mattedi (2010). Seção Científica de Matemática do CECIBA: propostas e atividades para renovação do ensino secundário de matemática (1965–1969) [Scientific Section of Mathematics of CECIBA: proposals and activities for the renewal of secondary education in mathematics (1965–1969)]. *Bolema, 23*(35B), 363–386.

Furinghetti, Fulvia (2014). History of international cooperation in mathematics education. In A. Karp & G. Schubring (Eds.), *Handbook on the history of mathematics education* (pp. 543–564). New York: Springer.

GEEM (Grupo de Estudos em Ensino de Matemática) (1965). Assuntos mínimos para um moderno programa de matemática para o ginásio [Minimum Topics in a Modern Mathematics Curriculum for *Ginásio* and *Colégio*]. In GEEM. *Matemática moderna para o ensino secundário* (pp. 89–95). São Paulo: IBECC.

Gimeno Sacristán, José (1998). *O currículo: uma reflexão sobre a prática.* Porto Alegre: Artes Médicas.

Gispert, Hélène (2010). Rénover l'enseignement des mathématiques, la dynamique internationale des années 1950 [Renovating mathematics education, the international dynamics of the 1950s]. In R. d'Enfert & P. Kahn (Eds.), *En attendant la réforme: disciplines scolaires et politiques éducatives sous la Quatrième République* (pp. 115–129). Grenoble: Presses Universitaires de Grenoble.

Gispert, Hélène (2014). Mathematics education in France: 1800–1980. In A. Karp & G. Schubring (Eds.), *Handbook on the History of Mathematics Education* (pp. 229–240). New York: Springer.

Goodson, Ivor F. (1988). Studying curriculum. In I. F. Goodson, *The making of curriculum: collected essays* (pp. 12–24). London; New York; Philadelphia: Falmer Press.

Gouveia, Aparecida Joly, & Havighurst, Robert James (1969). *Ensino médio e desenvolvimento* [High school and development]. São Paulo: Melhoramentos.

Howson, Albert Geoffrey (1973). A congress survey. In A. G. Howson (Ed.), *Developments in Mathematical Education*, Proceedings of the Second International Congress on Mathematical Education (pp. 1–74). Cambridge: University Press.

IBGE (Instituto Brasileiro de Geografia e Estatística). *Anuário estatístico do Brasil – 1972* [Statistical Yearbook of Brazil – 1972]. Rio de Janeiro: 1972.

Karp, Alexander, & Schubring, Gert (Eds.) (2014). Handbook on the history of mathematics education. New York: Springer.

Kilpatrick, Jeremy (2012). The new math as an international phenomenon. *ZDM - The International Journal on Mathematics Education,* 44(4), 563–571.

Kline, Morris (1973). *Why Johnny can't add: the failure of the New Math.* New York: St. Martin's Press.

Lamparelli, Lydia Condé, Canton, Adolpho Walter, Morettin, Pedro Alberto, & Indiani, Dalva Fontes (1971). *Matemática para o ginásio* [Mathematics for *Ginásio*]. Vol. 3. São Paulo: Edart.

Marques, Alex Sandro (2005). *Tempos pré-modernos: a matemática escolar dos anos 1950* [Premodern times: the 1950s school mathematics. Master thesis, São Paulo: Pontifícia Universidade Católica de São Paulo.

MEC (Ministério da Educação e Saúde) (1952). Portaria nº 1.045, de 14 de dezembro de 1951 [Ordinance No. 1,045, of December 14, 1951]. *Diário Oficial, 45,* 6–9.

MEC (Ministério da Educação) (1998). *Parâmetros curriculares nacionais*: Matemática [National Curricular Standards: Mathematics]. Brasília: MEC.

MEN (Ministère de l'Éducation nationale) (1958). Arrêté du 31 juillet 1958 [Order of 31 July 1958]. *Bulletin Officiel de l'Éducation nationale, 32,* 2513–2518.

MEN (Ministère de l'Éducation nationale) (1960). Arrêté du 20 juillet 1960 [Order of 20 July 1960]. *Bulletin Officiel de l'Éducation nationale, 29,* 2297–2308.

MEN (Ministère de l'Éducation nationale) (1971a). Arrêté du 22 juillet 1971 [Order of 22 July 1971]. *Bulletin Officiel de l'Éducation nationale, 30,* 1872–1878.

MEN (Ministère de l'Éducation nationale) (1971b). Circulaire nº 71-370 du 22 novembre 1971 [Circular No. 71-370 of 22 November 1971]. *Bulletin Officiel de l'Éducation nationale, 45,* 2867–2916.

MEN (Ministère de l'Éducation nationale) (1972). Circulaire nº 72-220 du 30 mai 1972 [Circular No. 72-220 of 30 May 1972]. *Bulletin Officiel de l'Éducation nationale, 23,* 1562–1564.

MEN (Ministère de l'Éducation nationale) (1973). Circulaire nº 73-087 du 19 février 1973 [Circular No. 73-087 of 19 February 1973]. *Bulletin Officiel de l'Éducation nationale, 8,* 625–634.

MEN (Ministére de l'Éducation Nationale) (1978). Circulaire nº 78-392 du 16 novembre 1978 [Circular No. 78-392 of 16 November 1978]. Enseignement des Mathématiques dans les classes de quatrième et de troisième des collèges. *Bulletin officiel de l'Éducation nationale*, spécial 1, 79–82.

Morin, Christiane (1971). Construction de l'ensemble R des nombres réels [Construction of the set R of real numbers]. *Bulletin de l'APMEP, 279*, 303–316.

Nunes, Clarice (2000). O "velho" e "bom" ensino secundário: momentos decisivos [The "old" and "good" secondary education: decisive moments]. *Revista Brasileira de Educação, 14*, 35–60.

OECD (Organization for Economic Cooperation and Development). (1961). *Synopses for modern secondary school mathematics*. Paris: OCDE.

Papy, Georges (1964). Méthodes et techniques de présentation des nouveaux concepts de mathématiques dans les classes du premier cycle de l'enseignement secondaire [Methods and techniques for the presentation of new concepts in mathematics in lower secondary education]. In Howard F. Fehr (Ed.), *Mathématiques modernes – guide pour enseignants* (pp. 79–119). Paris: OECD.

Papy, Georges (1967). La géometrie dans l'enseignement des mathématiques [Geometry in mathematics education]. In UNESCO, *New trends in mathematics teaching, vol. 1. Prepared by the International Commission on Mathematical Instruction* (pp. 283–302). Paris: UNESCO.

Pierro Netto, Scipione di (1970). *Matemática na escola renovada. Curso Ginasial.* [Mathematics in renewed school. Junior high school]. Vol. 3. São Paulo: Saraiva.

Pinto, José Marcelino de Rezende, Brant, Liliane Lúcia Nunes de Aranha Oliveira, Sampaio, Carlos Eduardo Moreno, & Pascom, Ana Roberta Pati (2000). Um olhar sobre os indicadores de analfabetismo no Brasil [A look at the indicators of illiteracy in Brazil]. *Revista Brasileira de Estudos Pedagógicos, 81*(199), 511–524.

Prost, Antoine (2004). *Histoire de l'enseignement et de l'éducation en France* [History of teaching and education in France]. Tomme IV. L'École et la Famille dans une societé en mutation (depuis 1930). Paris: Perrin.

Revuz, André (1972). Avant-propos [Foreword]. In INRDP. *Mathématiques en 3e: bilan d'une expérimentation dans le 1er cycle* (pp. 5–7). Paris: INRDP.

Sangiorgi, Osvaldo (1968). *Matemática: curso moderno* [Mathematics - Modern Course]. Vol. 3. São Paulo: Nacional.

São Paulo. Secretaria da Educação. Centro de Recursos Humanos e Pesquisas Educacionais "Prof. Laerte Ramos de Carvalho". (1973). *Guias Curriculares Propostos para as Matérias do Núcleo Comum do Ensino do 1° Grau* [Proposed Curricular Guidelines for School Common Subjects]. São Paulo.

São Paulo. Secretaria da Educação. Coordenadoria de Estudos e Normas Pedagógicas (1992). *Proposta curricular para o ensino de Matemática – 1° grau* [Curriculum Proposal for the Teaching of Mathematics]. São Paulo: SE/CENP.

Servais, Willy, & Varga, Tamás (1971). *Teaching school mathematics*. Paris: UNESCO.

SMSG (School Mathematics Study Group) (1959). *Mathematics for high school.* First course in Algebra (part I). New Haven: Yale University Press.

Soares, Flávia dos Santos, Dassie, Bruno Alves, & Rocha, José Lourenço da (2004). Ensino de matemática no século XX – da Reforma Francisco Campos à Matemática Moderna [Mathematics teaching in the twentieth century - from the Francisco Campos Reform to Modern Mathematics]. *Horizontes,* 22(1), 7–15.

Thom, René (1973). Modern mathematics: does it exist? In A. G. Howson (Ed.), *Developments in Mathematical Education*, Proceedings of the Second International Congress on Mathematical Education (pp. 194–209). Cambridge: University Press.

Trabal, Patrick (1998). La réforme des mathématiques modernes, discours, polémiques et réalités [The reform of modern mathematics, discourses, polemics and realities]. In B. Belhoste, H. Gispert & N. Hulin (Eds.), *Les Sciences au lycée*: Un siècle de réformes des mathématiques et de la physique en France et à l'étranger (pp. 181–195). Paris: Vuibert-INRP.

Valente, Wagner Rodrigues (2004). Euclides Roxo e o movimento internacional de modernização da matemática escolar [Euclides Roxo and the international movement for the modernization of school mathematics]. In W. R. Valente (Ed.), *Euclides Roxo e a modernização do ensino da matemática no Brasil* (pp. 45–83). Brasília: UnB.

Valente, Wagner Rodrigues (2008). Osvaldo Sangiorgi, um best-seller [Osvaldo Sangiorgi, a best-selling author]. In W. R. Valente (Ed.), *Osvaldo Sangiorgi: um professor moderno* (pp. 13–41). São Paulo: Annablume, Brasília: CNPq, Osasco: GHEMAT.
Vanpaemel, Geert, De Bock, Dirk, & Verschaffel, Lieven (2012). Defining modern mathematics: Willy Servais (1913–1979) and mathematical curriculum reform in Belgium. In K. Bjarnadóttir, F. Furinghetti, J. M. Matos, & G. Schubring (Eds.), *"Dig where you stand" 2. Proceedings of the second "International Conference on the History of Mathematics Education"* (pp. 485–505). Caparica: UIED.

Author Biography

Elisabete Zardo Búrigo is Associate Professor of mathematics education at the Mathematics and Statistics Institute of the Federal University of Rio Grande do Sul, in Brazil. She graduated with a mathematics degree from the same university and received her Ph.D. in Education from the University of São Paulo in 2004. Her research interests include the history of mathematical education in primary and secondary schools, curricular policies and practices, and teacher training. In 2011, she undertook a postdoctoral internship at the then existing Service on History of Education in Paris, France. She is a member of GHEMAT (Research Group on the History of Mathematical Education in Brazil) and participates in a CAPES-COFECUB project of cooperation between France and Brazil.

The Revolution in Mathematics Education During the Meiji Era (1868–1912): A Study of the Textbooks Used to Teach Computation, Geometry and Algebra

Marion Cousin

Abstract During the Edo period, Japan was relatively closed, although there was an important intensification of cultural and scientific activities. In 1853, Commodore Perry and his black ships landed on Edo bay to force Japan to open its trade road: authorities had to soften the border control and to face the realities of a new international configuration. As a consequence, the Meiji period (1868–1912) witnessed a nationwide modernization program in Japan, including sweeping reforms in the education system. This paper offers a general view of the reform of mathematics curricula by presenting several case studies: in elementary schools, where traditional teachings based on the manipulation of computation devices (abacus or counting rods) were abandoned; in middle schools, where Euclidean geometry and its argumentative language were introduced; and in higher level education, where traditional *tenzan* 点竄 algebra was replaced with Western algebra.

Keywords History · Mathematics education · Japan · Meiji era

1 Introduction

The Edo period (1600–1868) saw an unprecedented intensification of scientific activities, in the wake of a general increase in economic and cultural activities. In this context, mathematics developed rapidly (Horiuchi 2010, pp. 3–5; Cousin 2008, pp. 6–9). As had been done before,[1] Japanese scholars turned to Chinese texts and, from the very beginning of the 17th century, mathematicians wrote textbooks on the

[1]For example, we know that the *Suanjing shishu* 算経十書 (Ten Classics of Calculation), a compilation of textbooks gathered in China during the Tang dynasty, was adopted in mathematical education intended for the imperial administration after the Taika reform (645). See Horiuchi (2010, p. 6) and Keller and Volkov (2014, pp. 64–65).

M. Cousin (✉)
Lyons Institute of East Asian Studies (ENS Lyon), Lyon, France
e-mail: cousin_marion@yahoo.fr

F. Furinghetti and A. Karp (eds.), *Researching the History of Mathematics Education*, ICME-13 Monographs,
https://doi.org/10.1007/978-3-319-68294-5_3

matter and developed several original methods.[2] By the end of the Edo period, *wasan* 和算 (Japanese mathematics) had gained great influence in several classes of Japanese society and was taught from elementary schools (*terakoya* 寺子屋) to high-level private schools (*shijuku* 私塾).

But in the second half of the 19th century, after the United States forced Japan to open its trade roads, the government decided to launch a nationwide modernization program. The country needed an advance in technology and military power in order to renegotiate the unequal treaties signed at the end of the Edo period (on historical context, see Jansen (1989)). In this context, the school system, educational contents and teaching methods were radically changed. More specifically, there was a profound transformation in the teaching of mathematics. Formal education now introduced students to Western mathematics, and the authorities ordered that *wasan* teaching be abandoned.

As *wasan* was very prominent at that time, there were no Japanese translations of texts dealing with the new Western mathematics. Following the Decree on Education (*Gakusei* 学制 1872), teachers were required to use exclusively Western knowledge, meaning that they had to integrate new types of mathematical texts and objects, new methods, and new pedagogic strategies into their teaching. Thus, they had to switch from individual teaching methods to "lecture-type" classes, or "simultaneous teaching method"—*issei kyōjuhō* 一斉教授法: the teacher addressed the entire class rather than each student in turn. According to Galan (1999, p. 217), this pedagogical method was one of the first innovations that were systematically implemented in every classroom. It also implied changes in pedagogic methods and materials in mathematics classroom.[3]

In this paper, by relying on studies I led on mathematics textbooks published during this period, I intend to illustrate that this transfer of knowledge from Europe and the United States to Japan should not be considered as a simple technical assimilation. Rather, I will highlight the complexity of this transfer by showing that its characteristics differed according to mathematical domain and educational level. To this end, I will present three case studies:

In new elementary schools (*shōgakkō* 小学校, for students between 6 and 14 years old), teachers had to abandon use of the abacus (*soroban* 算盤) in the teaching of elementary operations for the exclusive use of computation on paper ("brush computation" *hissan* 筆算).

In middle schools (*chūgakkō* 中学校, for students between 14 and 19 years old), the teaching of Western geometry was introduced, even though Japanese

---

[2]Horiuchi's body of work gives a general view of how wasan developed. For the period until the first half of the 18th century, see Horiuchi (2010) and for an overview of the later period, see Mikami and Smith (1914, pp. 163–253), and Horiuchi (1998).

[3]For general analysis of mathematics education during the Meiji era, see Ogura (1974) and Matsubara (1987). In English, Ueno (2012) gives a good overview of the situation and the reader can also refer to the report on mathematical education given by Kikuchi (1909, see paragraph 2.2) at the University of London.

mathematicians had never reasoned in such a way: deductive reasoning and argumentative discourse were completely new for *wasan* mathematicians.

At higher education levels, although there was a long *wasan* tradition in algebraic practices, teachers had to introduce students to new methods and their notations imported from the West.

To perform the three case studies mentioned above, we relied upon Japanese studies describing the curricula of schools (Matsubara 1987; Neoi 1997) and established a representative corpus of textbooks used in mathematics teaching. We investigated the context in which these textbooks were written and used (mathematical education and careers of authors, sources used to create the content, reception in the community of teachers, tools used to teach them, etc.), and analyzed their form and content. Regarding elementary schools, we especially focused on the question of computation devices used in the classroom and, for the other two case studies, we took a special interest in language. In this paper, I will underline the characteristics of each case and the evolution of the situation until the beginning of the 20th century.

2 The Teaching of Computation in Elementary Schools

Before the Decree on Education, computation techniques taught in schools were based on *wasan* methods from the elementary level (in the *terakoya*) up to top-level private schools (*shijuku*). These methods favored the use of the abacus, especially in elementary and every day computation, and counting rods, especially in *tenzan* 点竄 computation, methods based on the Chinese *tianyuan* 天元 [celestial element, pronounced *tengen* in Japanese] method, that would today be classified as algebraic (on this research field of the *wasan*, see Horiuchi (2010), and on the *tianyuan* method, see for example Martzloff (2006, pp. 258–271). As part of a plan to westernize scientific education, the devices used in the *wasan* tradition were completely abolished by the decree, as all textbooks had to introduce Western computation. Moreover, in all algebra textbooks written after 1872 (see the last part of this paper), there is no trace of counting rods, a sophisticated device especially used in algebraic problem-solving.

The *Jinkōki* 塵劫記 was written by Yoshida Mitsuyoshi 吉田光由 (1598–1672) and published for the first time at the very beginning of the Edo period, in 1627. It gives a good idea of the content of mathematics teaching in *terakoya*, as it was a successful textbook used in most of these schools. During the Edo period, the government had no official control on the curricula of schools but historians have shown that the *Jinkōki* content is representative of the content of elementary teaching (Horiuchi 2014, pp. 166–169). Its content reveals that abacus computation was used to teach the rudiments of mathematics learning, such as elementary operations (addition, subtraction, multiplication and division). Students had to memorize the tables and practice computing on the abacus, but there was no attempt to explain to them why they had to perform operations this way.

Regarding the first Japanese textbooks of Western arithmetic, just before the Meiji Restoration, an author introduced students to Western arithmetic (*sanjutsu* 算術) by using *wasan* devices for computation: in *Yōsan yōhō* 洋算用法 [Rules for the use of Western mathematics], written by Yanagawa Shunsan 柳河春三 (1832–1870) in 1857, methods and notations for paper computation are explained using *wasan* devices. For example, *tenzan* notations are employed to translate Western basic operations. On Fig. 1, division is explained using *tenzan* notations: the Japanese kanji that signifies "division", *jo* 除 is followed by the sign "÷", and its pronunciation in Dutch "verdeling" is transcribed into *katakana.*[4] Then follows its inscription using *tenzan* notation (a representation of a vertical rod, with the kanji *jitsu* 実 —operand, dividend—on its right, and the kanji *hō* 法 —operator, divisor—on its left). Finally, the sentence "*Jitsu* is divided by *hō*." is given as an explanation for the process. In other words, this "definition" is a complete translation of the Dutch term and symbol into Japanese, employing traditional *wasan* notation. Following the Decree on Education, modernizers especially criticized the use of the abacus because of its dogmatic form of learning: procedures were applied automatically by students, and this didn't fit with Westernized ideas of mathematical education that had to be introduced in Meiji schools. Thus was the use of the abacus abolished at that time: in the textbooks designated in government directives,[5] *Hissan kunmō* 筆算訓蒙 [Elementary learning of brush computation][6] written by Tsukamoto Neikai 塚本寧海 (1833–1885) in 1869 and *Yōsan hayamanabi* 洋算早学 [Elementary study of Western mathematics] written by Yoshida Yōtoku 吉田庸徳 (1844–1880) in 1872, only Western arithmetic was to be taught and traditional devices and notations had disappeared from mathematical texts.

Nevertheless, teachers (who were, at first, teachers of the former *terakoya*) had little training in Western arithmetic and were still using the abacus in classrooms. Furthermore, the traditional device remained used by the population. As a consequence, the authorities decided to officially reinstate its use in elementary schools. Fujisawa Rikitarō 藤沢利喜太郎 (1861–1933), the mathematician whose textbooks stabilized arithmetical teaching, argued that arithmetic should be taught through problem-solving, as part of a Japanese cultural heritage, but that this teaching didn't have to be completely dogmatic. The studies I led on elementary textbooks showed that the use and shape of the abacus evolved during the Meiji era in order to integrate it into modernized mathematics education: for example, the *Jinkōki* (the most popular mathematics textbook of the Edo period mentioned

[4]The *kanji* 漢字 are Chinese ideograms that have several pronunciations in Japanese, and there are two types of phonetic characters, the syllabary *hiragana* ひらがな (used for Japanese terms), and the syllabary *katakana* カタカナ (used for foreign terms).

[5]See the "Teaching directives for elementary schools" (*Shōgaku kyōsoku* 小学教則) of 1872 reproduced in Ogura (1974, pp. 231–232).

[6]"Brush computation", as opposed to computation with devices, was a term used to name the computation methods imported from the West, executed on paper with a brush.

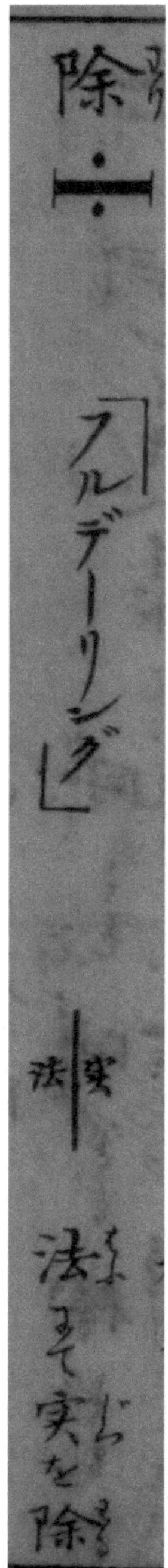

Fig. 1 Extract of Yanagawa (1857)

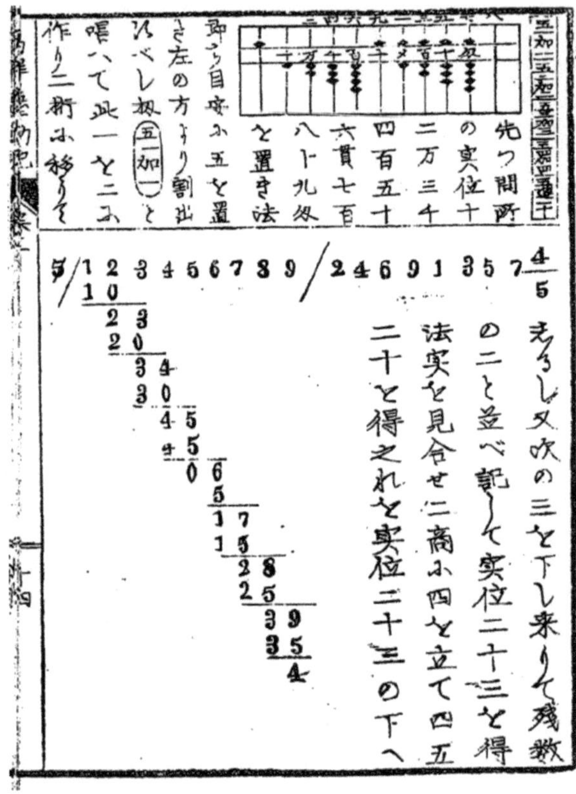

Fig. 2 Extract of Matsui (1872)

earlier) was reedited, but computation methods on the abacus were presented alongside Western methods (see Fig. 2: a page of the *Seiyō Jinkōki* 西洋塵劫記 where the division of 123 456 789 by 5 is explained with both methods).

3 Algebra and Geometry Textbooks During the Meiji Era

The situations in algebra and geometry show interesting contrast. On the one hand, in algebra, Japanese mathematicians found broad similarity between Western texts and those developed on the basis of *tenzan* research and, as a consequence, some of them wanted to use traditional terminology to translate Western texts. On the other hand, in geometry, although geometrical research was led by *wasan* scholars, deductive reasoning and argumentative discourse were completely new in Japanese mathematical culture, and the authors had to create a whole new mathematical language fitting those types of reasoning.

3.1 Studying the Mathematical Language to Understand the Contrast Between Algebra and Geometry Cases

The studies on geometry and algebra textbooks presented here allowed us to show that the creation of a modern mathematical language in Japan was a complex process that involved a reform of the Japanese language itself. Mathematical symbols, syntax, terminology and language characteristics (for example its accuracy) had to be renovated and adapted to a new vision of mathematics.

On the one hand, in algebra, some aspects of the mathematical language (for example its symbolic notations) were settled relatively fast and integrated into Japanese texts. Some historians argued that this was due to the existence of similar traditional practices. In particular, Horiuchi suggested that the "notational algebra" developed during the Edo period[7] must have "inspired and guided" Japanese translators (Horiuchi 2004, pp. 51–52). In the studies presented here, the case of algebra allowed us to show that the characteristics of mathematical language created during the introduction of Western mathematics into education and their relationship with traditional mathematics were dependent upon the domain and level of mathematical teaching: while traditional practices and terminology were preserved in elementary school arithmetical teaching, Japanese mathematicians did not try to maintain traditional notations in algebra, and the struggle of *wasanka* 和算家 (experts in traditional Japanese mathematics—*wasan*) for the preservation of traditional terminology linked with *tenzan* method did not succeed. For example, in a famous discussion about the translation of the term "algebra", the *wasanka* Kawakita Tomochika 川北朝鄰 (1840–1919) argued that the traditional term *tenzan* had an old history in Japanese culture and that its (late) meaning, "reveal the secrets" (the earliest meaning of *tenzan* being probably unknown by this *wasanka*) was closer to the original meaning of "algebra" than that of the term *daisūgaku* 代数学 ("science for the replacement of numbers"), literal translation originating from more recent Chinese translations. But as was often the case in algebra, the recent Chinese translation was chosen, mostly because it had already spread in Japanese texts.[8]

On the other hand, in geometry, the fundamental concepts of Euclidian construction (for example *theorem, definition* or *proof*) were completely new to Japanese mathematical culture, as axiomatic-deductive reasoning and argumentative discourse had no equivalents in *wasan*. As a consequence, these concepts were hard to integrate into Japanese culture: the studies I led on the terminology used in geometry textbooks showed that, during the 1870s, there were at least 7 terms used to translate *theorem* and that the terminology associated with fundamental concepts of Euclidian architecture was not set yet at the end of the 1880s (Cousin 2013, pp. 253–263).

---

[7]Horiuchi described the research associated to the *tenzan* method as a "notational algebra" based on the "positional algebra" developed by Chinese mathematicians (*tianyuan* method).

[8]Regarding the debates about the translation of "algebra", see Horiuchi (2003, pp. 247–248), Ōya (1966, pp. 294–298) and the reports on these debates in issue 43 of *Tōkyō sūgaku kaisha zasshi* 東京数学会社雑誌 [Journal of Tokyo Mathematical Society], TSK (1882).

3.2 *The Evolution of Geometry Teaching in Middle Schools of the Meiji Era: A Study of the Textbooks*

At the beginning of the Meiji era, Japanese authors had little training in Western mathematics: several of them were *wasanka* who discovered Western texts just before writing their textbooks. In this context, most of the textbooks were translations of American sources, principally because the instructors who were sent to Japan to create a new curriculum for middle schools were American. Studies of these textbooks show that authors ran into difficulties when transcribing the argumentative discourse in geometric texts.

For example, in *Shōgaku kikayōhō* 小学幾何用法 [Rules of use for elementary geometry] written by Nakamura Rokusaburō 中村六三郎 (1841–1907) in 1873, the Japanese author used a complex language that hid Euclidean structures or the logical links between some assertions. Nakamura translated and published in 1870 *Elements of geometry and trigonometry with applications in mensuration* written by Charles Davies (1798–1876), which was itself a translation of the geometry of Adrien-Marie Legendre (1752–1833) (Cousin 2013, pp. 147–190)). In the American textbook, the terms "point", "line", "surface" and "solid" were defined using the same sentence structure, to highlight the logical parallel between these concepts: "The [term] is that which has [properties possessed by the object], but not [properties not possessed by the object]". But when the Japanese author translated this section, he removed the parallel: first, the point is defined with a completely different structure ("3. One calls a point that which is extremely small, (that which is) hard to measure.", Nakamura (1873, Vol. 1, p. 1a)[9]) and second, probably to avoid repetition,[10] he used three different words to translate the English verb "to have" (Cousin 2013, p. 173).

Moreover, our studies of the mathematical content of *Shōgaku kikayōhō* (see Cousin (2013, pp. 157–172)) showed that the author often broke deductive chains that were in the original source. For example, in Davies' textbook, the 15th theorem of the first volume ("THEOREM XV. All straight lines which are parallel to the same line, are parallel to each other." Davies (1870, p. 28)) appeared as a direct consequence of the corollary ("*Cor.* If a line be perpendicular to one of two parallel lines, it will also be perpendicular to the other." Davies 1870, p. 26) of the 12th theorem ("If a straight line intersect two parallel lines, the alternate angles will be equal.", Davies (1870, p. 26), as we can see in the proof (Fig. 3):

"Let the lines *AB* and *CD* be each parallel to *EF*: then will they be parallel to each other.

---

[9]Regarding the textbooks printed traditionally (*fukuro toji* 袋綴じ method), as pagination was done using the folio system (where folded leafs are counted, not pages), the first page of leaf *n* is labeled *n*a and its second page is labeled *n*b.

[10]In China already, the repetitive aspect of Euclidean geometry had been criticized. See Martzloff (2006, pp. 112–118).

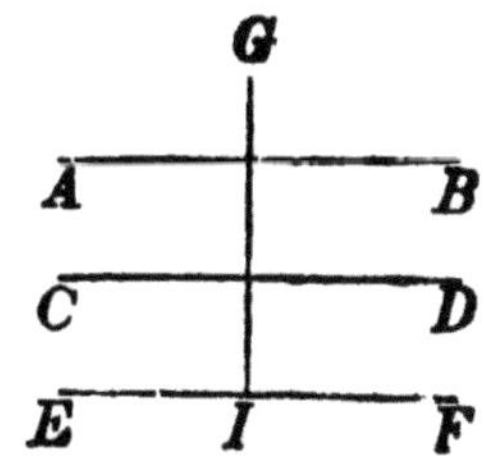

Fig. 3 Illustration of the 15th theorem of Davies (1870)

For, let the line GI be drawn perpendicular to the parallels AB, CD (**Th xii Cor.**). Then since the lines AB and CD are perpendicular to the line GI, they will be parallel to each other (**Th xii Cor.**).", Davies (1870, pp. 28–29).[11]

In his preface, Nakamura announced that, in his translation, he suppressed several parts of Davies' textbook when "sentences are […] complicated and when explanations are repeated" (Cousin 2013, pp. 147–153, where a translation of the preface is given). In this abridged translation of the American textbook, Nakamura removed the corollary of the 12th theorem and, when he translated the proof of the 15th theorem, he simply suppressed the reference to that corollary. Thus, he broke the deductive chain and made it difficult for a teacher to understand what result is used to prove the 15th theorem:

"Theorem 15. All straight lines that are parallel to one same line are equal to each other.[12]
When the lines *RoI*[13] and *NiHa* are parallel to the line *HeHo*, they should be equal to each other.[14]
Proof: When a line *RiTo* is drawn perpendicularly to *HeHo*, this line *RiTo* should also be perpendicular to *RoI* and *NiHa*. This is why the previous lines should be parallel to each other.", Nakamura (1873, Vol. 1, pp. 16b–17a) (Fig. 4).

Studying Nakamura's textbook (which was recommended by the ministry for mathematics teaching) shows that this author probably didn't understand some important aspects of argumentation in geometry, for example the fact that, in a proof, he cannot state an assertion without proving it. But this is linked to the fact that this author had a poor education in Western mathematics (he only learned mathematics in a military context) and that he had never taught this subject before writing his mathematics textbooks (see Cousin 2013, pp. 147–148).

[11]The formatting (especially the use of bold lettering) is the one of the original text.

[12]Here, it seems that Nakamura made a mistake. He should have translated "*parallel* to each other".

[13]The *katakana* (see Footnote 4) used to name the figures are transcribed with a phonetic transcription and with a capital letter at the beginning of each *katakana* transcription.

[14]See Footnote 12.

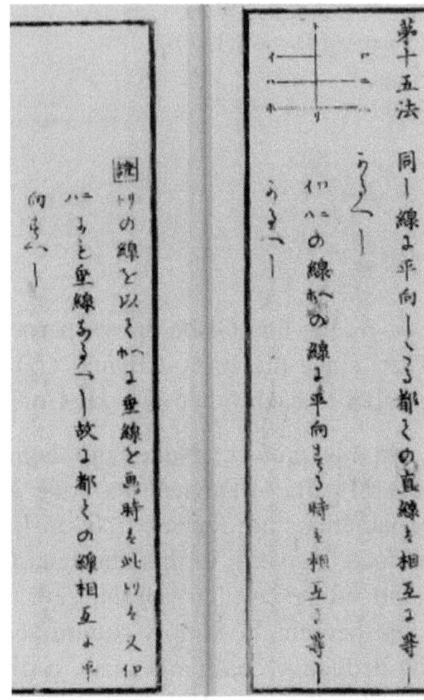

Fig. 4 Extract of Nakamura (1873)

During the 1880s, textbook authors had a better knowledge of Western mathematics and, as there were more and more discussions in the community of mathematicians on introducing Western mathematics into Japanese education, they had a better view and personal opinions on the subject. Therefore, they gained perspective on the sources they were using and started to reflect on the best ways to introduce deductive reasoning in their own textbooks. In particular, to write their textbooks, authors of the 1880s often used several books imported from America and England (see Cousin 2013, pp. 270–352). Moreover, since the end of the 1870s, authors had often tried to simplify Japanese mathematical language and to make the argumentative discourse more explicit and understandable in Japanese. For example, while some authors of the 1870s were reluctant to use repetitions (as was the convention in traditional texts, see for example Nakamura 1873), in *Kika kyōkasho* 幾何教科書 [Geometry textbook, 1882], a textbook widely used in middle schools, Tanaka Naonori 田中矢徳 (1853–?) used mathematical symbols and repetitions to simplify and normalize the mathematical language, as well as to underline logical structures. This author, who had started his studies in a *wasan* school, had a good knowledge of Western mathematics: from the beginning of the Meiji era, he studied mathematics in Kondō Makoto 近藤真琴 (1831–1886)'s school, *Kōgyokusha* 攻玉社, which had an important part in translating the Western textbooks. Moreover, he had been teaching this subject for ten years when he wrote his series of textbooks.[15]

[15]On Tanaka and his textbooks, see Cousin (2013, pp. 116–117 and pp. 276–311).

For example, let us look at the setting-out and specification of theorems in these textbooks. The setting-out (echtesis, Ἔκθεσις in the commentary of Proclus, 5th century) is the part of the proposition where the hypotheses are expressed while assigning names to the elements of the geometric figure. The specification (diorismos, Διορισμος) is the part where the conclusions of the proposition are given with those names. For example, the setting-out of the Pythagorean Theorem in the Davies' textbook is: "Let *BAC* be a right angled triangle, right angled at *A*:" (Davies 1870, p. 93); and the specification is: "then will the square described on the hypothenuse *BC,* be equivalent to the two squares described on *BA* and *AC*." Davies (1870, p. 93).

Though all the authors of the Western textbooks that were translated in Japan used a fixed sentence structure to write these parts of the text, in order to highlight the hypotheses and conclusions of theorems, most of the Japanese authors of the 1870s didn't reproduce this process. In fact, in our corpus for the 1870s, we found only one textbook, Miyagawa (1876), where the author used a fixed sentence structure for setting-out and specification. But Tanaka used a very simple and explicit structure as well as symbolic expressions to highlight the hypotheses and conclusions of a theorem: "[hypotheses expressed with symbolic expressions] -*eba*[16] [conclusions expressed with symbolic expressions] *nari*[17]" "… -eバ…ナリ" (which can be presented as: "If […], [then] […]"). For example, in proposition 4 (translation: "Theorem: If two sides and the angles they delimit are equal to each other, the other sides and the opposite angles of the equal angles are equal to each other. These two figures have the same area.", Tanaka (1882, p. 18)), Tanaka enounced setting-out and specification as follows (Fig. 5):

Explanation: "In the two triangles *ABC*, *DEF*, if $AB = DE$, $AC = DF$, $\angle BAC = \angle EDF$, [then] $BC = EF$, $\angle ABC = \angle DEF$, $\angle ACB = \angle DFE$, $\Delta ABC = \Delta EDF$." (Tanaka 1882, p. 19).

However, the first author who established real strategies with respect to geometry teaching and the introduction of logical argumentation in Japanese mathematics teaching was Kikuchi Dairoku 菊池大麓 (1855–1917), a mathematician who had a great knowledge of Western mathematics. Kikuchi studied mathematics in Cambridge University and was aware of the problems related to the introduction of Western mathematics into Japanese culture, as he participated in the establishment of a new modernized curriculum in Tokyo University for several years. Moreover, he occupied many high positions (among others, president of Tokyo University from 1898 to 1901 and Minister of Education from 1901 to 1903) that allowed him to play an important role in educational reforms. For geometry, he thought that language should be simplified and normalized to underline the logical structure but, contrary to Tanaka, he believed that mathematical symbols should not be used to

[16]The *–eba* form (conditional form) is used at the end of the setting-out.

[17]*Nari* is the conjunctive form of the verb *naru* (be, become).

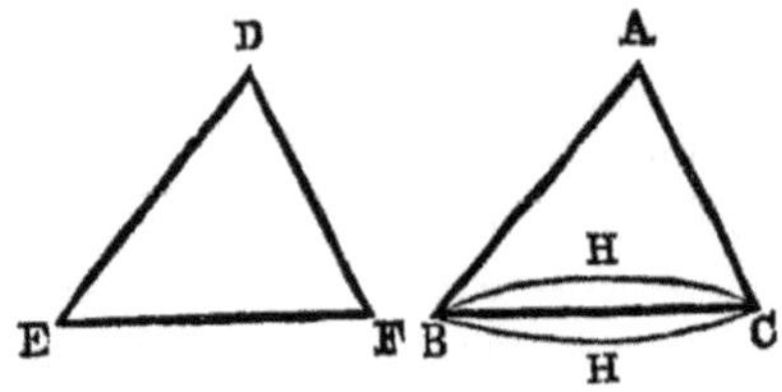

Fig. 5 Figure associated to proposition 4 in Tanaka (1882)

that end, and that a normalized Japanese literary language had to be established. Furthermore, he highlighted the fact that the distance between the spoken and written languages was a real problem in mathematics, where accuracy of the language is very important: students were not able to exactly write down what the teacher was saying. As a consequence, in *Shotō kikagaku kyōkasho. Heimen kikagaku* 初等幾何学教科書. 平面幾何学 [Textbook for elementary geometry. Plane geometry, 1889], which was used in almost all middle schools beyond the end of Meiji era, he created new grammatical structures and expressions in order to devise a precise logical language that, for the first time in Japanese, unified oral and written expressions.[18] For example, for each proposition, the setting-out and specification where expressed thanks to the sentence "… *to seyo: … beshi.*" ...トセヨ: …可シ ("Let us assume that…: in that case, …"), and Kikuchi specified that these sentences had to be pronounced exactly as they were written.

4 The Installation of Algebra Notations and Curricula in Japanese Universities

When Western algebraic works were studied for the first time by Japanese scholars, comparisons were made between the polynomial notations of the *wasan* tradition (*tenzan* research) and Western mathematical symbols, and traditional notations were used to explain the new writing. There were also debates in the community of mathematicians on the compared values of both traditions, in particular at the Tokyo Mathematics Society (*Tōkyō sūgaku kaisha* 東京数学会社), founded in 1877. Nevertheless, mathematicians were convinced that Western mathematics was the base of all scientific and technical knowledge (Horiuchi 2004, p. 49). As a consequence, in the minds of Japanese mathematicians, even while defending the *wasan* tradition, knowledge of the theories and notations of Western mathematics was necessary to access other scientific and technical works. Therefore, they had to be adopted and taught in all schools.

This is probably the reason why the position of Japanese authors regarding Western mathematical symbols was very different from that of Chinese authors who

[18]For an analysis of Kikuchi's language in his geometry textbooks, see Cousin (2013, pp. 539–570).

wrote works on Western algebra. In China, for the translations of European algebraic treatises realized during the 19th century by Protestant missionaries and Chinese mathematicians, authors used *tianyuan* notations and developed series of Chinese symbols to transcribe the symbols used by European mathematicians. Moreover, Latin letters were replaced by special ideograms used in Chinese mathematical tradition. Western symbols were transformed to take on a "Chinese aspect". On the contrary, in Japan, all authors decided to integrate Western mathematical symbols in Japanese textbooks as they were, and almost all of them used Latin letters in mathematical texts: in the whole corpus of textbooks that we considered, only two authors of geometry textbooks used Japanese characters to transcribe Latin letters.[19]

Even if Western symbols were instantly incorporated into mathematical text, it took a long time for them to blend into Japanese texts, because of the characteristics of the traditional layout. In particular, formulas didn't fit with vertical writing. Authors wrote formulas apart from the text, included them in the vertical text horizontally, or at a 90-degree angle that forced readers to turn the book to read them. Moreover, book history and printing history had an impact there, as typography progressively replaced xylography during that time. For example, in xylographic texts, the authors could choose any of the solutions quoted above. But in typographic texts, which became more and more common during the 1880s, it was harder to include formulas horizontally in the vertical text, as they had to be treated as pictures. Regarding the layout of typographic mathematical texts showed that, during this period, most authors chose to turn the formulas by 90 degrees, probably because publishers encouraged them to do so (as in xylographic texts, this was still not the standard). At the end of the 1880s, to solve these problems, mathematicians decided to write all mathematical text horizontally.[20] If these questions could seem insignificant, the evolution of the shape of mathematical text was important for the comprehension of textbooks and it determined the shape of modern mathematical (and scientific) text: for example, horizontal writing is the general standard for scientific texts since that period.

In 1877, the University of Tokyo was created by merging two institutions (mainly devoted to Western learning) where foreign teachers taught high level classes centered on sciences, in English.[21] As a consequence, the first teachers in Tokyo University were mainly foreigners (the few Japanese professors hired being

[19]In Nakamura (1873), Latin letters were transcribed using the *katakana* syllabary (see Nakamura's text studied above) and in Endō (1883), they were replaced by special *kanji* (see Footnote 9).

[20]On the apparition of horizontal writing in Japanese mathematics books, see Ogura (1974, pp. 251–253) and Cousin (2013, pp. 456–459).

[21]Concerning the curricula of these institutions and the establishment of the University of Tokyo (especially its mathematics curriculum), see Cousin (2013, pp. 381–389).

those who had studied abroad) and the classes were dispensed in English. For mathematics, Kikuchi was initially supported by two foreign professors[22] to establish the first curricula, but from 1879, the young scholar (he was only twenty-two when the University was created and had just come back from England) had to manage the teaching of mathematics on his own. Thus, the first mathematics curriculum in Tokyo University was highly inspired by English education, even though Kikuchi didn't exactly follow the pedagogic approach used in Cambridge (where he had studied previously). The innovative mathematical research developed in continental Europe had no impact on this teaching; for example, there were almost no classes on algebra in the curriculum established by Kikuchi. But in the 1880s, the foreign staff was rapidly replaced by Japanese students coming back from their studies abroad. In particular, the return of Fujisawa (quoted in the first section of this article) from Germany marked the beginning of a transformation of the mathematics department (and especially its algebraic curriculum), which would later become an international research centre.[23]

5 Conclusion

Studying the evolution of teaching in several mathematical domains during the Meiji era shows the complexity of the educational evolution that occurred during this period. It is necessary to contextualize the evolution of textbooks with the cultural situation (for example, with the fact that traditional computation devices were very important in Japanese mathematical culture), with the evolution in the mathematical education of authors, with the evolution of the school system, and with the changes in teaching tools and classrooms. Moreover, comparing the situations in various fields of mathematics and at different levels of education showed that, depending on field and level, textbook authors and teachers made heterogeneous choices regarding the use of traditional teaching, the foreign models adopted as references, or the writing of mathematics. The problems encountered by authors and the questioning that emerged while writing these new mathematics textbooks show that this transfer of knowledge from Europe and the United States to Japan cannot be seen as a "simple translation" or "technical assimilation" from one well-defined culture to another.

[22]William E. Parson (1845–1905), American, and Stéphane Mangeot (?–?), French.

[23]On the emergence of a "Research tradition of Western mathematics" in Tokyo University, see Sasaki (1994, pp. 181–184).

References

Cousin, Marion (2008). *Les premiers visages du* wasan. *Etudes du* Jinkōki *(1627) de Yoshida Mitsuyoshi et du* Jugairoku *(1639) d'Imamura Tomoaki* [The first faces of the *wasan*. Studies of the *Jinkōki* (1627) by Yoshida Mitsuyoshi and of the *Jugairoku* (1639) by Imamura Tomoaki]. University Lyon I: Master thesis.

Cousin, Marion (2013). *La " révolution" de l'enseignement de la géométrie dans le Japon de l'ère Meiji (1868–1912): Une étude de l'évolution des manuels de géométrie élémentaire* [The "revolution" in Japanese geometrical teaching during Meiji Era (1868–1912): A study of the evolution of textbooks on elementary geometry]. University Lyon I: PhD thesis.

Davies, Charles (1870). *Elements of geometry and trigonometry with applications in mensuration*. New York: A.S. Barnes & Burr (first edition: 1851).

Endō, Toshisada 遠藤利貞 (1883). *Shōgaku kikagaku* 小学幾何学 [Elementary geometry]. Tōkyō: Bungakusha.

Galan, Christian (1999). Les manuels de langue au lendemain de la Restauration de Meiji. Les innovations de la période du décret sur l'éducation [Japanese language textbooks after the Meiji Restoration. Innovations during the period of the Decree on Education]. *Cipango: Cahiers d'études japonaises*, *8*, 215–257.

Horiuchi, Annick (1998). Les mathématiques peuvent-elles n'être que pur divertissement? Une analyse des tablettes votives de mathématiques à l'époque d'Edo [Can mathematics be pure entertainment? An analysis of the mathematical votive tablets of the Edo period]. *Extrême-Orient, Extrême-Occident*, *20*, 135–156.

Horiuchi, Annick (2003). Kikuchi Dairoku (1855–1917), un mathématicien à l'heure de la modernization [Kikuchi Dairoku, a mathematician in the age of modernization]. *Daruma*, *12/13* (Autumn 2002/Spring 2003), 233–261.

Horiuchi, Annick (2004). Langues mathématiques de Meiji: à la recherche du consensus? [Mathematical languages of the Meiji era: in search for a consensus?] In Pascal Crozet & Annick Horiuchi (Ed.) *Traduire, transposer, naturaliser: la formation d'une langue scientifique hors des frontières de l'Europe au XIXe siècle* (pp. 43–70). Paris: L'Harmattan.

Horiuchi, Annick (2010). *Japanese Mathematics in the Edo Period (1600–1868). A study of the works of Seki Takakazu (?–1708) and Takebe Katahiro (1664–1739)* (trans. Wimmer-Zagier, S.). Basel: Birkhäuser.

Horiuchi, Annick (2014). History of mathematics education in Japan. In Alexander Karp & Gert Schubring (Eds.), *Handbook on the history of mathematics education* (pp. 166–174). New York: Springer-Verlag.

Jansen, Marius B. (1989). The Meiji Restoration. In *The Cambridge History of Japan. Vol. 5: The Nineteenth Century*, chapter 5 (pp. 308–366). New York: Cambridge University Press.

Keller, Agathe & Volkov, Alexei (2014). Mathematics education in Oriental Antiquity and Middle Ages. In Alexander Karp & Gert Schubring (Eds.), *Handbook on the history of mathematics education* (pp. 55–83), New York: Springer-Verlag.

Kikuchi, Dairoku 菊池大麓 (1889). *Shotō kikagaku kyōkasho. Heimen Kikagaku* 初等幾何学教科書. 平面幾何学 [Textbook for elementary geometry. Plane geometry]. Tokyo: Monbushō henshūkyoku.

Kikuchi, Dairoku (1909). *Japanese Education. Lectures delivered in the University of London*. London: John Murray.

Martzloff, Jean-Claude (2006). *A history of Chinese mathematics*. Verlag-Berlin-Heidelberg: Springer.

Matsubara, Gen'ichi 松原元一 (1987). *Nihon sūgaku kyōikushi* 日本数学教育史 [History of mathematical teaching in Japan]. 4 Vols. (1982, 1983, 1985, 1987), Tokyo: Kazama shobō.

Matsui, Koretoshi 松井惟利 (1872). *Seiyō Jinkōki* 西洋塵劫記 (Western Jinkōki). Tokyo: Aoyamu seikichi.

Mikami, Yoshio & Smith David E. (1914). *Japanese mathematics*. Chicago: The Open Court Publishing Company.

Miyagawa, Honzen 宮川保全 (1876). *Kika Shinron* 幾何新論 (New discourse on geometry). Tokyo: Ōmura chōei.

Nakamura, Rokusaburō 中村六三郎 (1873). *Shōgaku kikayōhō* 小学幾何用法 [Rules of use for elementary geometry]. Tokyo: Chūgai dōbotsuda.

Neoi, Makoto 根生誠 (1997). Meiji ki chūtō gakkō no sūgaku kyōkasho ni tsuite 明治期中等学校の数学教科書について [On the mathematical textbooks used in the middle schools of the Meiji era]. *Sūgakushi kenkyū* 数学史研究, 152, 26–48.

Ogura, Kinnosuke 小倉金之助 (1974). Sūgaku kyōiku no rekishi 数学教育の歴史 [History of mathematical education). In *Ogura Kinnosuke chosakushū.* 小倉金之助著作集 [Selected works of Ogura Kinnosuke], Vol. 6. Tokyo: Keisō shobō.

Ōya, Shin'ichi 大矢真一 (1966). Meiji jidai ni okeru sūgaku yōgo shū no kenkyū 明治時代における数学用語集の研究 [Research on mathematics vocabulary during the Meiji period). *Fuji ronsō* 富士論叢, (Vol. 11, pp. 289–313]. Tokyo: Fuji tanki daigaku shuppanbu.

Sasaki, Chikara (1994). The adoption of Western mathematics in Meiji Japan, 1853–1903. In C. Sasaki, S. Mitsuo, & J. W. Dauben (Eds.), *The intersection of history and mathematics* (pp. 165–186). Basel-Boston-Berlin: Birkhäuser Verlag.

Tanaka, Naonori 田中矢徳 (1882). *Kika kyōkasho* 幾何教科書 [Geometry textbook]. Tokyo: Shirai Renichi.

Tōkyō sūgaku kaisha東京数学会社 (TSK, Tokyo Mathematical Society) (Ed.). (1882). *Tōkyō sūgaku kaisha zasshi* 東京数学会社雑誌 [Journal of Tokyo Mathematical Society]. Tokyo: Tsuchiya Chūbei, Number 43.

Tsukamoto, Neikai 塚本寧海. (1869). *Hissan kunmō* 筆算訓蒙 [Introduction to brush computation] Numazu: Numazu gakkō.

Ueno, Kenji (2012). Mathematics teaching before and after the Meiji Restoration. *ZDM The International Journal on Mathematics Education*, *44*(4), 473–481.

Yanagawa, Shunsan 柳河春三. (1857). *Yōsan yōhō* 洋算用法 [Rules for the use of western computation]. Tokyo: Yamatoya.

Yoshida, Yōtoku 吉田庸徳. (1872). *Yōsan hayamanabi* 洋算早学 [Elementary study of Western mathematics]. Tokyo: Kobayashi Kiemon.

Author Biography

Marion Cousin is a post-doctoral researcher supported by the D. Kim foundation. She did a Ph.D. on history of mathematics in Japan. In her thesis, she focused on geometry textbooks written during the Meiji period (1868–1912) to highlight the complexity of the modernization process that characterized this period. After finishing her Ph.D., she obtained three fellowships to carry on her studies on the history of mathematics teaching during the Meiji period and to use her research in mathematics education.

Early Experiments with Modern Mathematics in Belgium. Advanced Mathematics Taught from Childhood?

Dirk De Bock and Geert Vanpaemel

Abstract During the early phases of the *New Math* reform, several series of classroom experiments were set up in Belgium by leading figures of the reform movement. Although there are no primary data documenting the experiments, some information did appear in scholarly publications. Moreover, the experiments were often referred to as powerful arguments in favour of the reform. We discuss the experiments set up by the structuralist school of Georges Papy, and their reception among Belgian and international reviewers. These experiments are compared to the very different approach advocated by Paul Libois. We conclude with some observations on the role of these experiments in the *New Math* reform in Belgium.

Keywords Belgian experiments · Georges Papy · Modern mathematics
New Math · Paul Libois

1 Introduction

New Math or *modern mathematics*, as it was commonly called in Europe, refers to a drastic change in the way mathematics was taught in most European countries, in the U.S. and in various other parts of the world. The peak of *modern mathematics* is typically situated in the 1960s, but the first reform ideas were already shaped from the mid-1950s on. A leading idea was to update school mathematics to the "needs of modern society". The main model for this operation, at least in Europe, was the academic discipline of mathematics, as re-constructed and formalized from the late 1930s on by the French structuralist school, better known under the name *Nicolas Bourbaki*. Initially the reform efforts were directed to secondary school mathematics, that had to prepare students better for further higher studies in mathematics, science or engineering, but in some countries, also the primary level soon came in sight. *Modern mathematics* not only changed the content of school mathematics,

D. De Bock (✉) · G. Vanpaemel
Faculty of Economics and Business, KU Leuven, Leuven, Belgium
e-mail: dirk.debock@kuleuven.be

F. Furinghetti and A. Karp (eds.), *Researching the History of Mathematics Education*, ICME-13 Monographs,
https://doi.org/10.1007/978-3-319-68294-5_4

but also led to a modernisation of the pedagogy of mathematics. At the international level, this reform movement was officially launched at the Royaumont Seminar (1959) (OEEC 1961a), organized by the *Organisation for European Economic Cooperation* (OEEC; later joined by nations outside Europe to form the *Organisation for Economic Co-operation and Development*, OECD), and further elaborated at other international conferences, including two OEEC/OECD follow-up sessions, one in Dubrovnik (1960) and the other in Athens (1963), focusing, respectively, on the reform of mathematics curricula (OEEC 1961b) and on teacher education (OECD 1964). These meetings provided valuable recommendations and guidelines for the actual implementation of *modern mathematics*, but individual countries, having different educational traditions and systems, had to develop their own new programs and instructional materials, including textbooks, and had to re-educate all their mathematics teachers.

Although the *New Math* phenomenon was the subject of many recent studies (see, e.g., Barbin 2012; Bjarnadóttir 2013; Kilpatrick 2012; Phillips 2015), dealing with various aspects of the reform, the way different countries prepared themselves to its coming in the classroom still remains underexposed in the literature on the history of mathematics education. This is particularly the case for Belgium which was, as well as France and the U.S., one of the forerunning and leading countries in this international movement. Already at Royaumont, the Belgian Willy Servais referred to ongoing experiments in his country, enabling him to present, with some authority, a quite detailed program proposal for the teaching of algebra in a unified mathematics curriculum (OEEC 1961a; Servais 1959). At the Athens meeting, Servais came up and provoked lively discussion with a complete program for *modern mathematics* for the scientific stream of the secondary school, though it was still a tentative one and in the experimental stage (Servais 1964). However, the most dominant Belgian mathematician and mathematics educator of the *New Math* era was Georges Papy. Papy started experimenting in the late 1950s and his experiments lasted for more than ten years. As president of the *Commission Internationale pour l'Étude et l'Amélioration de l'Enseignement des Mathématiques* (CIEAEM)/*International Commission for the Study and Improvement of Mathematics Teaching* during the mid-1960s and as author of *Mathématique Moderne* (Papy 1963–1967), a groundbreaking textbook series of which the first volume was published in 1963 and since then translated in many languages, Papy was also a prominent personality at the international forum. At the Athens meeting, Papy was invited to extensively report his experimental design of teaching *modern mathematics* which he had already elaborated in great detail (Papy 1964). Despite Papy's incontestable influence on the *New Math* reform in Belgium and abroad, his ideas and strategies to modernise mathematics education in a quite radical way, were not the subject of recent scientific analyses.

In this paper we attempt to fill, at least partly, this gap in the research literature. First, our goal is to understand the early Belgian experiments better. What did they consist of and how were they related to the debates on the reform of mathematics education? Was it a homogeneous endeavour, with all actors on the same wavelength, or can we distinguish different tendencies and phases? Also, how were the experiments framed as positive demonstrations of previously defined programmes? To clarify these issues, we re-construct, primarily on the basis of documents from that period, the different *New Math* experiments of the late 1950s and early 1960s in Belgium, and discuss their context, goals and design characteristics. Our second research question relates to the reception of these experiments, first, in the international community of mathematics educators, and second, by Belgian school psychologists. Which actors were enthusiastic and which were not, and what were these actors' main arguments?

2 A First Experiment with *Modern Mathematics* in Two Schools for Future Kindergarten Teachers

The idea to start experimenting with *New Math* ideas in schools, at least in Europe, seems to have originated within the CIEAEM community. This may not surprise as several "founding members" of the CIEAEM (for example, Gustave Choquet, Jean Dieudonné, André Lichnerowicz and Willy Servais) were members of, or were strongly influenced by the French structuralist school (Bernet and Jaquet 1998). Very early, the notion of "experiment" was used in connection to the formation of a new curriculum. In 1955, in a book written by six CIEAEM founders (Piaget et al. 1955), Lichnerowicz asked "for an experimental effort to introduce something of the spirit of modern mathematics into arithmetic, algebra, elementary geometry, by pointing out the structures of ring, group, and vector space and by adapting the language" (Furinghetti et al. 2008, p. 135). In August 1958, in the margin of the 12th CIEAEM meeting in Saint Andrews (Scotland, UK), the Belgians Willy Servais and Frédérique Lenger compiled the draft of a concrete program for the teaching of *modern mathematics*, which they intended to test in an "experiment" (Noël 2013).

Frédérique Lenger and Madeleine Lepropre obtained permission for already running the Servais-Lenger program in the school year 1958–1959 in two schools for future kindergarten teachers in the French speaking part of Belgium (one in Arlon and the other in Liège). In Belgium, future kindergarten teachers were trained in a special type of secondary schools, so the participants of the experiment were 15–16-year-old girls and certainly not belonging to the top streams of education for mathematics. It was however an important experiment because, as Lenger and Lepropre (1958–1959) stated:

> Not only for teachers at teacher training colleges, this experiment will be of interest, but for all mathematics teachers, because it is likely to influence, in the relatively near future, mathematics teaching in all sections of secondary education (p. 70).[1]

[1]All translations were made by the authors.

Lenger's and Lepropre's course (three hours of mathematics per week) started with fundamental notions from set theory, related to the genesis of natural numbers, and from topology as a basis for the study of geometry (Lenger and Lepropre 1959). According to Lenger and Lepropre, these notions were, at least to a certain extent, innate and already very young children are naturally inclined towards abstraction. To exemplify this, they reported a scene in which one of their students asked a four-years-old girl to draw her house, and the child's drawing was a *closed curve* (very similar to a Venn diagram) and the windows of the house were also *closed curves* within that first curve (see Fig. 1). So, as they argued, students' own daily-life experiences and imagination could be used as starting points for *modern mathematics* education. The role of the teacher mainly consisted in further stimulating the mental processes of the students and guiding and orienting what was after all a quite natural process of progressive abstraction.

Reasons to start with sets were both mathematical—according to Lenger and Lepropre (1959), their role in modern mathematics is beyond any discussion—but also didactical. The notion of set is an "empty" notion which can then, gradually, from the first lesson on, be "filled" with various examples of sets suggested by the students themselves. After introducing the notion of set, the course continued with set-theoretical topics such as the relations of inclusion and equality of sets, the main operations on sets (intersection, union and Cartesian product) and correspondences between sets (e.g. one-to-one correspondences). Examples were basically taken from students' experiences and from school life. The second part of the course included an arithmetic and a geometric track (treated in parallel in, respectively, one and two hours per week). In arithmetic, operations with natural numbers and properties of these operations were discussed from a set-theoretical perspective, as well as their application in the decimal system. Geometry started with some intuitive topological notions (e.g. open and closed figures, interior and exterior of a closed curve) and culminated in a study of the basic plane figures. This study was primarily based on geometrical transformations and led to the concepts of symmetry, congruence and similarity.

We found no evidence of a systematic or independent evaluation of this experiment, but for Lenger and Lepropre the main result was the encouraging lively and active response from students to the new material. If the experiment did prove one thing, it was that the *New Math* did not put off students, but on the contrary had inspired them with a "taste for mathematics". They also invited other teachers to share their experiences, to find still better ways to make the teaching methods more active.

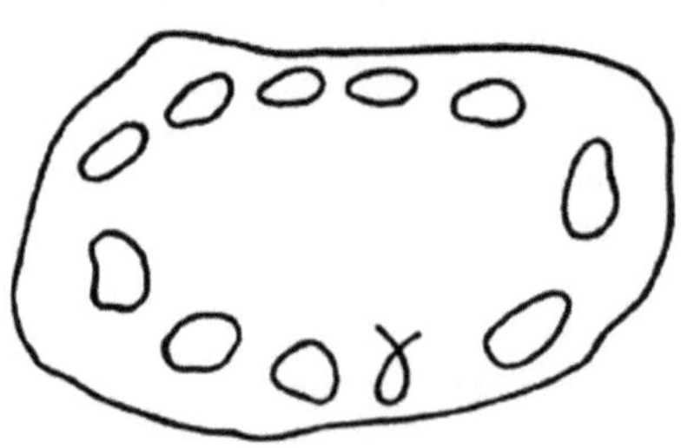

Fig. 1 Four-year-old girl's drawing of a house (Lenger and Lepropre 1959)

The experiment by Lenger and Lepropre received a lot of attention in the Belgian community of mathematics teachers. Already in July 1959, the *Société Belge de Professeurs de Mathématiques* (SBPM) [Belgian Society of Mathematics Teachers] organized the first of the *Journées d'Arlon* [Days of Arlon]—the start of a series of in-service teacher education courses aimed at mathematically and didactically preparing mathematics teachers for the coming of *modern mathematics* in their classrooms—and on that occasion, the experiment based on the Servais-Lenger program was presented and largely discussed (Papy 1959; Terfve 1968). The experiment was also brought to the attention of the international communities of mathematicians and mathematics educators. At the famous Royaumont Seminar (November 23 to December 5, 1959) Servais was invited to present his ideas about a modern and coherent approach to algebra in a unified mathematics curriculum (De Bock and Vanpaemel 2015). Servais' lecture was not published in full in the official report of the Seminar (OEEC 1961a), but we found the full text (in English) of his lecture in his personal archives (Servais 1959). Servais not only presented a detailed syllabus for a modernised approach to algebra for 12–18 year olds, but also positively referred to "some isolated tests". However, he acknowledges that further experimentation would be necessary to give the syllabus its final form:

> The syllabus proposed has not so far been tried out on any extensive scale, but isolated tests have shown that it is by no means utopian and that young people when faced with modern ideas show much greater aptitude than their reactions to traditional teaching might lead teachers to suppose. Before such a syllabus can be given its final form, it will have to be tested under real classroom conditions. Actual teaching practice will show which subjects can be assimilated quickly and thoroughly and which will need to be put off to a later date (Servais 1959, p. 4).

3 Papy's Experiments: Towards a Generalized Introduction of *Modern Mathematics* in Belgium

Servais and Lenger realized that they needed the assistance of an academic mathematician to take charge of the mathematical problems that would arise with the development of new teaching materials for follow-up experiments. Georges Papy, professor of algebra at the *Université libre de Bruxelles* (ULB), was asked to collaborate (Noël 1993). Although Papy was skeptical at first (Papy 1968), he accepted the challenge and played the game so convincingly that he soon became the uncontested leader and figurehead of the *modern mathematics* reform movement in Belgium. Papy's experiments covered a period of more than ten years and finally led to a generalized introduction of *modern mathematics* in Belgian secondary schools (from September 1968 on) and, about ten years later, in Belgian primary schools.

Papy's first classroom experiments built on Lenger's and Lepropre's work. In the school years 1959–60 and 1960–61, Papy accepted to teach himself two experimental classes in "*Berkendael*", a school for kindergarten teachers in Brussels (Fig. 2). It was Papy's first encounter with the teaching of mathematics at the secondary level (to students who moreover did not belong to the top streams of education for mathematics). One may ask why again it was chosen to conduct the experiment with 15–16-year-old future kindergarten teachers. In an article from that time, likely based on an interview with Papy, De Latil (1960) provides three arguments for the setup, whether or not premeditated: (1) These students are not likely to face competitions and examinations in which they need classical mathematics; also in their life and profession, they will have no need of the materials that were taught, it only exposed them to a "gymnastics of the mind" (an argument that also convinced the authorities of the Belgian Ministry of Education to provide the necessary permissions); (2) In these conditions, if the experiment would be successful, it would be highly convincing; (3) A future generation of kindergarten teachers is already made familiar with the new material which is, in an appropriate form, suitable for a mathematical initiation of 3–6-year olds. In "*Berkendael*", Lenger became Papy's partner, both professionally and personally (they married on October 1, 1960). Georges and Frédérique Papy (or: Frédérique Papy-Lenger) formed a complementary team, permeated by the same ideas and with a firm determination to actually implement the reform (Noël 1993).

For an active and talented university mathematician like Papy (Dieudonné 1992), teaching at "*Berkendael*" must have been a culture shock. Indeed, Papy's course (Papy 1960) looks like a tough university course for future mathematicians, rather than a course for 15–16-year-old future kindergarten teachers. Papy also started with "set theory", but although he illustrates the new concepts and methods with daily-life examples, these examples are quickly discarded and his approach becomes "formal", in the style of Bourbaki (1939), with emphasis on precise

Fig. 2 Papy and one of his students experimenting with *modern mathematics* (De Latil 1960)

definitions, axioms, logic and proofs, including proofs of "evident" properties (e.g. the reflexive, symmetric and transitive character of parallelism), and with a strong emphasis on the learning of an abstract and symbolic mathematical language based on set theory. For Papy, set theory was the ideal framework for developing a unified mathematical building and in the different chapters, he shows the unifying and illuminating role of sets in arithmetic, in Euclidean and projective (!) geometry, and in topology. The intrinsic difficulty level of Papy's course is very high, also because he consciously leaves little room for intuition (in its common sense). His approach is often deliberately counter-intuitive (e.g. straight lines are represented by points in a Venn diagram and Euclid's parallel postulate is reformulated as "The direction of every straight line is a partition of the plane").

However, Papy's structural and abstract view on mathematics was embedded in an innovative pedagogical approach. In his "*Berkendael*" experiments, Papy laid the foundation of the pedagogy that would become his trademark: colourful graphs, green-red conventions, playful drawings, "visual proofs" by means of drawings of film fragments, comics, …, all strongly appealing to the aesthetic and affective side of learning mathematics. Along with Papy's (and Lenger's) talents as teachers, this attractive and refreshing pedagogical approach undoubtedly contributed to the success of the experiment (at least according to its promoters because, once again, we did not find any evidence of a serious independent evaluation):

> Education was provided in these classes in an atmosphere of happiness. The hostility of the students towards mathematics had completely disappeared. We saw vividly that today's children are in resonance with the mathematics that is currently in use (Papy 1968, p. 29).

Papy was convinced that traditional mathematics before the age of 15 had already conditioned the students in a direction that was opposite to the spirit of *modern mathematics*. It was therefore necessary to start the reform efforts at an earlier age (Papy 1966). In May 1961, Papy published his *Suggestions pour un nouveau programme de mathématique dans la classe de sixième* (Papy 1961) and already in September 1961 a new and large-scale experiment was started in the first year of secondary school (and from then on gradually in the two subsequent years). To organise and to meticulously develop these and other upcoming reform initiatives, Papy had founded on May 24, 1961, with the cooperation of several personalities of secondary and university education, the *Centre Belge de Pédagogie de la Mathématique* (CBPM) [Belgian Center for Mathematics Pedagogy] of which he also became the chairman. The *Centre*'s goal was formulated in its *Articles* as "the study, the improvement and the reform of mathematics teaching. In particular, it will contribute to the promotion, the development and the diffusion of the teaching of modern mathematics" (Papy and Holvoet 1968, p. 133). During the school year 1963–1964 the *Centre* coordinated an experimental program of *modern mathematics* for a teachers' college, training teachers for the lower secondary level (12–15-year olds). The following school year, the experiments in the first three years of the secondary school were extended to the last three years (15–18-year olds) of the scientific study streams (and in 1967 also to the "non-scientific" streams), leading to a generalized introduction of *modern mathematics* in the first year of all Belgian secondary schools from September 1968 on (and from then on gradually in the subsequent years). But Papy was disappointed about the way "his" reform was implemented at the secondary school level and, together with Frédérique, he re-directed his interests towards the primary level in which the CBPM already had started in September 1967 an extensive experiment with two classes of six-year-old children (Papy 1970). Only in September 1976 and with little or no direct involvement of the Papy's, *modern mathematics* was introduced in the Belgian primary schools of the catholic network and two years later in the publicly run schools.

4 Libois's Experiments at the *École Decroly*: A World of Difference

In the late 1950s, Paul Libois, professor of geometry and Papy's nineteen years older colleague at the ULB, showed a renewed interest for the teaching of mathematics at the secondary level. Libois had been involved in the reform of the mathematics curriculum in secondary schools since 1945, but here we only discuss Paul Libois's involvement in mathematical classroom experiments; for more details about Libois and his earlier activities we refer to Bruffaerts (1994) or to Vanpaemel and De Bock (2017). Probably, Libois's interest at that time was rekindled by the current debate about the new "modern" mathematics programs, launched at the Royaumont Seminar (1959). That same year, Libois started a seminar on elementary mathematics at the ULB. His ideas were shaped and put into practice at the *École Decroly*, a so-called "method school" in Brussels for 3–18 year olds that followed the child-centred pedagogy of its founder Ovide Decroly (1871–1932). This pedagogy started from the child's own spontaneous activity in observing real nature in all its complexity, a process Decroly called "globalization" (Decroly 1928). The Decroly pedagogy later also strongly influenced Hans Freudenthal (1973). At the *École*, education was primarily based on an interdisciplinary approach, on societal involvement, and on active learning processes (Fig. 3). Libois worked closely together with the *École*'s secondary section, both by advising the teachers and acquainting them with new mathematical ideas, as by helping the students preparing their annual exhibitions on mathematical themes (Trompler 2009). Libois's ideas about a modernization of mathematics teaching were completely different from those of Papy. While Papy was strongly influenced by *Bourbaki*, Libois had studied in Italy under Frederigo Enriques and Guido Castelnuovo and had inherited from them a predilection for intuitive thinking in mathematics (Castelnuovo 1997; Menghini 2015). Rooted in this tradition, Libois proposed an intuitive approach to mathematics teaching, based on students' concrete experiences and actions, opposing a one-sided reliance on logic and the introduction from the beginning of an abstract theoretical framework (in this case

Fig. 3 Libois at the *École Decroly*, early 1960s (Centre d'Études decrolyennes)

the theory and language of sets), or, as formulated by his wife Lucie Libois-Fonteyne, mathematics teacher at the *École* (and later its director):

> [Our] approach is contrary to a dogmatic instruction, which under the pretension of rigour departs from "exact" definitions and "general" principles that are insufficiently supported by concrete experiences. Such an instruction leads to verbal training; it eliminates the intuitive aspects, and develops neither the intelligence nor true knowledge (Libois 1971, p. 183).

For Libois, not the abstract for its own sake, but the transition from the concrete to the abstract and the return to the concrete, were most important. The processes of idealization, abstraction and concretizing should be repeated as many times as necessary to ensure thorough understanding of mathematics (Libois 1955–6). In fact Libois's ideas were related to the advocacy of intuitive geometry (Vanpaemel and De Bock 2017), a movement in geometry education that flourished in Belgium during the 1950s and that was characterized by the role played by visualisations, perceptions, concrete materials and mental images in the generation of knowledge. The intuitive teaching methods encountered a strong opposition among Belgian mathematics teachers, who deplored the loss of Euclidean rigour. In the post-Royaumont era, the influence of the intuitive approach started to wane, at the same time when the structuralist approach of Papy was gaining supporters. It is therefore most surprising that Libois, and not e.g. Papy, was delegated to the Dubrovnik Seminar (1960) (OEEC 1961b), the sequel to Royaumont. Libois could not convince his colleagues: His proposals, which basically implied the gradual introduction and specification of the modern concepts of "set, structure and linearity" within the current programs (Libois 1960), were only partially reflected in the *Synopses for Modern Secondary School Mathematics* (OEEC 1961b)—in general, the views of the "*Bourbakist*" Gustave Choquet would be decisive for the future international development of the reform.

Still, Libois and his collaborators at the *École Decroly* continued their independent experimental efforts. They implemented a highly individual interpretation of the *Zagreb-Dubrovnik program* (OEEC 1961b) that they judged as completely in line with Decroly's pedagogical principles, principles they already applied for over fifty years for general education (Libois 1962). Basic reform ideas, such as the unity of all mathematics, the coordination between mathematics and physics teaching, and the stimulation of student's personal activity were put into practice at the *École* by organizing the mathematics curriculum around foci of interest: projects arising from reality or actuality, which evoke real problems that can be accessed by students from the age of 12 on from different perspectives and with various methods specific to different branches of mathematics, including statistics, and to other scientific disciplines. These foci of interest could vary each year depending on their appropriateness for a specific group of students. The role of the teacher in this was not to teach in the classical sense, but to assist students in constructing their understanding. The teacher participated in a joint investigation, managed the team, provided coordination and guidance, and ensured a successful completion of the project. The alternative "experiments" of Libois continued to attract the attention for several years of a number of Belgian and foreign circles—so, e.g., until the mid-1970s, Emma Castelnuovo and many young Italian mathematics teachers came regularly to Brussels to visit the mathematical exhibitions prepared by Libois's students at the *École* and at the ULB (Furinghetti and Menghini 2014; Menghini 2015) —, but "Papy'ists" qualified his approach as outdated and "waste of time".

5 Commentaries on the Belgian Experiments…

The Belgian experiments definitely attracted the attention of the international community of mathematics educators.

Modern mathematics, in particular Papy's approach was welcomed with enthusiasm in France, especially in circles of the *Association des Professeurs de Mathématiques de l'Enseignement Public* (APMEP) [Association of Mathematics Teachers of Public Education] (see also Barbin 2012). In the *Bulletin de l'APMEP*, Gilbert Walusinski evaluated Papy's experiments in glowing terms (Walusinski 1963). For him, these experiments demonstrated that it is really possible to teach *modern mathematics* in an active and non-dogmatic way, in real classes of 12–13-year olds, including students who are not necessarily gifted for mathematics. Walusinski characterised Papy's experiments as encouragements and inspiring examples for future developments in France:

> The time will come when a reconstruction of our public schools will be possible; it will not be a time of rest, but one of an action that should be fast and efficient. We must prepare for it from now. Papy is helping us. Very friendly, I want to thank him (p. 126).

In the report of a conference of the British *Association for Teaching Aids in Mathematics* (ATAM; now *Association of Teachers of Mathematics*, ATM) at which Papy had given a lecture, David Fielker was surprised about young children's receptivity for quite advanced mathematics in the spirit of *Bourbaki*. He wrote:

> Prof. Papy had taught sets to children from eight to twenty-five years old, and it was more difficult with the twenty-five year olds! Undergraduates were conditioned by the bad habits of traditional mathematics (…). Children of eight or ten were not so conditioned, and most success transpired with some fifteen-year-olds so poor in mathematics that they were uninfluenced by previous courses! (Fielker 1961, p. 48).

At that conference, Papy also put his experimental approach into practice with a "demonstration lesson" for a class of first-year grammar school girls. In that lesson, drawn from his Brussels course, the idea of *relation* as a set of pairs of points was developed. To present this abstract idea in a concrete context, the girls were represented by dots on a blackboard and various relationships between them were to be represented by arrows between these dots. Soon, the blackboard became covered with arrows, reverse arrows and loops in various colours (as shown in Fig. 2). The ideas of reflexivity, symmetry and transitivity were introduced with examples and counterexamples using this type of diagrams. But the girls were primarily struck by the strangeness of the situation and, although Papy brought his message with confidence and humour, he could not convince the audience of British teachers. At the end of the conference, David Wheeler, the secretary of ATAM, concluded with some irony:

> We had liked prof. Papy's lesson, but what came next, where did it fit in, and how could it be developed? It stood in its own right as something which educated children, but we had to learn, teach and carry out our own researches in order to see its place in the rest of mathematics. It was not necessarily dangerous to teach modern mathematics in the wrong way – we could do no worse than we were already doing! (reported by Fielker 1961, p. 52)

Of particular interest is a report by Emma Castelnuovo and a group of Italian mathematics teachers who assisted to and commented on some lessons at the *École Decroly* and at "*Berkendael*" on the occasion of a study visit to Brussels in 1964 (Castelnuovo 1965). Castelnuovo expressed little sympathy for Papy's experiment at "*Berkendael*". Castelnuovo explained that for Papy, the fundamental objective of secondary school mathematics is to develop the logical faculties of the student. Therefore, she observed, Papy always tried to escape the concrete, precisely to avoid that perception by the senses would prevail over reason. Castelnuovo raised serious doubts, both about Papy's goals ("training specialized mathematicians") and methods ("Are logical faculties really sharpened by diminishing the power of the senses?"). On the other hand, Castelnuovo showed herself quite excited about Libois's experiments at the *École*:

> Despite my concern for objectivity, I cannot hide my enthusiasm for the experiment at the *École Decroly*; this is also due to the fact that the method adopted by our working group, although it may not fit strictly within this school whose type is really unique, however has, with its methodology, many points in common and some issues are today studied simultaneously by them and by us (Castelnuovo 1965, p. 8).

6 Conclusions and Discussion

The Belgian mathematicians, especially those affiliated to Papy's CBPM, prided themselves on the performance of highly successful experiments. Yet, the experiments conducted by Belgian mathematicians at that time, were by no means "controlled" experiments (in the experimental-psychological sense). Nothing was objectively "measured". Although the methodology for experimental educational research was well known in Belgium and several pedagogical institutions carried out research programs in mathematics education, the teachers and mathematicians involved in the construction of the new curriculum did not care much for methodological concerns. The teacher-mathematician conducting the experiment took at best some notes and judged for himself whether his approach worked well or badly. The goal of the experiments was to show that teaching *modern mathematics* was possible and that students could understand the new concepts. Only in a few cases, students were also challenged to reinvest the newly learned concepts in problem solving activities. Most often, mathematicians simply pointed to the enthusiasm of the students as evidence of their experimental success. It seems as if the widespread fear of mathematics was a main concern of the teachers and the replacement of this fear by a more positive attitude was all that was required from the experiments:

> They all bring their contribution with good will, some with enthusiasm and a spirit of invention and abstraction which is extremely encouraging. [...] Only the future will be able to tell, [...], if we have engaged them on the path of love and light (Lenger and Lepropre 1959, p. 24).

From a didactical point of view, several important characteristics of the experiments should be highlighted. First, although the experimenters gave a lot of room for imagination and discovery, and for free and spontaneous discussion between the teacher and the students, and also between students, these discussions were always guided towards the mathematical concepts and representations the teacher had in mind. Students' own problem representations or visualizations, that not fitted within the prefabricated framework of *modern mathematics*, were not encouraged. Second, and this was especially the case for Lenger's and Lepropre's experiment: Their approach offered a "second" view on mathematical topics the 15–16-year-old students were already, at least to a certain extent, familiar with. Though Papy regarded this as an "infection" by traditional mathematics, the course likely functioned as an enrichment, a deepening or reframing of what these students already knew more or less.

Not surprisingly, critical comments on the "audacious" experiments carried out by Papy and his collaborators, aimed at introducing set theory and topological notions in several teachers' training colleges, were also formulated by leading school psychologists of that time (Hotyat 1960, 1961). They pointed to the lack of control and of objective measurement, necessary to build a truly experimental science. The personal experiences of the mathematicians led to a form of "experiential" (or experience-based) psychology, but did not contribute to "experimental psychology". Their proposals remained mere hypotheses as long as they were not supported by objective verification.

> Mathematics teachers have proposed many initiatives intended to improve the performance of their teaching. Their practical experience is irreplaceable because it results from the combination of a thorough knowledge of the matter and lessons suggested by their daily contact with students. But their projects are still at the stage of hypotheses as long as they have not undergone the test of objective verification (Hotyat 1961, p. 244).

However, as far as we know, there was little or no direct collaboration between the communities of mathematicians and school psychologists at that time, likely because that latter community could not offer usable alternatives. Nachtergaele (1964) formulated the reason for this difficult cooperation as follows:

> Efforts of psychologists related to the learning of mathematics have been very conscientious and often very detailed. But their conclusions necessitate significant reservations: the experiences commence from the mathematical knowledge acquired by a child of a certain age, within a particular pedagogy. So their observations give an account of the results of that pedagogy, about that specific body of knowledge.
>
> But to know what a child of 12 can assimilate, for example, about sets and relations, it should have received a pedagogically valid education on these subjects. The psychologist can then take suitable tests, prepared in collaboration with mathematicians, to measure the results obtained.
>
> But until now, studies that meet those requirements are missing. So, we must be temporarily satisfied with empirical evidence, gleaned from the teaching itself, and at the same time hope for a closer collaboration between psychologists and mathematicians (p. 84).

Finally, it should be noted that both Papy and Lenger were described by contemporaries as being exceptionally good teachers (see, e.g., Vredenduin 1967). They often gave demonstrations of experimental classes before audiences of colleagues, who never failed to be impressed by the ease with which they could motivate students of all ages and backgrounds. Their experiments could and did convince teachers that the *New Math* curriculum would make their job easier, or at least, more pleasant. It is to be doubted, however, how many of the Belgian mathematics teachers could ever duplicate their experimental success.

Sources

Personal archives Willy Servais, Morlanwelz.

Servais, Willy (1959). Fundamental concepts and their development in the school curriculum: Thoughts on the teaching of algebra in secondary schools (typewritten text). Paris: OEEC (OSTP).

References

Barbin, Evelyne (2012). The role of the French Association of Mathematics Teachers APMEP in the introduction of modern mathematics in France (1956–1972). In *Proceedings of the ICME-12 Satellite Meeting of HPM (History and Pedagogy of Mathematics) 2012, July 16–20* (pp. 597–605). DCC, Daejeon, Korea. Retrieved October 15, 2016, from http://hpm2012.onpcs.com/Proceeding/OT6/T6-01.pdf.

Bernet, Théo, & Jaquet, François (1998). *La CIEAEM au travers de ses 50 premières rencontres* [The CIEAEM through its first 50 meetings]. Neuchâtel: CIEAEM.

Bjarnadóttir, Kristín (2013). The implementation of the 'New Math' in Iceland: Comparison with neighbouring countries. *International Journal for the History of Mathematics Education, 8*(1), 1–18.

Bourbaki, Nicolas (1939). *Éléments de mathématique: théorie des ensembles* [Elements of mathematics: set theory]. Paris: Hermann.

Bruffaerts, Xavier (1994). *Paul Libois. Brouillon projet d'une biographie* [Paul Libois. Draft of a biography]. Unpublished master's thesis, Université libre de Bruxelles, Belgium. Retrieved January 14, 2017, from http://dev.ulb.ac.be/urem/Paul-Libois-brouillon-projet-d-une.

Castelnuovo, Emma (1965). Belgique: Mathématique d'avant-garde [Belgium: Pioneering mathematics] (French translation of Castelnuovo, Emma (1965). Matematica d'avanguardia. *Riforma della Scuola, 11*, 3–19). Retrieved October 15, 2016, from http://www.science.unitn.it/~fontanar/EMMA/riforma_scuola_maggio_1965.pdf.

Castelnuovo, Emma (1997). Federigo Enriques et Guido Castelnuovo [Federigo Enriques and Guido Castelnuovo] (French translation by Simone Trompler of Castelnuovo, Emma (1997). Federigo Enriques e Guido Castelnuovo. *Bollettino dell'Unione Matematica Italiana, 7*(11-A), 227–235). Retrieved October 15, 2016, from http://dev.ulb.ac.be/urem/Frederigo-Enriques-e-Guido.

De Bock, Dirk, & Vanpaemel, Geert (2015). Modern mathematics at the 1959 OEEC Seminar at Royaumont. In K. Bjarnadóttir, F. Furinghetti, J. Prytz, & G. Schubring (Eds.), *"Dig where you stand" 3. Proceedings of the Third International Conference on the History of Mathematics Education* (pp. 151–168). Uppsala: Uppsala University, Department of Education.

Decroly, Ovide (1928). *La function de globalisation et l'enseignement* [The function of globalisation in education]. Brussels: Maurice Lamertin.

De Latil, Pierre (1960). Sensationnelles expériences en Belgique: Les "Mathématiques modernes" enseignées dès l'enfance [Sensational experiments in Belgium: "Modern Mathematics" taught from childhood]. *Sciences et Avenir, 164*, 539–544. 553.

Dieudonné, Jean (1992). Les travaux mathématiques de Georges Papy [The mathematical works of Georges Papy]. *Bulletin de la Société Mathématique de Belgique (Série A), 44*(2), 149–151.

Fielker, David (1961). Developments in the teaching of mathematics. Report of the Easter Conference, London, 1961. *Mathematics Teaching, 16*, 32–52.

Freudenthal, Hans (1973). *Mathematics as an educational task*. Dordrecht: Reidel.

Furinghetti, Fulvia, & Menghini, Marta (2014). The role of concrete materials in Emma Castelnuovo's view of mathematics teaching. *Educational Studies in Mathematics, 87*, 1–6.

Furinghetti, Fulvia, Menghini, Marta, Arzarello, Ferdinando, & Giacardi, Livia (2008). ICMI Renaissance: The emergence of new issues in mathematics education. In M. Menghini, F. Furinghetti, L. Giacardi, & F. Arzarello (Eds.), *The first century of the International Commission on Mathematical Instruction (1908–2008). Reflecting and shaping the world of mathematics education* (pp. 131–147). Rome, Italy: Istituto della Enciclopedia Italiana.

Hotyat, Fernand (1960). Le 8ème colloque de pédagogie expérimentale de langue française, Morlanwelz, Belgique, avril 1960 [The 8th French language colloquium on experimental psychology, Morlanwelz, Belgium, April 1960]. *International Review of Education/Internationale Zeitschrift für Erziehungswissenschaft/Revue Internationale de l'Éducation, 6*(3), 353–354.

Hotyat, Fernand (1961). L'Enseignement des mathématiques au niveau secondaire. Les recherches dans les pays européens de langue française [Mathematics teaching at the secondary level. Researches in the French-speaking European countries]. *International Review of Education/Internationale Zeitschrift für Erziehungswissenschaft/Revue Internationale de l'Éducation, 7*(2), 235–247.

Kilpatrick, Jeremy (2012). The new math as an international phenomenon. *ZDM–The International Journal on Mathematics Education, 44*, 563–571.

Lenger, Frédérique, & Lepropre, Madeleine (1958–1959). Le programme B des écoles normales gardiennes [The program B of the schools for future kindergarten teachers]. *Mathematica & Paedagogia, 16*, 70–75.

Lenger, Frédérique, & Lepropre, Madeleine (1959). Initiation aux mathématiques fondamentales dans une classe de première normale gardienne [Initiation to fundamental mathematics in a first year class of training of kindergarten teachers]. In *Journée d'études 8: L'initiation aux mathématiques – Problèmes psycho-pédagogiques* (pp. 17–24). Bruxelles: Ministère de l'enseignement public, Secrétariat général de la réforme de l'enseignement moyen et normal.

Libois, Lucie (1962). Introduction [Introduction]. In *Journée d'études 22: Mathématiques. Programme O.E.C.E. Journées pédagogiques internationales tenues à l'École Decroly 9–12 avril 1962* (pp. 5–10). Brussels: Ministère de l'éducation nationale et de la culture, Secrétariat général de la réforme de l'enseignement moyen et normal.

Libois, Lucie (1971). Mathematics in the Decroly School. *Educational Studies in Mathematics, 4*, 182–186.

Libois, Paul (1955–6). L'Enseignement de la géométrie et la réalité [The teaching of geometry and the reality]. *Mathematica & Paedagogia, 7,* 30–34.

Libois, Paul (1960). Programme de Dubrovnik [Program of Dubrovnik] (Typewritten text of a lecture by Paul Libois on November 9th, 1960). In X. Bruffaerts (1994), *Paul Libois, Brouillon projet d'une biographie* (pp. 50–57), Mémoire de licence en mathématiques non publié, Université libre de Bruxelles, Bruxelles. Retrieved October 15, 2016, from http://dev.ulb.ac.be/urem/IMG/pdf/Paul_Libois_-_Projet_d_une_biographie_X.Bruffaerts_.pdf.Bruffaerts_.pdf.

Menghini, Marta (2015). Emma Castelnuovo's commitment to creating a new generation of mathematics teachers. In K. Bjarnadóttir, F. Furinghetti, J. Prytz, & G. Schubring (Eds.), *"Dig where you stand" 3. Proceedings of the Third International Conference on the History of Mathematics Education* (pp. 263–278). Uppsala: Uppsala University, Department of Education.

Nachtergaele, Jean (1964). XVIIème rencontre international de professeurs de mathématiques [XVIIth international meeting of teachers of mathematics]. *Mathematica & Paedagogia, 25*, 82–86.

Noël, Guy (1993). La réforme des maths moderne en Belgique [The reform of modern mathematics in Belgium]. *Mathématique et Pédagogie, 91*, 55–73.

Noël, Guy (2013). *1945–1960: 15 ans d'enseignement des mathématiques, y compris la création de la CIEAEM et de la SBPM* [1945-1960: 15 years mathematics teaching, including the creation of the CIEAEM and the SBPM]. Plenary lecture given at the 39th annual congress of the Société Belge des Professeurs de Mathématique d'expression française, Auderghem, Belgium. Retrieved October 15, 2016, from http://www.sbpm.be/wp-content/uploads/2013/06/Guy-Noel.pdf.

OECD (1964). *Mathematics to-day. A guide for teachers*. Paris: OECD.

OEEC (1961a). *New thinking in school mathematics*. Paris: OEEC.

OEEC (1961b). *Synopses for modern secondary school mathematics*. Paris: OEEC.

Papy, Frédérique (1968). Une expérience pédagogique de dix ans en Belgique [A pedagogical experiment of ten years in Belgium]. In G. Papy (in collaboration with P. R. Burgraeve, R. Holvoet, F. Papy, & A. Terfve) (Ed.), *Arlon 10* (pp. 25–89). Bruxelles: Centre Belge de Pédagogie de la Mathématique.

Papy, Frédérique (1970). L'expérience Belge à l'école primaire [The Belgian experiment in primary school]. In A. Revuz (Ed.), *New trends in mathematics teaching – Tendances nouvelles de l'enseignement des mathématiques*, Vol. II (pp. 95–111). Paris: UNESCO.

Papy, Georges (1959). *Aperçu général de la théorie des ensembles. Quelques notions essentielles de topologie* [General outline of set theory. Some basic notions of topology] (Report of Arlon 1, July 4–6, 1959). Bruxelles: Presses Universitaires de Bruxelles.

Papy, Georges (1960). *Premiers éléments de mathématique moderne* [First elements of modern mathematics]. Brussels: Author.

Papy, Georges (1961). Suggestions pour un nouveau programme de mathématique dans la classe de sixième [Suggestions for a new mathematics program in the sixth class]. *Mathematica & Paedagogia, 20*, 20–29.

Papy, Georges (1963–1967). *Mathématique moderne* [Modern mathematics] (Vols. 1, 2, 3, 5, 6). Paris, Bruxelles, Montréal: Editions Didier.

Papy, Georges (1964). Methods and techniques of explaining new mathematical concepts in the lower forms. In H. F. Fehr (Ed.), *Mathematics to-day. A guide for teachers* (pp. 99–147). Paris: OECD.

Papy, Georges (1966). La géométrie dans l'enseignement moderne de la mathématique [Geometry in modern mathematics teaching]. *L'Enseignement Mathématique, 12*(3), 225–233.

Papy, Georges, & Holvoet, Roger (1968). Centre Belge de Pédagogie de la Mathématique à Bruxelles: Statuts [Belgian Center for Mathematics Pedagogy in Brussels: Articles]. In G. Papy (in collaboration with P. R. Burgraeve, R. Holvoet, F. Papy, & A. Terfve) (Ed.), *Arlon 10* (pp. 133–135) (Reprint from Act 2472 published in the annexes of the *Moniteur Belge* of June 8, 1961). Bruxelles: Centre Belge de Pédagogie de la Mathématique.

Phillips, Christopher J. (2015). *The New Math: A political history*. Chicago: University of Chicago Press.

Piaget, Jean, Beth, Evert Willem, Dieudonné, Jean, Lichnerowicz, André, Choquet, Gustave, & Gattegno, Caleb (1955). *L'enseignement des mathématiques* [The teaching of mathematics]. Neuchâtel: Delachaux et Niestlé.

Servais, Willy (1964). A modern secondary school syllabus in mathematics for the scientific stream. In H. F. Fehr, *Mathematics to-day. A guide for teachers* (pp. 43–73). Paris: OECD.

Terfve, Andrée (1968). En suivant les cours du Centre… [By following the courses of the Centre…]. In G. Papy (in collaboration with P. R. Burgraeve, R. Holvoet, F. Papy, & A. Terfve) (Ed.), *Arlon 10* (pp. 13–22). Bruxelles: Centre Belge de Pédagogie de la Mathématique.

Trompler, Simone (2009). *Libois professeur* [Libois professor]. Unpublished paper based on a lecture given at "Une journée sur Paul Libois", Brussels, Belgium.

Vanpaemel, Geert, & De Bock, Dirk (2017). Marxism and mathematics. Paul Libois and intuitive geometry in Belgium. In K. Bjarnadóttir, F. Furinghetti, M. Menghini, J. Prytz, & G. Schubring (Eds.), *"Dig where you stand" 4. Proceedings of the Fourth International Conference on the History of Mathematics Education* (pp. 383–398). Rome: Edizioni Nuova Cultura.

Vredenduin, Piet J. G. (1967). Het experiment Papy [The experiment Papy]. *Euclides, 42*(6), 167–172.

Walusinski, Gilbert (1963). La réforme est en acte [The reform is in act]. *Bulletin de l'Association des Professeurs de Mathématiques de l'Enseignement Public, 233*, 121–126.

Author Biographies

Dirk De Bock is professor of mathematics at the Faculty of Economics and Business of the University of Leuven (Belgium). His major research interests are history of mathematics education, psychological aspects of teaching and learning mathematics, and the role of mathematics in economics and finance. His research in the field of history of mathematics education currently focuses on the role of Belgian mathematicians and mathematics teachers in the international New Math reform movement of the 1960s.

Geert Vanpaemel is professor in history of science at the University of Leuven (Belgium). His research concerns the history of science in Belgium since 1500, the history of universities and education, and the popularization of science. He published on the mathematical culture of the seventeenth century, in particular with respect to the Jesuits, the history of statistics during the nineteenth century and the introduction of modern mathematics in secondary schools. He is currently involved in a research project on scientific expertise and the fine arts.

The Role of a Journal on Teaching Mathematics and Sciences Issued at the Beginning of the 20th Century in Professionalizing Italian Primary School Teachers

Fulvia Furinghetti and Annamaria Somaglia

Abstract This chapter examines the first non-ephemeral journal on teaching mathematics and sciences in primary school, which appeared at the beginning of the twentieth century in Italy. The project of publishing such a kind of journal is set in the context of the initiatives launched in the new Italian nation for creating a course of studies specifically addressed at educating primary school teachers. We outline the mission and vision of the journal by presenting the personality of its founder Alberto Conti (a leading character among the secondary school mathematics teachers of the period), and analyzing the content of the journal. The subjects treated in the journal were sciences and mathematics, but mathematics was the prevailing subject covered in the journal because of its importance in primary education and the strong mathematical background of Conti. Most contributors, including the editors, were mathematics teachers in Italian schools for prospective primary teachers (*Scuole Normali*); their contributions reflect firsthand experience in this context. To involve the readers the journal published questions addressed to primary school teachers and to students of *Scuole Normali*. It is remarkable that female students were very active in answering the questions.

Keywords Professionalization · Primary school teachers · Mathematics Journals

1 Introduction

After the Congress of Vienna in 1815, the Italian territory was composed of some ten states. Gradually, a process of unification developed so that in 1861 the unified nation of Italy was born. The process of unification was promoted and led by the

F. Furinghetti (✉) · A. Somaglia
DIMA—Dipartimento di Matematica, Genoa University, Genoa, Italy
e-mail: furinghetti@dima.unige.it

F. Furinghetti and A. Karp (eds.), *Researching the History of Mathematics Education*, ICME-13 Monographs,
https://doi.org/10.1007/978-3-319-68294-5_5

Kingdom of Sardinia, whose king became the king of the unified Italy.[1] Turin, the capital of the Kingdom of Sardinia, became the first capital of Italy. Rome was annexed to Italy *de facto* in 1870 and in 1871 *de jure* became the capital of the kingdom.

One of the main tasks of the government of the new state was the construction of a system of instruction (Canestri and Recuperati 1976; Dei 1998; Genovesi 2007). Mathematical instruction was one of the main issues in this construction (Giacardi 2006a; Giacardi and Scoth 2014). Before the unification, most paramount mathematicians of the nineteenth century participated in the Risorgimento, the Italian movement that promoted the unification of Italy. After the birth of the new state, mathematicians again had an active role in the construction of the system of instruction and also had official appointments to the Government: Luigi Cremona, for one, was senator, Minister of the Public Education, and member of the Board of Governors of Public Education (Giacardi 2006b, 2012). Meanwhile, Italian mathematical research was achieving an international reputation. In this period rich with excitement for initiatives, which lasted until the First World War, mathematics teaching in school was acquiring a defined status and the figure of the mathematics teacher was emerging. There was a link between the communities of mathematicians and mathematics teachers: some math teachers maintained contact with universities and carried out research; some mathematicians paid attention to the problems of schools, in particular, those of teacher education (Furinghetti and Giacardi 2012). This context is the background for the study reported in this chapter. With this study we continue the research on Italian education and school teacher education that explored the shaping of the professional identity of mathematics teachers in post-unity Italy (Furinghetti 2012, 2017). At the same time, our study continues the line of exploring mathematics education journals carried out by Furinghetti (2002, 2003, 2006, 2009) and Furinghetti and Somaglia (1992, 2005). In this sense, it follows and makes use of the methodology of analyzing their content, the personality of the editors and the main contributors, and their mission and vision. This methodology, which relies on journals to study aspects of the mathematical life (research or teaching), has proven itself fruitful (Despaux 2011). An international research project is gathering information on the circulation of mathematics by journals (which specialized in mathematics or published mathematical articles regularly).[2]

Specifically, the goal of our chapter is to shed light on the mathematics education of prospective and in-service primary school teachers. Although the primary school was the place where to begin the plans for creating a modern and homogeneous system of instruction in the newborn country and extending alphabetization to the

[1]After the Vienna Congress (1815) the Kingdom of Sardinia included Savoy, Piedmont, Liguria, the county of Nice, and Sardinia.

[2]http://cirmath.hypotheses.org/. Retrieved January 26 2016.

entire population,[3] the *Scuola Normale*,[4] the school for preparing primary school teachers, was neglected in the system of the secondary schools. The following passage taken from the reform proposal of 1895 shows that still, more than 30 years after the unification, the organization of this kind of school was far from being settled: "È debito del governo … per le condizioni attuali dell'istruzione primaria, promuoverla moltiplicando il numero dei maestri." (M.P.I. 1909, p. 65) ["It is a duty of the government … for the present conditions of primary school instruction, to boost it by multiplying the number of teachers"].

To reach our goal we report on a study about a journal on mathematics and science addressed to primary school teachers and to secondary school students attending the schools aimed at educating primary school teachers, *Il Bollettino di Matematiche e di Scienze Fisiche e Naturali. Giornale per la Coltura dei Maestri delle Scuole Elementari e degli Alunni delle Scuole Normali.* [The bulletin of mathematics and of physical and natural sciences. Journal for the culture of primary school teachers and students of the *Scuole Normali*], issued from 1899 to 1917. This journal was founded and edited by a secondary school mathematics teacher, the very great majority of papers were written by teachers, and contact with primary school (in-service or prospective) teachers was one of its missions. These specific characteristics make the journal a means suitable to exploring how the professionalization of primary school teachers was promoted by initiatives coming from outside the institutions, with particular involvement of secondary school teachers working in *Scuole Normali.*

2 The *Scuola Normale* (Normal School) for Educating Primary School Teachers

At the moment of unification Italy was a set of smaller and larger states with different dialects, (that were the languages used by the low classes) money, systems of measures, cultural backgrounds, political statuses, traditions, administrations, economies, and school systems. As for mathematical research, Volterra (1909) reports that the old rules running the schools and the low number of university chairs made it difficult to modernize the university programs, and that it was difficult ameliorating this situation because of the lack of easy communication and the hostility of those governments "pei quali l'ignoranza pubblica era valido sostegno al potere" [whose power was supported by public ignorance, p. 57]. Nevertheless,

[3]From 1861 a census of population was conducted (usually every 10 years). In 1861 the percentage of illiterates was 72 for men and 84 for women (Dei 1998; Genovesi 2007). A person was classified illiterate if not able to write or read.

[4]The adjective "normal" is inherited from the French "*École Normale,*" its meaning is not related to the meaning of the English adjective "normal" (average, generic, etc.). In the following we will use the Italian expression *Scuola Normale* (singular) and *Scuole Normali* (plural) when referring to the school(s) for educating primary school teachers.

in the scientific community we may identify an element unifying the different states and constituting one of the ideal kernels of the future nation: since 1839 the scientists of the Italian territory, already perceiving themselves as Italian (though Italy did not yet exist as a state), began the tradition of having an annual meeting called *Unione degli scienziati italiani* [Meeting of the Italian scientists], held in various different towns (Silla 1939–1940). In these meetings the scientists discussed their scientific studies and also the problems of organization of the Italian instructional system. In the fifth meeting, held in Lucca in 1843, for the first time a commission was created to compile the statistics of schools for children (*Atti della quinta unione degli scienziati italiani* 1844, p. 110) and in the seventh meeting of 1845, in Naples, a commission was established with the aim of gathering data on the instruction in Italy and investigating which methods were most suitable to spread this instruction in the country (*Atti della settima adunanza degli scienziati italiani* 1846, p. 434). In successive meetings these commissions provided data systematic, though partial. The state of instruction emerging from these reports was very poor.

After the unification the school organization was substantially based on the Casati Law of 1859[5] (M.P.I. 1909, p. 18). As for the education of primary school teachers, this law has inherited the *Scuola Normale* established by the Minister Giovanni Lanza (Law number 2878, June 20 1858), whose organization and rules were substantially confirmed after the Italian unification. It lasted three years, including practical training, and was split into schools for men and schools for women. At the age of 16 for men and 15 for women there was an examination for the admission to the *Scuola Normale* based on topics taught in primary school.[6] After the first two years, the students obtained a license for teaching in the first two years of primary school termed *Scuola Elementare Inferiore* [Lower Primary School]. A preliminary course of preparation was established in 1880 for women and in 1889 for men. In 1895 this course was termed *Scuola Complementare* [Complementary School] and became again only for women, but in 1896 it was also for men. Initially the *Scuola Complementare* lasted two years and, later on, three years. In 1909 the *Scuola Normale* became the same for men and women (Law number 525, July 19 1909, and Royal Decree number 278, April 10 1910). The program of mathematics included only arithmetic, elementary geometry, and accounting. Conti (1911b) writes that until 1897 there were partial reforms of the *Scuola Normale*, but we know from the text of the proposal of reform of 1895 that the conditions of primary school instruction remained unsatisfying and that there was the need of "multiplying" the number of teachers (M.P.I. 1909, p. 65). An interesting and organic project for *Scuola Secondaria Superiore* [Higher Secondary School] was worked out by the Commission of 1909, established by the Minister Leonardo Bianchi (Law number 525, July 19 1909, and Royal Decree number 278, April 10 1910) with Paolo Boselli as a chair. The Report of this commission is

[5]Royal Decree of the Kingdom of Sardinia number 3725, November 13, came into effect in 1860.

[6]For some years the programs were different for men and women, the latter being a bit less difficult (Covato 2012).

Table 1 Male and female attendance in *Scuole Normali*

School years	Men	Women
1861–1862	947	2795
1875–1876	1248	5227
1881–1882	1383	7482
1899–1900	1323	19,864

accurate and extensive; it contains a rich and deep historical survey by Giovanni Vailati, who claimed the need to outline a history of the school reforms in the first 50 years of Italy (M.P.I. 1909, p. 23). The Report lacked a discussion and elaboration concerning the *Scuola Normale*, which was linked to the problem of the instruction of women.

The Daneo-Credaro Law (number 487, June 4 1911) tried to improve the quality of the *Scuola Normale* through the institution of a course, *Corso Magistrale* [Teachers' Course], directly after the gymnasium (five year middle school). This attempt was not successful. The most innovative change happened with the Gentile Reform in 1923[7]: the *Scuola Normale* became *Istituto Magistrale* [Teachers' Institute] and the complete course (*Corso Complementare* and *Scuola Normale*) changed the duration from six to seven years (*Istituto Magistrale, Corso Inferiore e Superiore*). This reform has some important deficiencies since, as stressed by Vita (1986, p. 82), the *Istituto Magistrale* was a school of general culture without any forms of practical training and in which the humanities and pedagogy were not covered adequately. Even less adequate was the treatment of the scientific disciplines.

An important point of the Daneo-Credaro Law was the passage of the administration of primary schools from the local municipalities to the central state. For the large towns this passage was optional. This change avoided some of the problems of primary school that existed in the second half of the nineteenth century, such as the poverty of schools due to the municipalities' lack of money and the subservient relationship of teachers with the local authorities who were running the schools. The Daneo-Credaro Law also increased salaries.

In this difficult context, for some years there had been cases of teachers in primary school without specific qualifications. To be primary school teachers was a frustrating profession for a low salary (even lower for women) and a difficult work context (Bini 1981; Fort 2015; Santoni Rugiu 2006). Nevertheless, the *Scuola Normale* was attended by women because it offered the possibility to continue their studies after primary school and to enter a profession. Santoni Rugiu (1982, p. 15) provides the data in Table 1 about the attendance in *Scuola Normale*.

The data on the instruction in primary school and in *Scuola Normale* in the years 1863/64, 1875/76, 1881/82, 1901/02 provided by the *Ministero dell'*Agricoltura,

[7]This Law, based on a series of Royal Decrees, is due to the Italian neo-idealist philosopher Giovanni Gentile (1875–1944), Minister of Public Education in 1922–1924. It represents the most important reform of the whole school system after the Casati Law.

Table 2 Primary school teachers in Italy in the second half of the nineteenth century

School years	Men	Women	% of women
1863–1864	18,443	15,820	46.17
1875–1876	23,267	23,818	50.59
1881–1882	21,670	26,485	55.00
1901–1902	21,178	44,561	57.78

Industria e Commercio [Ministry of Agriculture, Industry and Trade][8] show the impressive growth of the female presence as teachers in primary school (Table 2).[9]

Many novels and stories of Italian literature of the period we are considering have primary school teachers (especially female teachers) as a chief character. This fact shows the interest in this new figure emerging in society. A famous journalist and writer, Matilde Serao (1856–1927), attended the *Scuola Normale Femminile* [Female *Scuola Normale*] in Naples and left firsthand evidence of the education of girls as primary school teachers in the novel *Scuola Normale Femminile* published in the collection of novels *Il romanzo della fanciulla* [The novel of the girl, Milano, Treves, 1886). In the same collection the novel *Come muoiono le maestre* [How the female teachers die] is about a female teacher who committed suicide because of the difficult conditions of her work. This story echoes the sad story of Italia Donati (1863–1886), which occurred in Tuscany.

3 The Journal *Il Bollettino di Matematiche e di Scienze Fisiche e Naturali*

3.1 *The Foundation of the Journal*

Chiosso (1992) lists 475 periodicals dealing with education and school issued from the Casati Law to 1899, the years when the Italian system of education was in the making. The main group of publications considered are: journals dealing with didactics and/or school policy addressed to teachers, the bulletins of societies and associations addressed also to schools, and pedagogical journals. The list does not

[8]In some years this Ministry was in charge of this type of school (Di Pol 1996). Further statistical data can be found at https://ebiblio.istat.it/SebinaOpac/.do?idDoc=0004137#2 (Retrieved April 9 2017).

[9]Despite the higher percentage of women in *Scuole Normali*, appointment as teachers still remained difficult for women due to many factors, such as the concern about the loss of employments for men, their supposed inadequacy for transmitting virile values (patriotism, a positive attitude towards manufacturing work), and, in general, the persistence of some societal stereotypes against women present in Italian society. The feminization of Italian teaching profession is discussed in Covato (2012) and Pak Kazdan (2012).

include other publications such as journals for *Scuole Popolari* [Popular Schools], recreational journals, and children's storybooks that, indeed, sometimes had references to school issues. Also, the list does not include Annual reports (*Annuari*) and Almanacs because they were not published regularly.

Among the subjects covered in these journals there was mathematics; Pizzarelli (2017) illustrates how this subject was present in the educational journals issued in Piedmont, even before the unification of Italy. The appearance of the first journals dedicated only to mathematics teaching (sometimes associated with science) occurred later on. Before that, when the community of mathematicians was established, the need to create a network of communication among Italian researchers emerged, then the first journals addressed to mathematical research were published; see Table 3. Also, mathematics teachers felt the need to acquire a specific professionalization that could be fostered by the communication of experiences and results (Furinghetti 2017). To this purpose, journals dedicated only to teaching mathematics (in a few cases associated with science) were founded starting in the 1870s; see Table 4.

The mathematical journals for teachers listed in Table 4 were mainly addressed to those at the secondary school level. As far as we know, the first attempt to issue a journal addressed specifically to prospective and in-service primary teachers was *Il Piccolo Pitagora* [The little Pythagoras], with the subtitle *Periodico di Matematica per gli Alunni delle Scuole Secondarie e pei Maestri Elementari* [Periodical of mathematics for pupils of secondary schools and for primary school teachers] founded in 1883. The editor was Alberto Cavezzali (1848–1922), who for many years was the principal in primary schools. The journal was printed in Novara, where a printing house was active in producing school publications (Callegari 1998; Furinghetti 2017). The first issue was published in March 15 1883; in 1884 the publication ceased.

At the end of the century, the professionalization and the identity of secondary school mathematics teachers was strengthened by important events. The journals of the early 1870s ceased, but a new non-ephemeral journal (*Periodico di Matematica*), still published today, began publication in 1886. In 1895 the association of Italian mathematics teachers *Mathesis* was founded and teachers began to meet in local and national meetings organized by this association. The main field of interest of the *Periodico* and the association was secondary school mathematics teaching, but they also dealt with some problems of mathematics in primary school. In the first congress of *Mathesis,* held in Turin, (September 9th to 14th 1898) Corrado Ciamberlini, a leading mathematics teacher of the period, complained about the scientific and didactical mistakes found in textbooks and expressed the need to make primary school teachers aware of the studies on mathematics teaching carried out by the members of the association. For Ciamberlini, the only means to reach so many (more than 50,000, according to him) primary school teachers was to have a journal of arithmetic and geometry addressed exclusively to primary school

Table 3 The first Italian journals on mathematics or on the history of mathematics

Title	First issue in	Period of publication
Annali di Scienze Matematiche e Fisiche founded by Barnaba Tortolini, continued as *Annali di Matematica Pura ed Applicata*edited by Enrico Betti, Francesco Brioschi, Angelo Genocchi, Barnaba Tortolini	Rome Rome	1850–1857, 1858–
Giornale di Matematiche ad Uso degli Studenti delle Università Italiane (aka *Giornale di Battaglini*) founded by Giuseppe Battaglini, Vincenzo Janni, Nicola Trudi	Naples	1863–1938/39, 1947/48–1965/67
Bullettino di Bibliografia e di Storia delle Scienze Matematiche e Fisiche founded by Baldassarre Boncompagni	Rome	1868–1887
Rendiconti del Circolo Matematico di Palermo founded by Giovanni Battista Guccia	Palermo	1887–1917, 1919–1941, 1952–
Bollettino di Storia e Bibliografia Matematica (*Supplemento al Giornale di Matematiche di Battaglini*) edited by Gino Loria *Bollettino di Bibliografia e Storia delle Scienze Matematiche* founded by Gino Loria, continued as *Sezione Storico-Bibliografica* edited by Loria in *Il Bollettino di Matematica* founded by Alberto Conti	Naples Turin Bologna	1897 1898–1919, 1922–1943, 1947–1948

Table 4 Early mathematical journals for teachers after the unification of Italy

Title	First issue in	Period of publication
Periodico di Scienze Matematiche e Naturali per l'Insegnamento Secondario edited by Angelo Armenante, Eugenio Bertini, Davide Besso, Enrico De Montel, Luigi Pinto, Francesco Rodriguez, Leone De Sanctis	Rome	1873–1875
Rivista di Matematica Elementare, founded by Giovanni Massa	Alba (Piedmont)	1874–1885
Il Piccolo Pitagora founded by Alberto Cavezzali	Novara	1883–1884
Periodico di Matematica, founded by Davide Besso title changed into *Periodico di Matematiche*	Rome	1886–1916 1921–1943, 1946–
Rivista di Matematica, founded by Giuseppe Peano	Turin	1891–1906
Bollettino dell'Associazione Mathesis, first editor Giovanni Frattini. It underwent changes in format, editorial line, and name	Rome	1896–

teachers (Ciamberlini 1899).[10] The congress approved, but in a following meeting of the Steering Committee of *Mathesis* (25 August 1899 in Leghorn) the proposal was rejected for financial reasons and because, according to another outstanding teacher, Giovanni Frattini, there were no chances to overcome primary school teachers' beliefs (Conti 1899–1900). But the discussion had not been worthless, because only a year later Alberto Conti, a young teacher of *R.* [Royal] *Scuola Normale Femminile Anna Morandi Manzolini* of Bologna founded the journal *Il Bollettino di Matematiche e di Scienze Fisiche e Naturali* [The bulletin of mathematics and of physical and natural sciences] in Bologna. The subtitle *Giornale per la Coltura dei Maestri delle Scuole Elementari e degli Alunni delle Scuole Normali* [Journal for the culture of primary school teachers and students of *Scuole Normali*] indicates that primary school teachers and secondary school students attending the schools for educating primary school teachers were the target. Conti was well informed of the problems of this type of school, since in those years he was teaching only in a *Scuola Normale*. The first Volume has the date 1900, but the first two issues are dated December 1899 in the cover; Conte (1936) indicates December 1st, 1899 as the date of the foundation of the journal. The last Volume was published in 1917. The initial publisher was Zanichelli[11] in Bologna (Fig. 1).

3.2 The Editors

Conti was the only editor and administrator until Volume 10 (1908–1909). In this school year he was appointed as a teacher in the *R. Scuola Normale Femminile Margherita di Savoia* in Rome and address of the journal was established in this town. In 1909–1910, (Volume 11) beside Conti, there was an Editorial Board composed of secondary school teachers: Bisson-Minio Ersilia, Chiari Augusto, Ghezzi Teodosio, Mossini Dirce, Quintili Pierina, Rubini Luisa (for the section of problems); Tenca Luigi (for the bibliographical review and answers to readers); the secondary school teachers Mani Mario, Neri Francesco (for physical and natural sciences), and the primary school teacher Mancinelli Gemma for *Didattica speciale* [Special didactics] in primary school. In 1910–1911 (Volume 12) Tenca joined Conti as a co-editor and remained as the only editor from 1911 to 1912 (Volume 13) until the definitive closure of the journal in 1917 (Volume 18), even when enrolled in the Italian army during the First World War.[12] Conti (1911–1912) explains that the change of the editorship was due to his many engagements, which made the issues of the bulletin irregular for some years previous. In 1902 Conti had

[10]The proceedings of the conference are published in *Periodico di Matematica*, 1899, s. 2, 1, pp. 1–124.

[11]Zanichelli has been and still is an important publishing house of school textbooks.

[12]In his autobiography Conti (1936) claims that in the Volume 17 (1915–1916) he was again the editor due to the absence of Tenca at the front. This information is in contrast with what we read in the cover of this volume.

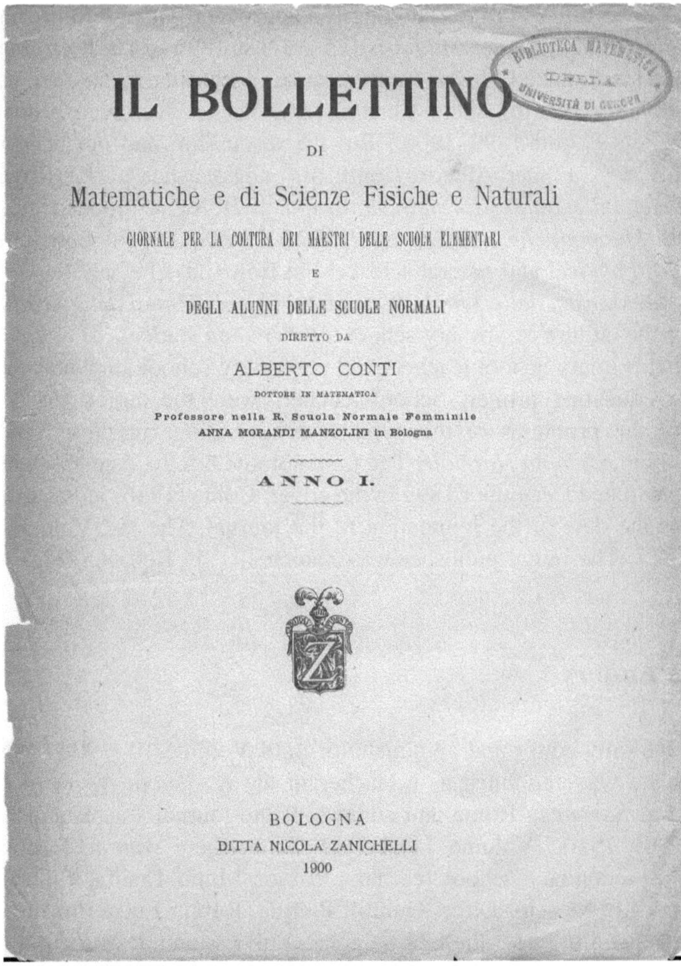
IL BOLLETTINO

DI

Matematiche e di Scienze Fisiche e Naturali

GIORNALE PER LA COLTURA DEI MAESTRI DELLE SCUOLE ELEMENTARI

E

DEGLI ALUNNI DELLE SCUOLE NORMALI

DIRETTO DA

ALBERTO CONTI

DOTTORE IN MATEMATICA

Professore nella R. Scuola Normale Femminile
ANNA MORANDI MANZOLINI in Bologna

ANNO I.

BOLOGNA
DITTA NICOLA ZANICHELLI
1900

Fig. 1 The cover of the first Volume of *Il Bollettino di Matematiche e di Scienze Fisiche e Naturali*

founded the journal *Il Bollettino di Matematica* [The bulletin of mathematics] mainly addressed to secondary school mathematics teachers. This journal, still published under the title of *Archimede* [Archimedes], became an important reference for the generations of mathematics teachers of the first half of the twentieth century (Furinghetti 2006). In spite of the changes in the editorial board, *Il Bollettino di Matematiche e di Scienze Fisiche e Naturali* can be considered a journal of Conti, reflecting his view of mathematics teaching and the profession of teaching. In his autobiography, Conti (1936, p. 89) defines it as the "papà" [dad] of his more famous *Il Bollettino di Matematica.*

Alberto Conti[13]

Alberto Conti was born in Florence (December 3, 1873) and died there (October 18,[14] 1940). He studied in the prestigious *Scuola Normale Superiore* (Higher *Scuola Normale*, a school initially established in Pisa with the aim of training secondary school teachers, but later on developed as a school aimed at training researchers in mathematics) and graduated with a degree in mathematics on November 12th, 1895, with a dissertation on the theory of connections. Among his professors were prominent Italian mathematicians such as Eugenio Bertini, Luigi Bianchi, and Ulisse Dini. He taught number theory for a few months to the students of the first year of *Scuola Normale Superiore*, afterwards he began down the career path of secondary school mathematics teacher that he followed all his life: in 1896 for four months in a technical school of Pausula, a village close to Macerata, in October of the same year in the Scuola Normale of Belluno, in October 1898 in *R. Scuola Normale Femminile Anna Morandi Manzolini* of Bologna, in October 1908 in *R. Scuola Normale Femminile Margherita di Savoia* of Rome, in 1918–1920 in *R. Collegio Militare* [Royal Military College] of Rome, in October 1920 in the *R. Istituto Tecnico* [Royal Technical Institute] of Florence, and from 1923 until his retirement in the *R. Liceo Ginnasio* [Royal Lyceum gymnasium] *Michelangelo* of Florence. As a consequence of the Gentile Reform that associated the teaching of mathematics and of physics in secondary school, from the school year 1926–27 he taught these two disciplines. Conti is an example of how teachers were a unifying element in the newborn country, since, thanks to the system of recruitment based on national *concorsi* [competitive examinations], he had positions in different places of North and central Italy before going back to his hometown (Florence) in 1920. Other events of his life, notably the foundation of didactic journals, contributed to establishing contacts in the country.

In his autobiography Conti (1936) mentions honors such as the title of knight of the Italian Crown, Ministry awards for his textbooks, promotions and upgrades of the salary for outstanding merit. His publications appeared in journals for mathematics teaching, in proceedings of congresses of the associations of teachers *Mathesis* and the *Federazione Nazionale Insegnanti Scuola Media* FNISM [National Federation of Middle School Teachers], and in school journals and bulletins. He delivered plenary talks in national congresses. Conti (1936) lists 18 works and mentions many others: teaching materials such as textbooks, didactic articles and surveys of the state of mathematics teaching, inquiries on textbooks, papers and interventions about school and university policy (recruitment, training, etc.). The

---

[13]This author has already studied the figure of Conti (Furinghetti 2012). My main sources for outlining the life and career of Conti are his autobiography and memories (Conti 1936, 1941), the obituaries (Anonymous 1941; La Direzione 1940; Rossi 1983), and Castelnuovo (1909). In Conti's autobiography there are some mistakes in the data about publications that I amended in this paper. Conti is listed among the mathematicians by Tricomi (1962).

[14]This is the correct date of the death reported in the official document of *Comune di Firenze* (*Anno* 1940, *Parte* 1, Vol. 04, *Atto* n° 2364).

mathematician Federigo Enriques asked Conti to collaborate on the collective work *Questioni riguardanti le matematiche elementari* [Questions concerning elementary mathematics], two volumes conceived to improve the preparation of mathematics teachers. He was interested in the political aspects of his profession and published articles in Italian newspapers as well. He was vice-president of FNISM in 1902–1904 and president of its Roman branch in 1910.

While he remained a schoolteacher all his life, Conti was in contact with mathematicians and also wrote didactic papers in collaboration with them. He participated in the International Congress of Mathematicians in Rome (1908) and in Bologna (1928). In Rome he contributed to the Fourth Section *Questioni filosofiche, storiche, didattiche* [Philosophical, Historical and Didactical Questions], (Conti 1907–1908b). In the closing plenary section (April 11) he presented a proposal for establishing an international association of mathematicians (Castelnuovo 1909, p. 33). The proposal was approved, though we know that it was only in 1920 that the International Mathematical Union was founded. In 1909 Enriques appointed Conti as a member of the Italian Delegation of the *International Commission on the Teaching of Mathematics* (the first incarnation of ICMI). For this Commission he wrote the reports (Conti 1911a, b), on mathematics teaching in *Scuole Normali*, in infant and primary schools.

Luigi Tenca[15]

Luigi Tenca (Gambara, close to Brescia, September 8th, 1877 – Florence, August 27,[16] 1960) graduated with a degree in mathematics at University of Pavia, where he was an assistant professor for four years. Afterwards he began his career as a secondary school teacher mainly in *Scuole Normali* for women. He had positions in various different towns in Italy. Afterwards he became a director of schools, a principal and a *Provveditore agli studi* [Commissioner of Education] Tenca gained promotions and upgrades for outstanding merit. He was a valiant soldier during the First World War, received six medals, and later was appointed a reserve brigadier general. He wrote a paper on the mathematicians who participated in World War I (Tenca 1959). When he became the editor of *Il Bollettino* he was a teacher in *Scuola Normale.*

Tenca is the author of about hundred publications, most on history of mathematics, but also on arithmetical and geometrical progressions from a higher standpoint and the theory of determinants.

---

[15]My main sources for outlining the life and career of Tenca are the obituaries (Procissi 1960; R. G. 1960). He is listed among the mathematicians by Tricomi (1962).

[16]This is the correct date of the death reported in the official document of *Comune di Firenze* (*Anno* 1960, *Parte* 2, *Serie* B, Vol. 2, *Atto* n° 1235), while the obituary (R. G. 1960) and the website *Edizione Nazionale Mathematica Italiana* report August 26.

3.3 The Mission and Vision of the Journal

Conti was a mathematics teacher, and though the journal gives space to articles on the sciences, the mission and vision of the journal emerge when mathematics is at play. In the words of Conti (1911–1912), the aim of the journal was to raise the prestige of primary school and of *Scuole Normali* to the level they deserved for their important mission in the life of the nation. Teacher education was the main means for reaching this aim. Conti, though a beginner in the profession of teaching, showed a precise and, we would say modern, idea of what knowledge for teaching should be. This emerges in the answer to a letter of Ciamberlini, one of the leading teachers of the period. In the circular letter announcing the birth of the journal and in the back cover of the first Volume the following contents of the journal were listed: articles of didactics, analysis of incorrect practices in classroom teaching, didactics (advice on how to teach), a column called "Intermediario" [Intermediary], which was aimed at promoting exchanges of opinions and problems among the readers, recreational columns, gleanings, variety, history, scientific articles, reviews of books, and questions for students. In his letter to the editor Ciamberlini wrote that only the first three types of content were of importance in a journal addressed to the target readership of the *Bollettino*. Conti's defense is a kind of manifesto of his view of mathematics teaching and teacher knowledge. He says that history contributes to enlarging the culture of teachers and students beyond what can be taught in school. Questions, which are specially addressed to students, are important because they offer the opportunity to apply what is seen in the classroom only from a theoretical point of view. Recreations are an element that could help to attract students. We see that Conti has an idea of mathematical knowledge required for teaching which goes beyond the knowledge of technical aspects, encourages practice that we would now call problem solving, and takes into account the public's attitude towards mathematics. All that is epitomized by the use, in the subtitle, of the expression "culture of teachers," which is not usual in literature of those times concerning school.

The foundation of the journals mentioned in Table 2 and the associations of mathematics teachers are events showing that communication and solidarity were perceived as values in the school milieu just as they were in some milieus of society. In accordance with these values, the journal encouraged contact with readers through correspondence, competitions on solving questions, and discussions. Teachers were asked to send reports on their experiences in the classroom in order to share the results and help their colleagues to improve their work. One initiative inspired by Conti's wish to promote the participation of the readers was the launch of a referendum based on 20 questions for teachers of *Scuole Normali* about the programs of scientific teaching in the *Scuola Complementare* and in the *Scuola Normale* (Conti et al. 1900–1901). Before the referendum was published, this issue had been repeatedly addressed in some readers' messages that had appeared in the first two volumes of the journal. Conti's intention was to present the majority's opinion at the second congress of *Mathesis* in Leghorn (17–22 August 1901). The questions concerned the choice of years to introduce the arithmetic operations, accounting, geometry, if it is advisable to have teachers of accounting with a specific qualification, and the role of froebelian courses for teachers of mathematics. In the following issues the answers from 16 readers were published with their names.

Conti's ideas underlying the mission and vision of the journal are clearly displayed in the talk delivered at the fourth International Congress of Mathematicians in Rome (1908) and published in the *Bollettino* (Conti 1907–1908b). At that time he was no longer a beginner teacher and showed a clear view of the problems of teaching at the primary school level and of educating primary school teachers. His opinions are those of a teacher who, in the modern terms of Schön (1983), could be described as a "reflective practitioner," a professional teacher who is able to look from the outside at his profession and, in certain cases, to interpret it in the light of theoretical aspects of mathematics education (intended as a discipline). He knew the work of Charles-Ange Laisant and other classical authors interested in education from ancient (Plato, Aristotle, Cicero, Quintillianus) to more recent (Vittorino da Feltre, Friedrich Wilhem August Froebel, Johann Heinrich Pestalozzi, Johann Friedrich Herbart, Louis-René de Caradeuc de La Chalotais). Conti took the basic ideas on mathematics teaching at the primary school level from Laisant (1905, 1906). He maintained that pupils have to follow a continuous path from simple situations that foster intuition to theoretical thinking and the acquisition of concepts. This claim was linked to the state of mathematics teaching in Italian infant and primary schools. Following Laisant, Conti suggested observation, action, and puzzles as good means for applying his ideas. To this aim the use of various types of manipulatives, drawing, paper folding, and the construction of objects was encouraged,[17] as well as the enlargement of the contents taught, for example by introducing geometrical transformations. Following Loria (1905), he also advocated introducing some new topics of mathematical research, such as functions, into secondary school teaching.

As for teacher education, the journal tried to overcome the difficulty, already discussed by Conti (1907–1908b), of combining general pedagogy with disciplinary (in our case mainly mathematical) pedagogy. In general, the emphasis given to the presentation and discussion of school programs reveals the strong belief that information is fundamental for teachers and that they have to be active deliberate actors in the school milieu.

[17]He also mentioned a toy imported from Britain called "The wood and cork worker" useful for constructing solids.

3.4 The Content of the Journal

The journal was founded on December 1st, 1899, and was published quite regularly until the last issues 5–6–7 (March, April, May) of 1917. Sometimes the volumes[18] are organized according to the school year, from fall to summer, and sometimes they follow the solar year from January to December. The volumes are divided into numbered issues, but the pages are numbered sequentially across the volumes. The number of issues varied from year to year, with at least eight issues per volume, with the exception of the last volume published in the hardest period of the First World War, which consisted of only seven issues. The number of pages varied around the range of 16–24. The journal was published by various printing houses in different towns, see Appendix 2.

The subscribers were mainly teachers and students of *Scuole Normali*. In the first Volume (a message to the founding subscribers published on an additional sheet) Conti writes that in the first year there were 500 subscribers, which included Commissioners of education, teachers of primary, secondary and upper secondary level classes, and students (he specifies "male and female").[19] Conti complains that just one of 200 Italian inspectors was among the subscribers. In Volume 3 (1902, p. 52) Conti acknowledges the support of the Ministry of Education, which has purchased subscriptions to the journal for the state *Scuole Normali* for men in 1901 and in 1902 did the same for female *Scuole Normali*. This fact evidences an openness of the Ministry of those times that is quite surprising if compared with the present situation in Italy, and also the acknowledgement of the role of the journal in teacher professionalization. The subscription was the condition for being allowed to send in answers to the proposed questions and, if correct, to receive a prize for good solutions. In the editorial published in Volume 13 Conti (1911–1912) claims that after some innovations introduced in 1902 the journal has increased spread of its readership more and more, especially among male and female students of *Scuole Normali* and that solving the proposed questions was becoming very popular. He complains that the journal was not so popular among teachers and that, unfortunately, many students dropped the subscription when they become teachers because they believed that their preparation was finished after their school years.

[18]The "volumes" are indicated as "years" and their numbers are represented with Roman numerals. Sometimes it happens that the articles have different titles in the table of contents and at the interior of the journal. Also, sometimes the number of pages is reported in the wrong way in the table of contents.

[19]In the short note dedicated to Sofia Kowalevski (1908, *9*, pp. 47–48) the anonymous author writes that it was decided to open a new column dedicated to women scientists because "the great majority" of the readers belongs to female gender. Already in previous volumes we find some notes on the history of women in sciences, e.g. "Le donne nella scienza [Women in science]" in the Volume 3 (1902), pp. 8–11, signed La Direzione.

The journal contains contributions and the following sections:

- Articles
- *Didattica per la scuola elementare* [Short notes of "Didactics for primary school"]
- *Rubrica Intermediario* [Column Intermediary]
- *Bibliografia* [Bibliography/Reviews]
- *Corrispondenza* [Correspondence]
- *Cronaca* [News]. Among the news there were reports on laws, programs, and Ministry regulations about primary school and *Scuola Normale*, and reports of meetings (of *Mathesis* and others)
- *Questioni da risolvere* [Questions to be solved] for teachers and for students
- *Ricreazioni scientifiche* [Scientific puzzles]
- *Concorsi a premi* [Contests with prizes for the readers]
- *Risoluzioni delle questioni* [Solutions of the questions proposed by the journal]
- *Risposte alle ricreazioni* [Answers to the proposed puzzles]
- *Curiosità, pensieri e massime* [Curiosities, thoughts, and sayings]
- *Necrologia* [Obituaries]
- *Temi di matematica proposti ad esame di concorso a posti di maestro nelle classi elementari* [Mathematical tasks proposed at competitions for positions as teachers in primary school]
- *Svolgimento dei temi proposti ad esami di concorso* [Solution of the tasks proposed at competitions]
- *Rubrica di storia* [Column of history]
- *Per la riforma delle Scuole Normali* [For the reforms of the *Scuole Normali*]. This column is a forum of proposals from teachers of the *Scuole Normali* and students of these schools.
- *Rubrica delle forme erronee* [Column of analysis of incorrect practices in classroom teaching]

In analyzing all issues of the journal we have identified 325 articles,[20] of which 203 concern mathematics. The others concern the sciences or general topics about education. Clearly, mathematics is the most frequent subject. In the last volumes it is almost the only subject covered in the articles. Given that, in general, questions to readers and the tasks proposed in national competitions for positions as teachers dealt with mathematics, we may say that the focus of the journal is on mathematics teaching. This, of course, depends on the emphasis given to mathematics in the programs of primary schools and *Scuole Normali* as well on the academic education of the editors, who had both degrees in mathematics and were in touch with the mathematical milieu.

The articles have an elementary approach. They may be divided in these main categories:

- treatment of theoretical aspects (mainly in the field of arithmetic and geometry)

[20]We have not included in this number the pieces signed *La Direzione* (The editor).

- teaching suggestions and caveats for teachers
- reports of experiences in classroom
- reports of activities and initiatives such as visits to museums and laboratories in schools.

Some of the articles discuss history (mainly of mathematics, but also of physics and the natural sciences), a choice that is consistent with the idea of the culture for teaching held by the founder. Moreover, we must remember that in those times the mathematicians Federigo Enriques and Gino Loria, strong supporters of the role of history in mathematics teaching and teacher education, were very active in education. The articles on history were aimed at giving some information about significant persons or about the development of concepts, without any suggestion of using history in the classroom. Conti was not a historian, nevertheless he gave history serious consideration. His main journal *Il Bollettino di Matematica* hosted a separate section of history and bibliography from 1922 to 1948, when Loria's the historical journal, entitled *Bollettino di Bibliografia e Storia delle Scienze Matematiche* [Bulletin of bibliograhy and history of mathematical sciences] ceased its publication. An interesting article on history of science is "Una breve esposizione della teoria di Darwin" [A short outline of the theory of Darwin] (1908, *9*, 10–14, 17–24, 58–60, 92–101) by Alpinolo Natucci (1883–1975). This author was a teacher who did research mainly on the history of mathematics, and also wrote papers on physics and other sciences (Maracchia 1975; Togliatti 1975).

It is remarkable that some articles contain references to foreign works, not only those translated into Italian, but also published only in foreign languages.

Usually the topics in the articles on science are various and without attempts at establishing a link with mathematics. In these articles the didactical reflection is rather absent. Some topics, such as agriculture, alcoholism, hygiene, and widespread diseases such as malaria, mirror the pre-industrialized status of Italy and its social problems. Other themes are linked to aspects of the scientific culture of the period: an example is the article by Alberto Alberti "Due razze umane in Europa" [Two human races in Europe] in the Volume 2 (1900–1901, 65–74; 100–105). Some early technology is also present, see the one page note on cinema ("Il cinematografo") in Volume 12 (1910–1911, 5–6).

The section of questions and problems presents a variety of situations. There are arithmetical problems proposed for the classroom or proposed in the national *concorsi* (competitive examinations for having permanent positions as teachers): often they have the form of word problems. Sometimes the proposed questions require proofs of simple theorems. The few science questions are often presented in a friendly and amusing form. Examples of tasks proposed to readers are:

1902, *3*, p. 41:

> 133. Why, when a bottle of beer is close to a fire, does the cap sometimes jump in the air?
> 142. Find the edge of a cube equivalent to a regular tetrahedron with the edge of 1.37 m.

1907, *8*, p. 29:

5. Prove that a triangle having two equal heights is isosceles.

1907–1908, *9*, p. 9

9. Prove in a arithmetic and geometric way … that:
(a) The product of two consecutive numbers is equal to the square of the lesser plus the lesser
(b) The difference of the squares of two consecutive numbers is equal to their sum

1907–1908, *9*, p. 42:

3. How would you deliver a lesson on the theme "Practical calculus of ordinary fractions?"

The column of wrong forms is a special feature of this journal. It reflects the concern of some contributors, mainly Ciamberlini and the editor, for the precision and the clarity of the language. Some examples are:

1906, *7*, p. 32:

The chain of false equalities
315 + 425 = 740 × 8 = 5920: 5 = 1184

1907–1908, *9*, p. 5:

Definition of cone: Solid with a circular basis, which goes gradually ending in a tip.

3.5 The Contributors

There are 98[21] credited authors of articles, of whom 60 write on mathematics. The authors of the science articles are mainly unknown teachers. The only famous name is Giuseppe Mercalli (1859–1914), seismologist and volcanologist, who developed the earthquake intensity scale named after him, Scala Mercalli, which was widely used in Italy. Accounting, though one of the subjects in the *Scuola Normale* programs, has little space, it appears only in some questions.

Mathematics is the main aspect characterizing the journal. Some authors of mathematics articles have a solid mathematics background. Some evidence of this fact comes from the comparison of the list of the authors with the list compiled by Tricomi (1962), which contains short biographical notes on 371 Italian mathematicians who died in the period of January 1st, 1861, to December 31st, 1960. In this list the qualification of mathematician has to be intended in a broad sense as researchers who produced at least a few papers on mathematics, so that researchers of various degrees of importance known for their mathematical research are mentioned. There are 10 authors of the journal on this list; some authors who would satisfy the conditions to be included in the list died after 1960. This fact reflects a general trend of those years: sometimes teaching in school was the first step for entering an academic career. Most authors are teachers in *Scuola Normale*. On one hand, this fact guarantees that the

[21]Some articles are signed *La Direzione* (The Editor).

contributors were well aware of the problems of this kind of school. On the other hand, it shows the low interest in this type of school by those on the outside, such as teachers of other schools and academic mathematicians.

The only contributors who were university professors are Enriques, Giovannni Garbieri, and Giuseppe Peano. Enriques and Peano are both first rank mathematicians (Enriques in algebraic geometry, Peano in logic) with a marked interest in mathematics education. Enriques (1871–1946) contributed with a short note on vision (1899–1900, *1*, 113–114) and a report on the first competition of the journal. Peano (1858–1932) wrote the short paper "Sul prodotto di grandezze" [On the product of quantities] (1914–1915, *16*, 99–100). Giovanni Garbieri (1849–1931) wrote a touching letter to Conti on December 18th, 1899, after having received the first two issues of the journal; some passages are published in the first Volume at page 33. He praises the initiative of founding the journal and the aims expressed by the editor. Garbieri was aware of the problems of primary school teachers' education and the teaching in primary schools, since he started his career as a primary school teacher (Gallarati 1999). As a full professor at the University of Genoa, he endeavored to establish the *Scuola di Magistero* (a university level institute for teacher training) in mathematics, which was officially approved by the Ministry of Public Education in 1906. He was one of the founders of the library of the School, which afterwards became the Library of the present Department of Mathematics of University of Genoa. He wrote textbooks whose effects were widespread and appreciated.

We have evidence that some research mathematicians were among the readers. The article on the history of literal calculus written by Omobono Buzzi (1899–1900, *1*, 19–23, 51–55, 114–117, 168–171, 209–212, 330–333, 361–366) received remarks by the historians Ettore Bortolotti (1899–1900, *1*, 375–377) and Giovanni Vacca (1899–1900, *1*, 377–378).[22]

There are 15 authors who contributed with at least four articles. Some of them, Artom, Ciamberlini, Conti, and Natucci, are chief characters in the history of mathematics teaching in Italy, but there are also forgotten contributors who are interesting figures. Giuseppe Scoto (1861–1908) is an example of a person who became a mathematics teacher in technical schools and *Scuole Normali*, and a science teacher in *Scuole Normali*, without any academic degrees. He belonged to a poor family and, with enormous difficulty, earned the diploma of a primary school teacher. According to the obituary by Conti (1907–1908a) he was an esteemed teacher and author of well-appreciated textbooks, which had good reviews.

---

[22]Buzzi wrote also a book linked to the the history of mathematics: Buzzi Omobono (1888). L'insegnamento dell'aritmetica nella scuola elementare, dedotto dalla storia della scienza dei numeri [The teaching of arithmetic in the primary school based on the history of the science of numbers]. Treviglio: Stab. Tip. Sociale.

Buzzi was a teacher of Italian in a *Scuola Normale* (http://lombardiabeniculturali.it/archivi/unita/MIUD0B10B8/. Retrieved 23 November 2016). He gave an interesting outline of aspects of *Scuole Normali* in the talk celebrating the 50 years of a *Scuola Normale*: Buzzi Omobono (1912). *La R. Scuola Normale promiscua "Candida Lena-Perpenti", commemorando il suo primo semisecolare giubileo* [The Royal *Scuola Normale* for boys and girls "*Candida Lena-Perpenti*", celebrating its first 50 years]. Tirano: Tip. Fiorentini e C.

An example of the deep concern for mathematics education is offered by the contributions of Emilio Artom (1888–1952). He studied in Turin, where a school of algebraic geometry was flourishing around Corrado Segre. After receiving a degree in mathematics he began an academic career as a teaching assistant of Enriques, but soon decided to become a teacher and at the age of 23 he got a position in the *Scuola Normale* of Aosta. In 1920 he passed to the Technical Institute and after the Gentile Reform to the *Liceo scientifico Galileo Ferraris* in Turin. His academic career was halted because his refusal to affiliate with the national fascist party[23] made it impossible for him to earn the *libera docenza*, a special qualification delivered on the basis of an oral examination and scientific publications that allowed the holder to teach at a university. The Racial Law against Jews promulgated by the Italian state in 1938 obliged him to leave the state schools; he became a teacher in a Jewish school. The memories of persons who met Artom described him as an exceptional teacher for his disciplinary competence, his human character, moral probity, and aspirations for social justice.[24]

Artom was author of textbooks and some 50 works of research in various aspects of mathematics and mathematics teaching. He collaborated to the collective work *Enciclopedia delle matematiche elementari e complementi* [Encyclopedia of elementary mathematics and complements]. For his work on the history of conics in elementary mathematics he was awarded with a national prize of *Mathesis*. The list of his publications in the *Bollettino* exemplifies the aspects that make a teacher a good teacher. Three articles concern some important mathematical concepts taught in primary school; in particular, one of them focuses on the passage from primary to secondary school. One article considers an element not inherent in the discipline that can affect learning mathematics, such as the role of memory. The article entitled "Sulla rappresentazione grafica delle coppie di grandezze dipendenti" [On the graphical representation of couples of dependent quantities] is remarkable (1913–1914, *15*, 200–202). Firstly, the interest lies in the fact that the subject, even not explicitly mentioned with its name, is functions, a concept not usually introduced in Italian schools. Secondly, the structure of the article resounds, though naively, with that of a modern article of mathematics education, consisting of a theoretical presentation followed by case studies to support the claims of the author. Artom's deep and open concern for problems of mathematics education is evidenced by his participation in the inquiry about the attitude towards mathematics launched by Gino Loria in the bulletin of Mathesis (Furinghetti 2002).

---

[23]Moscone (1965) reports that Artom was one of the three teachers with permanent tenure in Turin secondary schools who were never affiliated with the Fascist party.

[24]Useful information based on the documents and the Annual reports of the Liceo Scientifico "Galileo Ferraris" are provided by the text of a research carried out by two students of this institute (Alessandra Costa and Silvia Fanni) in the school year 2002–2003, as an activity in the historical laboratory of the Liceo Scientifico "Galileo Ferraris" of Torino (the teachers Barbara Garofani and Anna Maria Strumia were the supervisor), see http://share.dschola.it/galileoferraris/Attivit%20extracurricolari%20previste%20dal%20POF/Laboratorio%20storico/Laboratorio%20storico_Materiali%20e%20Testi/Profilo%20di%20Emilio%20Artom.pdf. Retrieved 23 November 2016.

We have identified 12 female authors: Bisson-Minio Ersilia, Capuzzo Adele, Crespi Isabella, Danesini Emilia, Frisone Rosetta, Giudici Maria Teresa, Maestro Ida, Magnani Teresa, Mancinelli Gemma, Quintili Pierina, [25]Vanni Elena, and Viriglio Luisa. They are primary school teachers or teachers in *Scuole Normali* with a degree in mathematics and, as in the case of Frisone, Quintili and Viriglio, doing research in collaboration with university mathematicians (Linguerri 2010; Luciano and Roero 2008). Their papers discuss elementary topics of arithmetic and geometry or teaching problems. Capuzzo and Mancinelli consider activities with concrete materials such as paper folding and cutting. Before the foundation of the *Bollettino,* Capuzzo contributed a note in the *Rivista di Matematica Elementare*, the first non-ephemeral journal addressed to mathematics teaching; she is a real pioneer in history of female engagement in math education. After being a teacher she became the *direttrice* [headmistress] in the *Regia Scuola Normale* of Grosseto.

The number of female authors could be a little higher, since some authors have not signed their articles with their full names,[26] in either case this number, though not reflecting the percentage of Table 2, is remarkable if compared with the numbers of female authors found in the mathematics journals addressed to school in Table 3 and in the study by Fenaroli et al. (1990) for women active in mathematical research. We add that the solutions of the questions and the problems, especially in the case of items addressed to students, are mainly due to women.

4 Conclusions

The journal *Il Bollettino di Matematiche e di Scienze Fisiche e Naturali* is a real window on the world of mathematics teaching in primary school. The substantial absence of academic mathematicians fostered spontaneity and freedom in fulfilling Conti's project and allowed a better adherence to the reality of schools.

Though at the moment of its founding, the main aim was to act on mathematical knowledge and to amend the frequent mistakes found in mathematics teaching in primary schools, the main achievement that we perceive in reading the journal is that it involved prospective and in-service teachers of primary school in the discussion of the programs and of the problems of mathematics teaching, thus making them protagonists of their prospective.

The analysis of the content shows that the journal was open to new methodologies, such as the use of manipulatives, and new topics, such as functions. Some ideas of the founder were (and still are) difficult to realize. One such case is overcoming the fragmentation of the teaching inherent in the programs of primary

[25]The articles are signed by Quintili P., but it is likely that they refer to Pierina Quintili, a pupil of Volterra. Crespi signed with the initial I. By crossing information, we can guess that the full name is Isabella.

[26]We have not found the full names of five authors of mathematics articles and six of sciences articles.

school, which contain many different topics. As discussed above, the journal was mainly a journal addressed to mathematics teaching, even if it contained contributions on the various branches of science (physics, geology, biology, etc.). Articles such that of Antonio Sartori on phyllotaxis, (1913–1914, *15*, 164–165, 196–199) which attempt to combine mathematics and sciences, are not usual in the journal.

The marked inclusiveness characterizing the spirit of the journal had an important side effect in encouraging women to collaborate as solvers of questions or as authors of contributions. This feature, which makes this journal so different from the main journal addressed to mathematics teaching of those years (*Periodico di Matematica*), mirrors the actual role of teachers in Italian primary schools.

Acknowledgements The authors thanks Maria Di Santo, Nicoletta Pavia and Antonio Salmeri for their invaluable help.

Appendix 1: The Editors and Their Schools During the Editorship

1899–1900, 1900–1901 (a. 2), 1902 (a. 3), 1902–1903 (a. 4), 1904 (a. 5), 1905 (a. 6), 1906 (a. 7), 1907 (a. 8), 1907–1908 (a. 9): Conti Alberto, R. Scuola Normale Femminile Anna Morandi Manzolini, Bologna.

1908–1909 (a. 10): Alberto Conti, R. Scuola Normale Femminile Margherita di Savoia in Rome.

1909–1910 (a. 11): Alberto Conti, R. Scuola Normale Femminile Margherita di Savoia in Rome with the cooperation of an editorial board.

1910–1911 (a. 12): Alberto Conti, R. Scuola Normale Femminile Margherita di Savoia in Rome, Lugi Tenca R. Scuola Normale Femminile Maffeo Vegio in Lodi.

1911–1912 (a. 13), 1912–1913 (a. 14), 1913–1914 (a. 15), 1914–1915 (a. 16), 1915–1916 (a. 17), 1916–1917 (a. 18): founded by Alberto Conti, R. Scuola Normale Femminile Margherita di Savoia in Rome and edited by Luigi Tenca, R. Scuola Normale Femminile Maffeo Vegio in Lodi.

Appendix 2: The Publishers/Printers

1899–1900 (a. 1), 1900–1901 (a. 2): Nicola Zanichelli, Bologna

1902 (a. 3), 1902–1903 (a. 4): Tipo-Litografia Luigi Pongetti Bologna

1904 (a. 5), 1905 (a. 6), 1906 (a. 7), 1907 (a. 8), 1907–1908 (a. 9): Tipografia di P. Cuppini succ. Cenerelli, Bologna

1908–1909 (a. 10): not indicated

1909–1910 (a. 11) Tipografia di P. Cuppini succ. Cenerelli, Bologna
1910–1911 (a. 12), 1911–1912 (a. 13), 1912–1913 (a. 14) Società tipografica succ. Wilmant, Lodi
1913–1914 (a. 15), 1914–1915 (a. 16), 1915–1916 (a. 17), 1916–1917 (a. 18) Premiata Tipografia successori Fratelli Fusi, Pavia

Appendix 3: Distribution of the Issues

1899–1900 (a.[27] 1), 1–12, from December 1899 to November 1900
1900–1901 (a. 2), 1–12, from December 1900 to November 1901
1902 (a. 3), 1–7, from January 1902 to July 1902
1902–03 (a. 4), 1–7, from December 1902 to June 1903
1904 (a. 5), 1–10, from January 1904 to December 1904
1905 (a. 6), 1–10, from January 1905 to December 1905
1906 (a. 7), 1–6, from January 1906 to December 1906
1907 (a. 8), 1–8, the months are not indicated in the cover, the issue 8 is indicated as the last issue of the year
1907–1908 (a. 9), 1–8, n. 1 has not month indicated, nn. 2 and 3 December 1907, from January1908 to the issue of April-May-June 1908 (indicated as the last issue of the year)
1908–1909 (a. 10), 1–8, from November 1908 to June 1909
1909–1910 (a. 11), 1–8, from November 1909 to June 1910
1910–1911 (a. 12), 1–8, from November 1910 to June 1911
1911–1912 (a. 13), 1–8, from November 1911 to June 1912
1912–1913 (a. 14), 1–8, from November 1912 to June 1913
1913–1914 (a. 15), 1–8, from November 1913 to June 1914
1914–1915 (a. 16), 1–8, from November 1914 to June 1915
1915–1916 (a. 17), 1–8, from November 1915 to June 1916
1916–1917 (a. 18), 1–7, from November 1916 to May 1917

References

Anonymous (1941). A. Conti. *Bollettino dell'Unione Matematica Italiana, s. 2, 3*, 350–351.
Atti della quinta unione degli scienziati italiani tenuta in Lucca nel settembre del MDCCCXLIII [Proceedings of the fifth meeting of the Italian scientists held in Lucca in September 1843) (1844). Lucca: Tipografia Giusti.

[27]a. stays for year.

Atti della settima adunanza degli scienziati italiani tenuta in Napoli dal 20 di settembre a' 5 di ottobre del MDCCCXLV [Proceedings of the seventh meeting of the Italian scientists held in Naples from September 20 until October 5 1845]. (1846). Napoli: Stamperia del Fibreno.

Bini, Giorgio (1981). Romanzi e realtà di maestri e maestre [Novels and reality of man and women teachers]. In C. Vivanti (Ed.), *Intellettuali e potere* [Intellectuals and power] (pp. 1195–1224). Torino: Einaudi.

Callegari, Carla (1998). *Alberto Cavezzali: un dirigente scolastico nella Treviso di fine Ottocento* [Alberto Cavezzali: a school director in Treviso in the late nineteenth century]. Ferrara: Ricerche Pedagogiche.

Canestri, Giorgio, & Recuperati, Giuseppe (1976). *La scuola in Italia dalla legge Casati a oggi* [School in Italy from Casati Law to today]. Torino: Loesher.

Castelnuovo, Guido (1909). *Atti del IV Congresso Internazionale dei Matematici* [Proceedings of the IV International Congress of Mathematicians] Vol. 1. Roma: Tipografia della R. Accademia dei Lincei.

Chiosso, Giorgio (Ed.) (1992). *I periodici scolastici nell'Italia del secondo Ottocento* [School periodical in Italy in the second half of the nineteenth century]. Brescia: La Scuola.

Ciamberlini, Corrado (1899). Libri di testo dal punto di vista scientifico e didattico. Errori che vi dominano, mezzi perché si limiti, per quanto si può, il danno che tali errori arrecano alla scuola [Textbooks from the scientific and didactic point of view. Mistakes in them, means for limiting, as far it is possible, the damage provoked by these mistakes in school]. *Periodico di Matematica, s. 2, 1*, 37–57.

Conti, Alberto (1899–1900). Due parole d'introduzione [Two words of introduction]. *Il Bollettino di Matematiche e di Scienze Fisiche e Naturali, 1*, 1–3.

Conti, Alberto (1902). L'insegnamento della Matematica elementare nelle scuole complementari e normali [The teaching of elementary Mathematics in the *scuole complementari* and *scuole normali*]. In *Atti del secondo congresso dei professori di matematica nelle scuole medie italiane) tenuto a Livorno nei giorni 17, 18, 19, 20, 21 e 22 agosto su iniziativa dell'Associazione Mathesis* [Proceedings of the second congress of mathematics teachers organized in Leghorn in August 17, 18, 19, 20, 21 and 22 by the Association *Mathesis*] (pp. 29–59). Livorno: R. Giusti.

Conti, Alberto (1907–1908a). Giuseppe Scoto. *Il Bollettino di Matematiche e di Scienze Fisiche e Naturali*, 9, 62–63.

Conti, Alberto (1907–1908b). Sull'iniziazione alle matematiche e sulla preparazione matematica dei maestri elementari in Italia [On the initiation to mathematics and on the mathematical preparation of primary school teachers in Italy]. *Il Bollettino di Matematiche e di Scienze Fisiche e Naturali*, 9, 81–91. Also in G. Castelnuovo (Ed.) (1909). *Atti del IV Congresso Internazionale dei Matematici* (Vol. 3, pp. 519–528). Roma: Tipografia della R. Accademia dei Lincei.

Conti, Alberto (1911a). L'insegnamento della matematica nelle Scuole infantili ed elementari [The teaching of mathematics in the infant and primary schools]. *Il Bollettino di Matematica, 10*, 213–248.

Conti, Alberto (1911b). L'insegnamento della Matematica nelle Scuole Normali [The teaching of mathematics in the *Scuole Normali*]. *Il Bollettino di Matematica, 10*, 249–318.

Conti, Alberto (1911–1912). Uno sguardo al passato e all'avvenire di questo "Bollettino" [Looking at the past and at the future of this "Bulletin"]. *Il Bollettino di Matematiche e di Scienze Fisiche e Naturali, 13*, 1–2.

Conti, Alberto (1936). I miei quaranta anni … [My first forty years …]. *Il Bollettino di Matematica, n. s., 15*, 89–96.

Conti, Alberto (1941†). Un quarantacinquennio di attività della "*Mathesis*" (1895–1940) [Forty-five years of activity of "*Mathesis*"]. *Atti della società italiana di scienze fisiche e matematiche "Mathesis", n. s., 4*(1), 9–24.

Conti, Alberto; Del Prete, Guelfo; Crepas, Emilio; & Mari, Mario (1900–1901). Discussione dei programmi per l'insegnamento scientifico nella Scuola Complementare e nella Scuola Normale

[Discussion on the programs for the scientific teaching in the *Scuola Complementare* and the *Scuola Normale*]. *Il Bollettino di Matematiche e di Scienze Fisiche e Naturali, 2*, 74–86.
Covato, Carmela (2012). Maestre d'Italia. Uno sguardo sull'età liberale [Women teachers of Italy. A view of the liberal age]. *Storia delle donne, 8*, 165–184.
Dei, Marcello (1998). *La scuola in Italia* [School in Italy]. Bologna: Il Mulino.
Despaux, Sloan E. (2011). Victorian mathematical journals and societies. In R. Flood, A. Rice, & R. Wilson (Eds.), *Mathematics in Victorian Britain* (pp. 155–176). Oxford: OUP.
Di Pol, Redi Sante (1996). *Scuola e popolo nel riformismo liberale d'inizio secolo* [School and people in the liberal reformism of the beginning of the century]. Torino: Sintagma Editrice.
Fenaroli, Giuseppina; Furinghetti, Fulvia; Garibaldi, Antonio C.; & Somaglia, Anna (1990). Women and mathematical research in Italy during the period 1887–1946. In L. Burton (Ed.), *Gender and mathematics. An international perspective* (pp. 144–155). London: Cassell.
Fort, Ester de (2015). Maestri e maestre in Italia dalla fine dell'Antico Regime alla salita al potere del fascismo. Nascita e sviluppo di una professione [Women and men teachers in Italy from the fall of the Old Regime to the rise of fascism. Birth and development of a profession]. *Historia y Memoria de la Educación, 1*, 131–201.
Furinghetti, Fulvia (2002). Il *Bollettino della Mathesis* dal 1909 al 1920: pulsioni tra temi didattici internazionali e nazionali [The *Bollettino della Mathesis* from 1909 to 1920: drives between national and international didactic themes]. *PRISTEM/Storia. Note di matematica, Storia, Cultura, n. 5*, 31–58.
Furinghetti, Fulvia (2003). Mathematical instruction in an international perspective: the contribution of the journal *L'Enseignement Mathématique*. In D. Coray, F. Furinghetti, H. Gispert, B. R. Hodgson, & G. Schubring (Eds.), *One hundred years of L'Enseignement Mathématique*, Monographie n. 39 de *L'Enseignement Mathématique*, 19–46.
Furinghetti, Fulvia (2006). Due giornali ponte tra ricerca e scuola: la *Rivista* di Peano e il *Bollettino* di Loria [Two journals bridging research and school: the *Rivista* of Peano and the *Bollettino* of Loria]. In L. Giacardi (Ed.), *Da Casati a Gentile. Momenti di storia dell'insegnamento secondario della matematica in Italia* [From Casati to Gentile. Moments of the history of mathematics teaching in the secondary schools of Italy] (pp. 181–237). Livorno: Agorà.
Furinghetti, Fulvia (2009). The evolution of the journal *L'Enseignement Mathématique* from its initial aims to new trends. In K. Bjarnadóttir, F. Furinghetti, & G. Schubring (Eds.), *"Dig where you stand". Proceedings of the conference on "On-going research in the History of Mathematics Education"* (pp. 31–46). Reykjavik: University of Iceland – School of Education.
Furinghetti, Fulvia (2012). Secondary teachers in the unified Italy: a group portrait with a zoom. In K. Bjarnadóttir, F. Furinghetti, J. M. Matos, & G. Schubring (Eds.), *"Dig where you stand" 2. Proceedings of the second "International Conference on the History of Mathematics Education"* (pp. 181–201). Caparica: UIED.
Furinghetti, Fulvia (2017). The mathematical journals for teachers and the shaping of teachers' professional identity in post-unity Italy. In K. Bjarnadóttir, F. Furinghetti, M. Menghini, J. Prytz, & G. Schubring (Eds.), *"Dig where you stand" 4. Proceedings of the fourth "International Conference on the History of Mathematics Education"* (pp. 101–116). Rome: Edizioni Nuova Cultura.
Furinghetti, Fulvia, & Giacardi, Livia (2012). Secondary school mathematics teachers and their training in pre- and post-unity Italy (1810–1920). *ZDM – The International Journal on Mathematics Education, 44*(4), 537–550.
Furinghetti, Fulvia, & Somaglia, Annamaria (1992). Giornalismo matematico 'a carattere elementare' nella seconda metà dell'Ottocento [Mathematical journalism 'of elementary type' in the second half of the nineteenth century]. *L'Insegnamento della Matematica e delle Scienze Integrate, 15*, 815–852.
Furinghetti, Fulvia, & Somaglia, Annamaria (2005). Emergenza della didattica della matematica nei primi giornali matematici italiani [The emergence of didactics of mathematics in the first Italian mathematical journals]. In D. Moreira & J. M. Matos (Eds.), *História do ensino da*

Matemática em Portugal [History of mathematics teaching in Portugal] (pp. 59–78). Beja: Sociedade Portuguesa de Ciências da Educaçao.

Gallarati, Dionisio (1999). Giovanni Garbieri. *Dizionario biografico degli italiani, 52.*

Genovesi, Giovanni (2007). *Storia della scuola in Italia dal Settecento a oggi* [History of the school in Italy from the eighteenth century to today]. Roma-Bari: Laterza.

Giacardi, Livia (Ed.) (2006a). L'insegnamento della matematica in Italia dall'Unità all'avvento del Fascismo [The teaching of mathematics in Italy from the Unification to the Fascism]. In *Da Casati a Gentile. Momenti di storia dell'insegnamento secondario della matematica in Italia* [From Casati to Gentile. Moments of the history of the mathematics teaching in the secondary schools of Italy] (pp. 1–63). Lugano: Lumières Internationales.

Giacardi, Livia (Ed.) (2006b). *Da Casati a Gentile. Momenti di storia dell'insegnamento secondario della matematica in Italia* [From Casati to Gentile. Moments of the history of the mathematics teaching in the secondary schools of Italy]. Lugano: Lumières Internationales.

Giacardi, Livia (2012). «Pel lustro della Scienza italiana e pel progresso dell'alto insegnamento». L'impegno dei matematici risorgimentali [For the glory of the Italian science and for the progress of higher education]. In A. Ferraresi & E. Signori (Eds.), *Le Università e l'Unità d'Italia (1848–1870)* [The universities and the unification of Italy (1848–1870)] (pp. 233–254). Bologna: CLUEB.

Giacardi, Livia, & Scoth, Roberto (2014). Secondary mathematics teaching from the early nineteenth century to the mid-twentieth century in Italy. In A. Karp & G. Schubring (Eds.), *Handbook on history of mathematics education* (pp. 201–228). New York - Heidelberg - Dordrecht - London: Springer.

La Direzione (1940). Alberto Conti. *Il Bollettino di Matematica, s. 4, 1*(1), 81–84.

Laisant, Charles-Ange (1905). *L'éducation fondée sur la science* [The education founded on science] (second edition). Paris: Félix Alcan.

Laisant, Charles-Ange (1906). *Initiation mathématique* [Mathematical initiation]. Paris: Librairie Hachette.

Linguerri, Sandra (2010). *Un matematico un po' speciale: Vito Volterra e le sue allieve* [A bit special mathematician: Vito Volterra and his female pupils]. Bologna: Pendragon.

Loria, Gino (1905). Programmi del passato e programmi per l'avvenire [Programs of the past and programs for the future]. *Bollettino della Associazione "Mathesis" fra gli Insegnanti di Matematica delle Scuole Medie, 9*, 71–80.

Luciano, Erika, & Roero, Chiara Silvia (Eds.) (2008). *Numeri, atomi e alambicchi. Donne e scienza in Piemonte dal 1840 al 1960* [Numbers, atoms, and alembics in Piedmont from 1840 to 1960], Parte I, Torino: Centro Studi e Documentazione Pensiero Femminile.

M.P.I. (1909). *Commissione reale per l'ordinamento degli studi secondari in Italia. Relazione* [Royal Commission for the regulations of secondary schools] (Vol. 1). Roma: Tipografia ditta Ludovico Cecchini.

Maracchia, Silvio (1975). Necrologio di Alpinolo Natucci [Obituary of Alpinolo Natucci]. *Archimede, 27*, 56–58.

Moscone, Francesca (1965). *Il D'Azeglio e la Resistenza* [The D'Azeglio and the Resistance]. Torino: Editoriale Rattero.

Pak, Julie Kazdan (2012). *Italy's primary teachers: The feminization of the Italian teaching profession*. Ph.D., History 0429, UCLA.

Pizzarelli, Chiara (2017). Teaching of mathematics in school periodicals: the case of the Piedmont (1849–1894). In K. Bjarnadóttir, F. Furinghetti, M. Menghini, J. Prytz, & G. Schubring (Eds.), *"Dig where you stand" 4. Proceedings of the fourth "International Conference on the History of Mathematics Education"* (pp. 293–308). Rome: Edizioni Nuova Cultura.

Procissi, Angiolo (1960). Luigi Tenca (1877–1960), *Bollettino dell'Unione Matematica Italiana, s. 3, 15*, 466–468.

R. G. (1960). La scomparsa del prof. Luigi Tenca [The death of prof. Luigi Tenca]. *Archimede, 12*, 268 (likely R. G. is Roberto Giannarelli).

Rossi, Francesco Saverio (1983). Alberto Conti. *Dizionario biografico degli italiani, 28.*

Santoni Rugiu, Antonio (1982). Orientamenti culturali, strumenti didattici, insegnanti e insegnamenti [Cultural orientations, didactic tools, teachers and teachings]. In A. Santoni Rugiu (Ed.), *Storia della scuola e storia d'Italia dall'Unità ad oggi* [History of the school and history of Italy from the unification to today] (pp. 9–23). Bari: De Donato.
Santoni Rugiu, Antonio (2006). *Maestre e maestri. La difficile storia degli insegnanti elementari* [Women and men teachers. The difficult history of primary school teachers]. Roma: Carocci.
Schön, Donald A. (1983). *The reflective practitioner. How professionals think in action*. New York: Basic Books.
Silla, Lucio (Ed.) (1939–1940). *Un secolo di progresso scientifico italiano 1839–1939* [A century of Italian scientific progress]. (Seven volumes). Roma: Società Italiana per il Progresso delle Scienze.
Tenca, Luigi (1959). Matematici combattenti. *Archimede, 11*(4), 186–190.
Togliatti, Eugenio (1975). Necrologio di Alpinolo Natucci [Obituary of Alpinolo Natucci]. *Bollettino UMI, s. 4, 12*, 425–426.
Tricomi, Francesco G. (1962). Matematici del primo secolo dello Stato unitario [Mathematicians of the first century of the unified State]. *Memorie dell'Accademia delle Scienze di Torino, Classe Scienze Fisiche, Matematiche, Naturali, s. 4, 1*, 1–120.
Vita, Vincenzo (1986). *I programmi di matematica per le scuole secondarie dall'Unità d'Italia al 1986* [The programs of mathematics for the secondary schools from the unification of Italy to 1986]. Bologna: Pitagora Editrice.
Volterra, Vito (1909). Le matematiche in Italia nella seconda metà del secolo XIX [Mathematics in Italy in the second half of the nineteenth century]. In G. Castelnuovo (Ed.), *Atti del IV Congresso Internazionale dei Matematici* (Vol. 1, pp. 55–65). Roma: Tipografia della R. Accademia dei Lincei.

Author Biographies

Fulvia Furinghetti is professor emerita of Mathematics Education in the Department of Mathematics (University of Genoa). Her research concerns: beliefs, images of mathematics in society, proof, problem solving, use of history of mathematics in teaching, teacher professional development, history of mathematics education. She has organized the celebrations of the Centenary of the journal *L'Enseignement Mathématique* and of ICMI and was one of the editors of the proceedings. She has developed the website on the history of the first hundred years of ICMI. She wrote the chapter "From mathematics and education to mathematics education" in the *Third International Handbook of Mathematics Education* and the chapter on the history of the international cooperation in mathematics education in the *Handbook on the history of mathematics education*. In 2000–2004 she chaired HPM (the International Study Group on the relations between History and Pedagogy of Mathematics affiliated to ICMI).

Annamaria Somaglia is a retired mathematics teacher in Italian *Scuola Secondaria Superiore* (students of ages 14–19). Also, she taught university courses for teacher training and professional development. Her research concerns history of mathematics, history of mathematics education, and the integration of history in mathematics teaching.

Russian Mathematics Teachers, 1830–1880: Toward a Group Portrait

Alexander Karp

Abstract This paper is a study of the lives of Russian mathematics teachers during the middle of the nineteenth century. The study of the life of "ordinary" people—as opposed to only renowned ones—has long ago began to attract researchers. Nor have "ordinary teachers" been ignored. But such studies (at least in Russia) have usually concerned teachers in general, without distinguishing them by subject, and the periods studied were also later ones, about which more evidence has survived. Meanwhile, the teachers of different subjects had their own differences, and the period to which this study is devoted also had its important distinctive features. This paper studies various aspects of the lives of teachers: their origins, education, workload, involvement in methodological work, material circumstances, and so on. The present study relies mainly on materials from St. Petersburg archives and focuses mainly on St. Petersburg teachers.

Keywords Mathematics teacher · Biographies
Mathematics education methodology · Russia

1 Introduction

This paper is a continuation of this author's study (Karp 2014), which in turn constituted a continuation of Prudnikov (1956). In order to understand the development of mathematics education, it is important to study the biographies of those who have been active in mathematics education, including not only prominent figures, such as the authors of the most popular textbooks, but average figures as well—ordinary teachers. In recent times, there have been several discussions of the biographies and daily lives of Russian teachers (see, for example, Gurkina 2014; Zubkov 2010). These studies, however, typically focus on somewhat later periods than the one addressed in this paper, and most importantly, they are devoted to

---

A. Karp (✉)
Teachers College, Columbia University, New York, USA
e-mail: apk16@columbia.edu

F. Furinghetti and A. Karp (eds.), *Researching the History of Mathematics Education*, ICME-13 Monographs,
https://doi.org/10.1007/978-3-319-68294-5_6

teachers in general, without distinguishing them by subject. Naturally, such an approach is also legitimate; the basic characteristics of a teacher's existence were common to all teachers, regardless of their subject. Nonetheless, by no means everything was the same in all cases: there were specific differences in the ways in which teachers acquired their education, in the requirements associated with their subjects, in the different positions of their subjects within the schools—and these differences merit discussion. This paper is devoted to the study of the biographies and everyday life specifically of mathematics teachers.

2 Some Remarks on Methodology

The study of collective biographies of relevant groups of persons (prosopography), has long become an accepted methodology in the history of science (Shapin and Thackray 1974). However, the relevant group that is being studied must be precisely defined. The teaching of mathematics took place at different levels, and although there were connections between these levels, the teaching of mathematics was often carried out by individuals of quite different types and social positions. We must therefore note at the outset that the present study relies first and foremost on materials from St. Petersburg's archives and consequently is devoted first and foremost to St. Petersburg's teachers, mainly teachers at gymnasia and *real schools* (*real'nye uchilischa*), that is, institutions that were attended by a quite small (although very important) segment of the population. Zubkov (2010, p. 299) estimates the number of gymnasium and real school teachers as constituting 9% of the overall number of teachers in Russian elementary and secondary schools by 1897. During the period discussed here, this percentage was unlikely significantly greater. Nonetheless, it was specifically this group of teachers that prepared what would become the most cultured segment of the Russian population.

The present article examines the period from 1830 to 1880, but the chosen time frame is not rigid, and the years 1831 and 1881 could have been chosen just as well. The end of the 1820s saw a certain regularization of mathematics education and of education in general. On the other hand, during the last decades of the nineteenth and the first decades of the twentieth centuries, the number of students being taught mathematics substantially increased, and this was accompanied by an increase of interest in methodology and a certain revolution in education in general (Karp 2015). The period in question encompasses the reigns of two czars almost in their entirety—that of Nicholas I and that of Alexander II. These reigns were very different in spirit and in the attitudes that prevailed in educated society. This article, however, focuses not on the differences within this period, but on its common features. And such features undoubtedly existed. A certain professional class was taking shape, whose members had similar origins, education, and careers.

In order to understand the distinctive nature of Russian education, it should be borne in mind that since the time of Peter I, the so-called *Table of Ranks* was in force: there were 14 grades for those serving in the army and navy, and in the civil service (and being granted even the lowest, fourteenth grade, was already an important step for very many). Titular counselors and infantry captains, for example, were in the ninth grade, while the eighth grade was granted to collegiate assessors and majors, and the fifth grade to state councillors and brigadiers (this title was later discontinued). Special distinction was given to "persons of the first four grades," that is, generals. Individuals who belonged to the lowest grades (up to the ninth) were referred to as junior officers (*ober-ofitsery*); in civil service, the junior officer rank did not endow an individual with hereditary nobility.

To repeat, this study is based on materials from St. Petersburg's archives, first and foremost the curricula vitae or so-called "formulary lists" of teachers, that is, official documents, filled out in a uniform manner for all civil servants, which summarized their biographies and all the details of their employment history. We have studied practically all surviving lists for teachers employed at the most privileged state-run public institutions for boys over the period in question (gymnasia and real schools) and the formulary lists of teachers from a number of other institutions (Commercial Academy, Navy Corps), although naturally the number of educational institutions which it has been possible to study falls far short of the total number of educational institutions even in St. Petersburg alone. Naturally, this study does not discuss all of the examined biographies, but only those which in our view are the most representative and permit us to sketch the contours of what was most typical. Certain other relevant archival and printed materials (for example, publications devoted to specific educational institutions) were also examined.

3 Beginnings

Mathematics education, which appeared in Russia effectively during the reign of Peter I, developed rather slowly. Neither in the eighteenth, nor even in the beginning of the nineteenth century, were there enough qualified and educated teachers of mathematics (indeed, the need for them was often not well recognized). We frequently encounter marginal figures among teachers, and teaching conditions themselves were far from what we usually expect (Karp 2007, 2014). The teachers of mathematics in the eighteenth and early nineteenth centuries were foreigners—beginning with the famous Scotsman Fargwarson; military men—ranging from the prototypes of the character of Tsifirkin in the famous play *The Minor* to far more educated graduates of military academies; and, finally, Russians, in gradually increasing numbers, who had received their mathematics education at civil educational institutions—universities and pedagogical institutes.

Gradually, the situation in education changed. In 1826, Emperor Nicholas I wrote:

> Reviewing with especial attention the organization of the educational institutions in which Russian youth is educated to serve the state, I regret to observe that they lack that necessary and indispensable uniformity which must be the foundation of both child rearing and instruction.[1] (Perepiska 1826–1828, p. 1)

Both the content and the administrative-organizational side of education began to be pushed toward an increasingly rigid uniformity. Far more attention than before also began to be devoted to who was doing the teaching. (An example of this is furnished by a correspondence, discovered by the author in the archives, which deals with the question of whether a certain foreigner named Schors was in possession of a certificate qualifying him to teach—for unknown reasons the emperor suddenly became concerned about this issue, and the head of the Gendarmerie, a minister, and other highly-placed officials rushed to determine how matters stood—Po otnosheniu grafa Benkendorfa 1831). This new period is the subject of the present paper.

4 Gymnasia and Mathematics in Gymnasia

Gymnasia occupy a key place in this paper, so a few words must be said about them. Within the complicated system of Russian school education, gymnasia were the main, and indeed the only form of non-military secondary educational institution that provided students with every necessary right to further education—even the so-called *real schools*, which appeared after a certain time, provided students with only partial rights (which varied at different stages). The number of gymnasia, by comparison with the size of the population, was very small. Even in 1915 in St. Petersburg, the capital, there were only 15 state-run (so-called governmental) gymnasia (Nikoltseva 2000), while the city's population was 2,315,000.

The curricula for teaching mathematics, as well as the number of hours allocated to mathematics classes, changed over the course of the nineteenth century. Nonetheless, the mathematics curriculum as a whole always remained intensive. To give an example, we will summarize the curriculum used at St. Petersburg's Gymnasium No. 1 in 1838, grade by grade (Soloviev 1880). There were six grades at that time.

In first grade, students were taught arithmetic, including fractions. In second grade, they got to proportions and the rule of three. In third grade, they completed the course in arithmetic, learning to find square roots and cube roots, and began algebra, which they covered up to algebraic fractions and first-degree equations (including equations with several variables). In fourth grade, they completed the course in algebra, covering quadratic equations, combinatorial mathematics, and

---

[1]All translations from Russian are by the author.

Newton's binomial, as well as progressions. In the same grade, they began studying plane geometry, covering material up to regular polygons and finding circumferences (and touching on the theory of limits). In fifth grade, they completed plane geometry and studied solid geometry. Finally, in sixth grade, the covered plane trigonometry, the theory of logarithms, and analytic geometry.

The classes were demanding, and unsatisfactory results were not infrequent. For example, an inspection carried out in 1865 yielded the following results for the fourth grade (in two sections): satisfactory grades—22; unsatisfactory grades—10; not questioned—18; and these results were deemed satisfactory (Soloviev 1880, p. 341).

5 Teachers at Gymnasia: Several Lives

Curricula vitae and other documents pertaining to the employment and education of teachers have been preserved in archives. One example is offered by the biography of Ivan Konovalov, a teacher at St. Petersburg Gymnasium No. 1 (Ob uchitele… 1838). The son of a work-assignment clerk at the court, Konovalov was admitted as a state-sponsored/independent (*kazenno-vol'nyi*) pupil at a second-order teachers college in 1820, and subsequently transferred to a St. Petersburg gymnasium, later called Gymnasium No. 3. In 1827, he enrolled as a student at St. Petersburg Imperial University, graduating from it with a candidate's degree (which was the highest possible degree that one could earn at the university and granted the graduate the right to a rank of grade 10 immediately upon starting work, by contrast with the also possible title of "actual student", whose holders received a rank of grade 12) from the physics-mathematics department in 1830. In 1831, Konovalov became a senior teacher of mathematics at Gymnasium No. 1, and in 1834 he began working at the Academy of Arts, becoming a professor there in 1836. In 1838, having attained the grade of collegiate assessor, he left the gymnasium for the Academy of Arts.

A later example is offered by the sad fate of Ivan Varaksin (O sluzhbe prepodavatelya matematiki Varaksina I.Z. 1882), whose career as a teacher at Gymnasium No. 3 was cut short. As a junior officer's child, Varaksin completed the Pskov gymnasium in 1857, and in 1862 he received a candidate's degree from the physics-mathematics department of St. Petersburg University. At this time, he went north, becoming a senior teacher at the Olonets gymnasium. Later, Varaksin taught at a women's institute; in 1866 he was transferred to the Kronstadt gymnasium; and in 1880, he was appointed inspector at the Pskov gymnasium. In 1867, he was promoted to collegiate assessor, in 1869 to court councillor, in 1873 to collegiate councillor, and finally in 1877 to state councillor (5th rank). Varaksin received various honors. In 1881, he was transferred as a mathematics teacher to St. Petersburg Gymnasium No. 3, but become ill and died in the same year, at the age of 42.

Another example is the case of the much more famous teacher Irakliy Petrovich Vereschagin. Vereschagin was famous mainly as an author of problem books in arithmetic and trigonometry (Vereschagin 1883, 1884, 1886). The renowned playwright and his former student Pyotr Gnedich (2000) spoke about him in quite ironic terms, referring to his near-sightedness and "face, which seemed covered in dust." The writer was especially amused, however, by Vereschagin's love for mathematics, which led Vereschagin to exclaim: "How can one not love mathematics?… It is everything! It is a revelation! Using mathematics, man becomes god" (p. 35). Vereschagin's problem books, however, remained consistently in demand, as may be judged for example by the numerous editions of collections of solutions to his problems (for example, Nazarin 1916).

Vereschagin was born in 1846 into a family that was described in his curriculum vitae as "low-middle" (*meschane*) In 1868, he graduated from St. Petersburg University, and in 1869 he began teaching at the Kronstadt boys' gymnasium—and in 1870 at the girls' gymnasium as well. (The authoritative pre-Revolutionary Brockhaus and Efron encyclopedic dictionary explains that he was urged to remain at the university "to prepare for the title of professor," but that he himself refused due to poor health.) The following year, Vereschagin was transferred to a commercial academy, and in 1873 he began teaching at St. Petersburg's Progymnasium No. 4 (the "progymnasium" was an educational institution consisting of the first grades of a gymnasium), where he again did not stay long, beginning in 1874 to work at St. Petersburg Gymnasium No. 1. Meanwhile, little by little he climbed through the ranks—grade 10 (collegiate secretary) was officially granted to him only in 1875, but effectively he was counted as having this rank since 1869 (because Vereschagin had graduated from the university with the degree of candidate). In 1876, he was promoted to titular councillor; in 1877, to collegiate assessor; and in 1879, he was nominated for the rank of court councillor. In 1877, Vereschagin was awarded the Order of St. Stanislaus, third degree, and three years later he was recommended for the Order of St. Anna, third degree. His work was not confined to Gymnasium No. 1: in 1875, he started working at the Foundling Home Academy (*Vospitatel'noe uchilische*), and in 1877 also at Gymnasium No. 6 (O prepodavatele matematiki Vereschagine 1884).

A mathematics teacher could, of course, spend his whole life working as a mathematics teacher, but he could also become an administrator (which opened the way to higher ranks, at the level of generals). One example of such a mathematician-administrator was Ivan Fyodorovich Knorring, who ultimately became the director of St. Petersburg Gymnasium No. 4 (the so-called Larinskaya Gymnasium). He was born in 1824 in the Tula gubernia into the family of a pharmacist, and completed a gymnasium in the North, in Olonets. In 1847, he received the title of candidate at St. Petersburg University, and in 1848 he was appointed senior teacher for the lower grades at the Olonets gymnasium. In 1848, he was transferred to the position of senior teacher for the upper grades at the same gymnasium. In 1852, Knorring became a titular councillor, and in 1854 a collegiate assessor. In 1853, he served as acting inspector due to the actual inspector's ill health, and he did so again in 1855–1856. Knorring received various kinds of support—a commendation from the region supervisor of schools, and in 1854 a commendation during the inspection of the gymnasium—and was the beneficiary of monetary gifts as well (which, however, were not rare in general). In 1856, he was transferred to teach in the capital, as a mathematics teacher at Gymnasium No. 1 (Ob uchitele matematiki Knorringe 1856). But he did not stay there long: in 1858, he began working as an inspector at the Larinskaya Gymnasium, and somewhat later he became acting director of the gymnasium. In 1858, he became a court councillor. He received monetary gifts in this position, too (now not the usual 150 rubles, as in Olonets, but 500), and subsequently was awarded orders as well: the Order of St. Anna, third degree, and the Order of St. Stanislaus, second degree. In 1862, Knorring became the director of Gymnasium No. 6, and in 1864 he was brought back to the Larinskaya Gymnasium, now as its official director. (O sluzhbe direktora 6 gimnazii Knorringa 1862–1864). In this position, he successfully attained the rank of active state councillor, heading various methodological committees in mathematics while serving as director of the gymnasium (O periodicheskikh sobraniyakh uchiteley odnorodnykh predmetov 1882).

The opposite example—in a certain sense—is presented by the life of Ivan Yankevich, a teacher at Gymnasium No. 3. Yankevich came from impoverished Lithuanian gentry (but he himself was already Russian Orthodox). His father's documents, which have been preserved in Yankevich's file, allow us to compare the possibilities for a career in different types of service. His father had entered the artillery arsenal as a gun carriage maker, and after years of diligent service—which won him various awards—he reached the rank of titular councillor in 1843. Born in 1833, Ivan graduated from the St. Petersburg Pedagogical Institute with a gold medal and stayed on to prepare for a master's title, but in 1856 he was appointed to Gymnasium No. 3 as a senior mathematics teacher. He also worked at the Nikolayevskaya Academy on a part-time basis. On track to receive the rank of titular councillor in the early 1860s, he became involved in various shady financial dealings. Letters of complaint about him began to appear—with allegations of failure to pay his debts—and most importantly he was implicated in a theft of treasury notes and arrested. The gymnasium supported Yankevich in various ways, appointing—as indeed it was required to do—a special representative to his case. This was Vladimir Stoyunin, a leading figure in Russian philological education. In the end, Yankevich was released, but the debts had to be paid: he continued working at the gymnasium, and they began to be deducted from his salary. In 1862, Yankevich died, leaving a mother and a sister without any means. In support of their petition for a pension, the director of the gymnasium wrote: "Personally, as a teacher, the late Yankevich was among the most capable and active of my colleagues, facilitating the development of our pupils. He devoted almost six of the best years of his life to... the Gymnasium" (O naznachenii g. Yankevicha 1856–1862, p. 92).

The information collected at the Larinskaya Gymnasium concerning the teachers who worked at the school from 1835 until 1863, the year when the document was prepared, is representative in several respects (Svedeniya o chinovnikakh... 1863). Of the mathematics teachers (of different categories, including so called *sverkhshtatnykh* teachers, that is those not officially included in the personnel) named in the document, four were of noble birth, three were children of junior officers, one was the son of a priest, one was described as having a low-middle class background (*meschane*), and one came from a family of *raznochintsy* ("persons of miscellaneous ranks"). Russian Orthodox teachers predominated—only two of those mentioned on the list were Protestants. The teachers received their higher education at the Pedagogical Institute and at St. Petersburg University (in about equal numbers), and one had studied at the University of Derpt.

Those seeking employment in schools, like other civil servants, were required to provide certain evidence of their political trustworthiness. Gymnasium employment files usually contain an oath of loyalty to the czar, signed by teachers (including mathematics teachers), with a declaration to the effect that they did not now and had not ever belonged to any secret society (see, for example, Ob opredelenii v gimnaziu raznykh chinovnikov 1853). Later, the ministry began putting out special lists of persons who were banned from pedagogical work for political reasons (see, for example, O litsakh, koim zapreschena pedagogicheskaya deyatel'nost' 1880). The reasons in such cases could range anywhere from disagreement with Christian doctrine to such "unfavorable conduct" as the intention to commit suicide; political unreliability, of course, could be cited as a reason as well, and individuals could be deemed politically unreliable for such actions, for example, as storing and distributing banned books. Such restrictions formed a general background to all pedagogical activity and were in no way specific to the teaching of mathematics.

6 Real Schools and Their Mathematics Teachers

The "real school" (*real'noe uchilische*) in Russia was based on the model of the German Realschule. The history of real schools in Russia should probably begin in the year 1864, when a statute concerning real gymnasia was passed, although educational institutions or classes with similar aims and objectives had already existed prior to this (Ganelin 1954). These schools went through various transformations, and were even designated in a number of different ways, sometimes as gymnasia, sometimes as academies. They were secondary educational institutions, similar in position to the classical gymnasia, but as has been noted above, somewhat inferior to the latter. Thus, although their graduates could enroll in all possible institutions of higher technical education, they could not enroll in universities—at one time, not at all, and later, only in certain departments. Their programs in mathematics were in certain respects more extensive than those in the gymnasia, but by and large they remained quite similar to the latter. What is most important for us is that there was practically no difference between mathematics teachers at state-run real schools and mathematics teachers at state-run gymnasia—it was not uncommon for a teacher to transfer from a gymnasium to a real school, or vice versa; to be employed simultaneously at different educational institutions; and so forth.

One example of a teacher who worked at a real school (albeit not for very long) is the well-known mathematician Matvey Tikhomandritsky (1844–1921). Of noble birth (thanks to his father, also a professor and a mathematician, who had been one of the top graduates from the Tver seminary, and then one of the top students at the St. Petersburg Main Pedagogical Institute), Matvey Tikhomandritsky received a candidate's degree from St. Petersburg University in 1865, began working as a so called *vospitatel'* (instructor responsible mainly for students behavior and similar matters) at the Progymnasium No. 2 in 1867, and in 1871 was transferred to the position of *nadziratel'* (class supervisor, again basically having mainly responsibilities not connected with any subject matter) teaching mathematics when the mathematics teacher was absent due to illness. In 1873, after passing the examinations for the master's degree, and being already a court councillor, he transferred to the newly opened Real School No. 2, where he spent several years working as a teacher of mathematics (Ob opredelenii Matveya Tikhomandritskogo 1879). Subsequently, Tikhomandritsky became a professor at Kharkov University. In general, it was not uncommon for future professors to work at a gymnasium or real school before obtaining a position at a university or other higher educational institution.

7 Teachers at Military Institutions

The system of educational institutions that existed before 1917 was very varied and heterogeneous. An important part of this system was made up of military secondary educational institutions, such as the cadet corps and so forth. Although we will not attempt to describe mathematics teachers at all existing military institutions, nonetheless a word should be said about the lives of teachers at one military institution: the Navy Cadet Corps, which had grown out of Russia's oldest school, the Mathematical and Navigational School. Already in the eighteenth century, this school to a certain extent prepared its own staff. A career path that was common at the time is described in the following report from 1784:

> Vasiliy Abrosimov and Fyodor Yakimov, while still classroom students, began to teach mathematics and through their great zeal in teaching achieved good results in their own pupils. In 1784, having been promoted to apprentices in the mathematical sciences, they continued to teach mathematics with as much enthusiasm and success as before. Finally, being deemed capable, they were assigned classes in Navigation, which they continue to teach with all manner of diligence and promise of success in their students. (Perepiska ob uchitelyakh 1783–1791, p. 73).

As a consequence, it was proposed that these two individuals should be promoted to full teachers.

This does not mean that all teachers at the Corps were even then prepared in only this way. The "chief inspector of classes," who made decisions such as the one just referred to, was Vasiliy Nikitin (Karp 2014), who had a degree from Oxford. Nonetheless, the practice of preparing its own staff endured at the Corps even in the nineteenth century.

One such teacher prepared at the Navy Cadet Corps, for example, was Vsevolod Chirkov, a junior officer's son originally admitted for study (in 1817) to the so-called third level of the Corps, who had by 1825 advanced to the first level, and who began to teach at the Corps in 1826. Only in 1830 was he granted a rank of the fourteenth grade, and only 15 years later (in 1845) did he become collegiate assessor—an official of the eighth grade (Formulyary 1849). Chirkov's career may be considered quite successful: far from all of his colleagues attained such ranks.

Another teacher working at the Corps during the same years as Chirkov was Nikolai Fus, a University of Derpt (currently, Tartu) graduate and one of the mathematics teachers about whom information was collected in the aforementioned Larinskaya Gymnasium (where he also taught). Fus was four years younger than Chirkov, but by 1849 he was two ranks above him, due primarily to the fact that he had obtained his first ranks much faster than Chirkov.

In general, the world of lower-rank teachers was markedly different from the more refined world of the higher ranks. Evidence to this effect may be found in a surviving letter of complaint to the director of the Corps, the famous Admiral Krusenstern, from a certain Lysina, the wife of an official of the ninth rank, against the same Chirkov, which reads as if it came out of the classic Russian literature of the time (O vydache spravok 1837). The cause for the complaint lay, naturally, not in the teaching of mathematics, but in a practical conflict: Lysina had begun to scold a certain Matryona Yakovleva, the wife of a carpenter who had previously resided in the house of her sister, when all of a sudden

> Cherkov [spelled incorrectly], a twelfth-grade teacher at the Navy Cadet Corps, jumped out of the side building, and with an expression of rage on his face and an unnatural appearance, forgetting about the honor and respect due to a well-born woman who was older than he was, began with vehemence to shout at me so loudly that passersby on the street stopped to watch this shameful scene; he used words that were most indecent and offensive to me; and he even wanted to spit in my face; and incidentally, he said that he was forever Matryona Yakovleva's faithful defender and protector (p. 39).

The insulted Lysina requested "an inquest into Chirkov's conduct, that judgment might be passed upon him, and that [she might] in this way receive lawful satisfaction" (p. 40).

An inquest was begun, and Chirkov provided an explanation, blaming everything on Lysina, who had allegedly subjected him to every manner of insolence, so that he even "desired to know who she was that she permitted herself to speak with him so insolently, seeing that I was wearing an officer's uniform?" (p. 42).

Rear Admiral Gorkovenko, the class inspector who investigated the matter, finally reported to Krusenstern that "Chirkov committed no transgression other than a certain indiscretion and impropriety in the way in which he expressed himself, and furthermore I have the honor to report that Chirkov's conduct has always been good and his work excellent" (p. 43). The scenes preserved in these documents can hardly be characterized as exemplars of academic decorum.

8 Teachers at Commercial Academies

Commercial academies were also considered secondary education institutions (Khutsieva 2003), but their graduates could enroll only in commercial and technical institutes. Mathematics education occupied an important place in their curricula. Consequently, among their teachers, one finds individuals who worked concurrently or during other periods in gymnasia and real schools (as, for example, Irakliy Vereschagin, who has already been repeatedly mentioned). The schools' orientation toward practical applications, however, opened their doors to individuals with a different kind of preparation as well.

A representative memorandum submitted by the director of an academy to its senior director in 1828 states: "Because teachers of more advanced areas of mathematics may not always be familiar with the multifarious applications and subtleties of arithmetic in merchants' affairs, it would be quite expedient for the academy to separate these branches" (Ob uvol'nenii 1828, p. 1). This is followed by a proposal to hire two teachers—one of whom, a St. Petersburg University candidate, would teach algebra and geometry; while the other, the presumed future teacher of arithmetic, would be a certain Mikhelson, whom the writer of the memorandum himself does not know well, but who has good qualifications, and above all a good recommendation from the pastor of the Reformed Church.

This plan, however, was not realized. The university candidate proved too busy and did not accept the position. What is more interesting, however, is the fact that, as the director writes, "I attempted to test Mr. Mikhelson's knowledge in the presence of Mr. Detenhofer, the teacher of accounting and the science of trade, and of Mr. Hauptorsel, Baron Stieglitz's chief accountant; but although he is quite familiar with the practical aspects of Arithmetic, yet for various reasons he is incapable of teaching it at this Institution" (Ob opredelenii v sie uchilische uchitelem 1828–1838, p. 1).

Although the "practical expert" was not hired in this instance, the history of the St. Petersburg Commercial Academy includes teachers of mathematics who combined teaching with work in accounting, which they taught as well. And former "practical experts" taught not only arithmetic. Thus, in 1838, Nikolai Nosov began to teach algebra at the academy (Ob opredelenii v sie uchilische uchitelem algebry 1838–1848).

Nosov was of noble birth and in 1827 he graduated from the Tsarskoye Selo Boarding School for Nobles (*Tsarskosel'skiy blagorodnyj pansion*, an educational institution that was for a time a part of the highly exclusive Tsarskoye Selo Lyceum, which enjoyed numerous privileges no less broad, and sometimes broader, than those of Universities; the boarding school was somewhat less prestigious). Subsequently, Nosov served in the Pavlovsky Guard Regiment and was decorated for maneuvers, as well as for organization and order in campaigns against Polish insurgents and for courage in battle. Discharged due to illness in 1832, Nosov was appointed to the customs office in order to "acquire knowledge of the practice." Meanwhile, his military rank was "renamed," as was said at the time, into the civilian rank of collegiate secretary, and in 1835 he was further promoted to titular councillor (ninth grade). Even prior to this, in 1834, he had entered the pedagogical field, leaving the customs office for the Gatchina Orphanage, where he worked as a class supervisor and soon also as a teacher of arithmetic. Nosov did not stay there long either, however, and in 1836 began to work at the Ministry of Foreign Affairs, doing work in "finances and accounts." This multifaceted expert evidently worked as a teacher of algebra until 1843, having opened a boarding school in 1842 that was dedicated to preparing students for the Commercial Academy; and he continued to work at the Academy until 1847 (pp. 4–11, 22).

9 Teachers' Workload and Methodological Work

The workload of a teacher at a school was comparatively modest—12 classes per week was a common norm (see, for example, O raspredelenii urokov 1881), although sometimes the number was higher, say, 18 (O predostavlenii godichnogo otcheta o sostoyanii Larinskoy gimnazii 1838); therefore, teachers sought supplemental income elsewhere, which could also open up additional career opportunities for them. In general, it may be argued that the existence and availability of such supplementary positions was a distinctive advantage enjoyed by teachers of mathematics (although of course such positions could be found by teachers of other subjects also). Teachers could also earn supplemental income by doing work at their own schools—as librarians, secretaries at teachers' meetings, class supervisors, and so on, which were all positions often held by teachers of mathematics. Such work was taken note of and encouraged. Thus, for example, Serebrovsky, a teacher at the Larinskaya Gymnasium, received acknowledgement in 1839 for his previous work as a teacher of mathematics and secretary at the Yaroslavl Seminary Board for "special diligence in the post of secretary and for his skill in his duties as preceptor" (Ob ob'yavlenii blagodarnosti 1839, p. 1).

The work of a teacher was quite carefully monitored: usually, various reports about students' progress and assimilation of the curriculum were required. At the same time, there was a certain amount of flexibility: teachers were permitted to use different textbooks (within certain bounds, of course), to deviate from them, and so on.

For example, in 1837 teachers at Larinskaya Gymnasium were given a questionnaire, which included the following questions: (1) What method of instruction do you use in teaching? (2) Which students perform exceptionally well? Also, which students are excellent and exemplary? And the same question also with regard to their behavior. (3) Which students exhibit poor behavior and perform poorly in your subject? (4) Do you engage in any scholarly work in addition to your teaching duties? Where has such work been submitted? What have you published? (1837).

Along with such questions, the teachers were also asked about their social background, religious affiliation, teaching at other educational institutions, education, salary, and whether they resided in state owned apartments.

The teacher Andrievsky answered the first question as follows (p.11):

> Using Bourdon's *Arithmetic and Algebra* and Scheglov's *Arithmetic* as textbooks, in my teaching I mainly use the method that was used by the famous Bourdon, deviating from it (1) where it would be difficult to understand, particularly for my youngest listeners; and (2) where new improvements or new discoveries have been made.

About his scholarly work, he answered in the following way:

> My personal work relates exclusively to developing certain parts of the subject that I teach. But in general my work is still unknown to the world and has not been brought to a stage of perfection that would permit me to demonstrate it.

His response to the question concerning students contains a long list indicating those who are stronger, those who are weaker, and so on.

The following year, the teachers were again given a similar, although not completely identical questionnaire (O predostavlenii godichnogo otcheta 1838). This time, Andrievsky confined himself to a short list of good students. However, he told in greater detail about the material that had been covered, and this information was later repeated in the gymnasium's year-end report (p. 33):

> In the fourth grade, in geometry, students covered line geometry as well as circles and angle measurement; in algebraic analysis, students covered sections on analyzing the quadratic equation and applying it to various problem, the characteristics of quadratic trinomials and applying them to questions about maxima and minima and to solving equations of the form $x^{2n} + px^n + q = 0$. In the third grade, students covered elementary algebra and four operations on algebraic quantities.

About the teaching of Fus (who worked at the Navy Corps as well), the report states:

> Mr. Fus, in teaching mathematics, used the course in *Arithmetic* by Busse, published by the Department of Public Education, and for exercising the students in various questions pertaining to arithmetic he employed arithmetical sheets composed by Guryev. In the second grade, he covered: decimal fractions, continued fractions, relations and proportions, and in the application of proportions the simple rule of three, and problem solving exercises pertaining to the simple direct and inverse rules of three. In the first grade, [he covered] computation and the first four rules on abstract numbers (p. 34).

As we see, what went on in the classroom was monitored both inside the school and by the regulatory authorities, to whom annual reports were submitted; but teachers were asked what textbooks they used precisely because there were no mandatory requirements. By this time, the number of available textbooks was quite large, although teachers continued using translated textbooks as well, and not only those written in Russia.

Such questionnaires as the one cited above were handed out in other educational institutions also, and scholarly research was encouraged. The authorities deplored the fact that teachers at some institutions failed to pursue scholarly work over and above their duties, adding that "teachers must be encouraged to pursue such activities" (Ob otchete po uchebnoy chasti 1854, p. 59).

Inquiries into teachers' aims and teaching methods were made on a regular basis. We are informed concerning the same Andrievsky—clearly based on his report (O predstoyaschem g. Popechiteliu otchete 1840)—that in the lowest grades he uses an *erotetic* approach, while employing an *acroamatic* approach in grades 4–5 (that is, an approach based on questions and an approach based on lectures, respectively). In a collection published in honor of the gymnasium's anniversary and based largely on surviving documents, we read:

> In terms of teaching method, [Andrievsky] complied with the requirements of the subject itself and sought to develop the mental abilities of the students and their practical assimilation of the subject.... geometry and arithmetic [were taught] mainly through a conversation with the students; in algebra, certain departures from this method of teaching [were observed], in accordance with the demands of the discipline; finally, for homework assignments in all the sections of mathematics covered at the gymnasium, students were constantly provided with problems to solve and questions to answer (Pyatidesyatiletie S.-Petersburgskoy Larinskoy gimnazii 1886, p. 13)

Problem solving could also play a role that would be rather unusual from today's perspective. We read the following about Gintovt, a teacher at St. Petersburg's Gymnasium No. 1:

> During almost every lesson, [Gintovt] gave students problems to solve at home. The teacher viewed the solving of these problems not as independent work, but as preparation for the following lesson, since most students, when working at home, usually do not solve problems on their own, but rely on someone else's help to do so; but this in the teacher's opinion was nonetheless not useless for them, if only they tried to understand and fully and consciously to grasp the methods for solving the problems. Later on, the teacher required students whose progress was mediocre to solve these problems in class on the blackboard and to go over the theories on which their solutions were based, while he himself made comments regarding the methods for solving them (Soloviev 1880, p. 345).

This description bears a very close resemblance to what the teacher himself wrote about his method. Let us continue the quote, now using his own report (O predmetakh prepodavaniya 1868, pp. 3–4):

> In my opinion, the notebooks with these problems [which were examined on the blackboard] should not have been taken away from the students. The students can be made to write these problems, after correcting them, in clean notebooks, and these notebooks can be immediately taken away for safekeeping in the library. From time to time, usually after the students complete sections in arithmetic and algebra, they are given problems for solving in class on separate sheets, to check whether they have fully assimilated the material of the section. I examine these sheets at home, correct the mistakes, and after putting grades on the sheets themselves and in my journal, I give them back to the students. It is possible not to return these sheets to the students, but it would be better, after returning them, to collect them again for keeping in the library. For independent work, stronger students are given problems during each lesson for solving in class on another blackboard. The solutions to these problems are of course examined at once in the classroom, and when I notice a mistake, I make the student solve the problem again and find the mistake; if he is unable to do so even after he has been shown the place where the mistake has been made, I give this question to another, and sometimes even to a third student.

In one way or another, problem solving occupied a large place in teaching, and thus in the work of the teacher. Belyaev, a teacher at St. Petersburg's Gymnasium No. 5, wrote in 1845: "In studying each arithmetical operation and in investigating algebraic quantities, the students will be given as many problems as possible, predominantly of a practical nature, and the questions posed will be as diverse as possible" (Ivanov 1895, p. 28).

From time to time, methodological committees (such as the one chaired by Knorring) were formed to address the work of gymnasia in general. During the period examined here, several reforms in mathematics education were carried out (on the whole, it may be said that, for all the differences in content—increased and decreased attention to various sections—educational curricula became more and more uniform, although of course none of this could be compared with the uniformity that would come later, in Soviet times). Teachers naturally became involved in these developments.

At meetings chaired by Knorring, for example, they were invited to present their own reports (O periodicheskikh sobraniyakh uchiteley 1882), and in 1845 gymnasia were invited to report their curricula to the authorities, "from which it might be seen which parts of mathematics are taught in each class and to what extent" (Po otnosheniu direktora 3 Sankt- Petersburgskoy gimnazii 1846, p. 1). Thus, teachers prepared curricula, which were sent up the channels (in this instance, to Fyodor Busse, the director of Gymnasium No. 3 and a leading figure in mathematics education) . Subsequently, the curricula were examined and discussed, and final drafts of the curricula were composed, printed, and sent back to the schools (p. 13).

In a number of cases, teachers also became involved in methodological work that went beyond their ordinary duties, for example, by reviewing or even writing new textbooks. It is noteworthy that the aforementioned Vereschagin requested that his textbook on trigonometry be examined at a teachers' meeting; this request was granted. Naturally, a great deal here depended on the teacher himself and on his environment.

The already mentioned Fyodor Busse was something like a consultant for the ministry on mathematics education issues; thus, he was regularly sent textbooks for review, which he then sent on to teachers at his own gymnasium. This, for example, is what he wrote to one teacher:

> Herewith included on official orders is one copy of the gymnasium course in algebra published by Lesnevskii, a senior instructor at the Novgorod Gymnasium, which I instruct you to examine and return to me along with your opinion of the merits of the book, the benefits of using it in educational institutions or of acquiring it for school libraries (O rassmotrenii knigi Lesnevskogo 1852, p. 2).

A similar official letter was sent to another teacher, too, about the book *Arithmetic on an Abacus* (Lesnevskii 1852). The teachers carried out Busse's instructions, indicating numerous slips, mistakes in the Russian language, and so on, in their reports. After this, Busse himself prepared a final report, which was rather favorable on the whole, but indicated that the abacus employed by Lesnevskii was significantly different from the type commonly used, which limited the application of his book.

Textbooks were written by many, although of course not all, teachers. Ivanov (1895) lists all mathematics teachers who worked at Gymnasium No. 5 (13 teachers up to 1881), and names only one (A.D. Dmitriev) as a published author, noting, however, that he listed only those teachers who had written to him. Among those not named we find the very famous textbook author Konstantin Kraevich, whose name became legendary, as well as Vasily Latyshev, a very prominent figure in mathematics education, although both of them wrote not a little during the years when they were working at the gymnasium, not to mention later periods. What is important, however, is that those who wrote and published included not only teachers who went on to author famous textbooks, but many others as well—"ordinary teachers."

In conclusion of this section, a word must be said about the methodological preparation of future teachers. Teachers' positions at gymnasia were held by graduates from the University or the Pedagogical Institute: the methodology of mathematics education received little attention there (although pedagogy as whole was, of course, discussed at the Pedagogical Institute; a special position was dedicated to it, which was occupied by Fyodor Busse; and short-term pedagogical courses were often completed by the university's graduates as well). The importance of learning the methodology of mathematics education, however, was recognized, although not everywhere and not by everyone to the same degree.

The following letter, for example, was sent by the Supervisor of St. Petersburg school region to Gymnasium No. 3 in support of the appointment of university graduate Wilhelm Busse to what today would be called an intern's position:

> with a view to submitting Busse's teaching ability to observation in practice, and hoping with guidance from experienced and enlightened mentors to acquaint him with at least the most important techniques of pedagogy, I ask you, Kind Sir, to make use of him for the purpose of teaching mathematics in one class or another at St. Petersburg's Gymnasium No. 3, at your discretion, designating to this end a time, two or three times weekly, and at such hours as you should find most convenient for this purpose" (O dopuschenii 1861).

It must be pointed out, of course, that the man on whose behalf this letter of recommendation was sent was the nephew of the same Fyodor Busse, who may be considered one of the founders of Russian mathematics methodology, and that the letter was addressed to the gymnasium of which Busse was the director until 1858. Nonetheless, the very language of the letter reveals a recognition of the importance of professional experience. Many teachers began by obtaining already mentioned *sverkhshtatnye* teaching positions—without a salary, but with a teacher's rights and privileges, and with the possibility of earning wages for separate lessons. In its way, this was also a path of professional education.

10 The Position of Mathematics Teachers

As can be seen even from the biographies described above, teachers' positions and opportunities for advancement could vary greatly. Teachers' material positions, especially in less privileged schools, were not ideal. The archives of the Commercial Academy include a correspondence from 1819, in which not only the directors participated, but also the empress herself, the patroness of the school (Ob opredelenii 1829). The problem was that the collegiate assessor Rozov, a teacher of mathematics (and subsequently of the Russian language), who had a salary of 600 rubles and a government apartment, due to a shortage of apartments in the building of the academy, occupied only one room, in which it was impossible to set up a kitchen. The school's previous director had arranged for Rozov to receive meals from the government kitchen, but upon the director's death this arrangement was terminated. Consequently, it was proposed (and ultimately agreed) that Rozov should be granted two portions of the meals to which he had a right, for his dinner and supper (p. 9). There is no need to point out that the picture of life that arises before us is not a wealthy one. Nonetheless, in a country in which almost half of the population were serfs, the position of teachers was comparatively enviable. One important aspect of it, for example, was the procurement of a pension, which after 25 years of service was granted in its entirety, and if a teacher obtained permission to continue working (and mathematics teachers usually did obtain such permissions), then after five more years, the pension would be increased by one fifth, and after 35 years of service by another fifth.

Although a comparison between the position of mathematics teachers and other teachers requires further research, one can argue that teachers of mathematics at the very least were not inferior to other teachers either in terms of their material well-being or in terms of status, but rather had advantages over them. It is enough to note that teachers of mathematics (and Latin), in accordance with a statute that remained for a long time in effect, received the title of senior teacher (with an annual salary of 643 rubles 31 kopecks in silver), while teachers of Russian were considered junior teachers with a salary of only 343 rubles 10 kopecks in silver. Educators fought against this injustice (Soloviev 1880, p. 184) , but the fact that it existed is telling in itself.

As has already been noted, mathematics teachers also derived advantage from the fact that they usually worked concurrently at several educational institutions (naturally, other teachers did this as well, but evidently to a lesser degree). Without going into the issue of the value of rubles during that period, which of course changed over time, or the exchange rate between silver rubles and assignation rubles, whose value differed by many multiples and also changed over the course of the period discussed here, we will cite a few facts about the earnings of the frequently aforementioned Fus and Andrievsky, as well as some of their colleagues, for the year 1837–1838 (O predstavlenii godichnogo otcheta 1838, p. 44). Fus received a principal annual salary of 2250 rubles from the gymnasium (as far as can be judged, in assignation rubles), and in addition 1500 rubles from the Navy Corps, and on top of that 470 rubles out of additional funds set aside for teachers (probably for additional work and classes). This made 4220 rubles altogether. Andrievsky received a principal annual salary of 2250 rubles at the gymnasium, plus 240 rubles from additional funds, for a total of 2490 for the year. Here it may be pointed out that, during the same year of 1838, for the same sum (2500 rubles), the freedom of the great Ukrainian poet Taras Shevchenko was purchased (Kratkaya literaturnaya entziklopediya 1975, p. 646), and in Shevchenko's case the price was exceptionally inflated—usually, a serf cost substantially less. It is more instructive to note, however, that a literature teacher in the same gymnasium earned 2850 rubles, an inspector 3300 rubles (and a government apartment), while a director—on top of a government apartment—made 12,300, of which 3000 rubles were his actual salary, 800 rubles came from additional funds, and the rest was earned for his work at the University and the Pedagogical Institute. The possibility of working at several places at once was thus quite important for those who wished to increase their earnings (although, of course, it is altogether impossible to compare the incomes of teachers with those of the social elite of the time).

One more source of income must be mentioned. As is clear, for example, from the report of the teacher from Gymnasium No. 1 quoted above, students were not even especially expected to work independently at home on home assignments. In a well-to-do family, gymnasium students would be helped, and probably not only by their parents. Galakhov (1999), who had graduated from a gymnasium as early as 1822, recalled how, unable to handle the intricacies of school mathematics on his own, he had availed himself of the assistance of his old arithmetic teacher. It is evident that such assistance was provided during later periods as well, although perhaps often not by gymnasium teachers themselves, but by teachers from less privileged educational institutions.

Finally, it must be said that a person's position during this period was by no means reducible to his or her financial circumstances. And in this respect, a teacher of any subject—as a state official, who wore "an officer's uniform," to use the words of Chirkov, the Navy Corps teacher cited above—was fundamentally different from the vast majority of the population. The authorities regularly and variously underlined a teacher's favored status: teachers were regularly awarded medals, and although these were not, of course, the highest honors, they were awarded not for any exceptional acts, but for everyday work; teachers were paid monetary awards, decorated for years of irreproachable service, and so on. This system survived in the future also (Zubkov 2010).

11 Discussion and Conclusion

"A cobbler's son who graduated from the university—that's what the Russian intelligentsia is," wrote the pre-revolutionary Russian journalist Doroshevich (1905, p. 188). Among the teachers mentioned above, almost none fit this profile. During the period discussed in this paper, for a cobbler's son to gain access to a university education was prohibitively difficult—that's the difference between this period and the one that followed, when such access became easier (for example, the Russian writer Fyodor Sologub, who was born in 1863 and taught mathematics for many years, was the son of a tailor). The teachers of the 1830s–1880s came either from poor noble and low-ranking officers' families, or from the families of priests, or from the top of low-middle class (*meschane*, *raznochintsy*); they were at best the grandsons of cobblers, whose fathers had already made some kind of way in the world.

The career of a teacher, however, proved to be a good path to advancement. The son of a junior officer, or even of a skilled craftsman or work-assignment clerk, could work up to a rank of the sixth or fifth grade (and in exceptional cases, the fourth grade, that of a general). This was true for all teachers, but teachers of mathematics also had their own distinctive characteristics.

The teachers of the preceding period were typically teachers in general, teaching whatever was needed. Indeed, even after 1830, one could find teachers in gymnasia who began by teaching literature and then transitioned to mathematics, or vice versa (one example of such a teacher was Fyodor Busse's brother Franz, see O sluzhbe Busse Frantsa 1837–1866), but nonetheless it was precisely during these years that a professional group developed and teachers who taught specifically mathematics began to be prepared and graduated on a relatively large scale. This professional group was connected to (and to a certain extent drew from) the world of universities and academic researchers, on the one hand, and the world of practical experts and skilled practitioners, on the other—bookkeepers, sailors, and other specialists. The connection with another world (including the military world, for which mathematics was a necessity) opened up additional possibilities during even earlier periods (Karp 2014).

Mathematics teachers were undoubtedly prominent and respected members of the professional group of teachers, certainly not inferior to others either in terms of their incomes or in terms of the prestige of their profession. The professional community of mathematics teachers was just beginning to develop at this time—only the rudiments of any kind of common professional methodological life can be observed—as mathematics teachers would begin to meet occasionally at various joint seminars, and infrequently to write or review various books aimed not just at their own educational institutions alone. Nonetheless, the very fact that teachers worked within the framework of a common system, which moreover was augmenting this uniformity during the period in question, facilitated a certain degree of professional unification. Furthermore, already at that time mathematics teachers found themselves connected to some degree (if only through textbooks) with what was taking place abroad as well.

Mathematics was inherently a less ideologically charged subject than many others. This did not mean, of course, that the same Busse brothers were engaged in any kind of struggle against the government (although it would be curious to inquire whether such a brilliant career as Fyodor Busse's would have been possible for a teacher of the humanities who acquired a Russian citizenship only 12 years before his death—O sluzhbe pri Sankt-Peterburgskoy gimnazii uchitelya Fyodora

Busse 1823–1859, p. 97). But even so, if history teachers in working with students were especially praised for following Ustryalov's course, which was supposed to "reinforce their loyalty to the Czar and the fatherland" (Ob otchete po uchebnoy chasti 1854, p. 60), then mathematics teachers were inevitably far more open to Western influences, again, if only because they relied much more heavily on foreign or translated literature.

Prokhorenkov (2010), who has studied the teaching staff (in all subjects) of St. Petersburg's Gymnasium No. 1 during the years 1850–1900, observes a certain degree of stability in adherence to the profession; in fact, very few people left the field of teaching, although of course such individuals existed. One such figure was Dmitry Peschurov, who later became a well-known Asian studies scholar, and who began his career as a gymnasium mathematics teacher (O sluzhbe kandidata Peschurova 1853–1856). But transitioning to a different field, that is, continuing one's career under the aegis of a different government agency, was not always easy—Peschurov himself, because he failed to remain at the gymnasium for the full term of service required by the conditions of his own education, needed to obtain permission from the czar in order to move to a position at the Ministry of Foreign Affairs.

Professional stability in terms of a teacher's place of work, however, was much less prevalent. Naturally, work at St. Petersburg gymnasia was prestigious, and once people obtained such positions, they usually did not hurry to leave them, unless the departure was accompanied by a promotion, as was the case with Andrievsky, who has already been repeatedly mentioned, who left for the position of school director in the Mogilev gubernia (Svedeniya o chinovnikakh 1863). But prior to finding a long-term position, teachers usually moved around quite intensively—to find a good position was not easy, as can be seen from the passages cited above; and sometimes special commissions were established to assess the professional aptitude of candidates (as with Mikhelson's appointment to the Commercial Academy).

The present study is intrinsically limited in scope, since it only peripherally touches on education outside the capital, while such education had its own distinctive characteristics. Ignatiy Lesnevskii, who has already been mentioned as the author of a reviewed textbook, is an example of such a provincial teacher: he spent several decades working at the Novgorod gymnasium for boys. Surviving evidence indicates that he played a large role in the city's cultural life, serving as the first head of the city's public library (http://news.novgorod.ru/news/122318/) and as the editor of its newspaper (http://www.novsu.ru/file/1110699). In general, it is natural to expect that a teacher in a small city would almost automatically become one of the focuses of local cultural life. Probably the most well-known Russian author of mathematics textbooks, Andrey Kiselev, who taught in the provincial city of Voronezh from 1876 on, became a member of the city duma there, and then a candidate—albeit unsuccessful—to the national parliament, the State Duma (Karp 2012). On the other hand, Fyodor Sologub painted grim portraits of provincial teachers, including teachers of mathematics, who were by no means always bearers of culture to the provinces. The study of the life and work of provincial mathematics teachers constitutes an important and interesting problem.

At the same time, even the limited group to which this article is devoted is of considerable interest. Before the First World War, a congress of mathematics teachers will take place in Russia, initiated in part by the work of the International Commission on Mathematical Instruction, which became a clear milestone that marked mathematics teachers' recognition of themselves as a separate professional group with its own traditions. This recognition developed gradually—with the appearance of special journals devoted to mathematics education, public discussions about teaching, and teachers' congresses (without distinction by subject, but nonetheless facilitating a growing sense of community among members of the group. All of this took place later, for the most part, but its foundations were laid and shaped precisely during the years examined above.

References

Doroshevich, Vlas M. (1905). *Sobranie sochinenii* [Collected works]. Vol. 2. Moscow: Tovarischestvo I.D. Sytina.

Formulyary grazhdanskikh chinovnikov [Curricula vitae of civil servants]. (1849). Russian State Naval Archive, f.406, op. 6, d. 62.

Galakhov, Alexey D. (1999). *Zapiski cheloveka* [The memoirs of a man]. Moscow: Novoe literaturnoe obozrenie.

Ganelin, Sholom I. (1954). *Ocherki po istorii sredney shkoly v Rossii vtoroy poloviny XIX veka* [Sketches in the history of Russian secondary schools during the second half of the nineteenth century]. Moscow: GosUchPedGiz.

Gnedich, Pyotr P. (2000). *Kniga zhizni. Vospominaniya* [Book of life. Memoirs]. Moscow: Agraf.

Gurkina, Nina K. (2014). *Provinsial'nyi uchitel' v dorevoliutsionnoy Rossii* [Provincial teachers in pre-Revolutionary Russia]. St. Petersburg: Poltorak.

Ivanov, Konstantin A. (1895). *Pyatidesyatiletie S.-Peterburgskoy pyatoy gimnazii 1845–1895* [Fifty years of St. Petersburg's Gymnasium No. 5, 1845–1895]. St. Petersburg: Obschestvennaya pol'za.

Karp, Alexander (2007). 'We all meandered through our schooling...' Notes on Russian mathematics education during the first third of the nineteenth century. *British Society for the History of Mathematics Bulletin, Vol. 22*, 104–119.

Karp, Alexander (2012). Andrey Kiselev: the life and the legend. *Educ. Matem. Pesq., São Paulo, 14*(3), *398–410.*

Karp, Alexander (2014). Russian mathematics teachers: Beginnings. *International Journal for the History of Mathematics Education, 9* (2), 15–24.

Karp, Alexander (2015). Problems in old Russian textbooks: How they were selected In K. Bjarnadottir et al. (Eds.), *"Dig where you stand" 3* (pp. 203–218). Uppsala: Uppsala University.

Khutsieva, Viktoriya V. (2003). *Istoriya Sankt-Peterburgskogo Kommercheskogo uchilischa: rol' v stanovlenii rossiiskogo kommersanta, 1772–1920-e gg.* [The History of the St. Petersburg Commercial Academy: Its role in the formation of the Russian businessman, 1772–1920]. Author's Dissertation Summary Submitted in partial fulfillment of the requirements for the degree of Candidate of the historical sciences, St. Petersburg: Herzen Pedagogical University.

Kratkaya literaturnaya entziklopediya [The Shorter literary encyclopedia] (1975). Shevchenko, T. G. v.8, p. 646. Moscow: Sovetskaya entsiklopediya.

Lesnevskii, Ignatiy V. (1852). *Arifmetika i prilozhenie pravil ee k vykladkam na schetakh* [Arithmetic and the application of its rules to calculations on the abacus]. Novgorod.

Nazarin, A.B. (1916). *Podrobnye resheniya i obyasneniya vsekh arifmeticheskikh zadach I. Vereschagina* [Detailed solutions and explanations for all problems in arithmetic by I. Vereschagin]. Odessa: M. A. Min'kovsky.

Nikoltseva, Natalya F. (2000). *Nachal'noe i srednee obrazovanie v Sankt-Peterburge XIX-nachalo XX veka.* [Elementary and secondary education in St. Petersburg during the nineteenth and early twentieth Centuries]. St. Petersburg: Liki Rossii.

Ob obyavlenii blagodarnosti uchiteliu Serebrovskomu [On awarding a certificate of acknowledgement to the teacher Serebrovsky]. (1839). Central State Historical Archive of St. Petersburg, f. 276, op. 1, d. 226.

Ob opredelenii Matveya Tikhomandritskogo prepodavatelem matematiki [On the appointment of Matvey Tikhomandritsky as teacher of mathematics]. (1873–1879). Central State Historical Archive of St. Petersburg, f. 388, op. 1, d. 4.

Ob opredelenii v gimnaziu raznykh chinovnikov [On appointing different civil servants to the gymnasium]. (1836–1853). Central State Historical Archive of St. Petersburg, f. 276, op. 1, d. 14.

Ob opredelenii v sie uchilische uchitelem algebry kollezhskogo assessora Nikolaya Nosova [On the appointment to this school of collegiate assessor Nikolai Nosov as teacher of algebra]. (1838–1848). Central State Historical Archive of St. Petersburg, f. 239, op. 1, d. 1167.

Ob opredelenii v sie uchilische uchitelem algebry kollezhskogo sekretarya Vasiliya Bruevicha [On the appointment to this school of collegiate secretary Vasily Bruevich as teacher of algebra]. (1828–1838). Central State Historical Archive of St. Petersburg, f. 239, op. 1, d. 789.

Ob opredelenii v sie uchilische uchitelem Arifmetiki, Algebry i Geometrii tituliarnogo sovetnika Rozova [On the Appointment to this Academy of titular councillor Rozov as a teacher of arithmetic, algebra, and geometry] (1829). Central State Historical Archive of St. Petersburg, f. 239, op. 1, d. 339.

Ob otchete po uchebnoy chasti [On the academic staff report]. (1854). Central State Historical Archive of St. Petersburg, f. 114, op. 1, d. 1581.

Ob uchitele matematiki Knorringe [On Knorring, teacher of mathematics] (1856). Central State Historical Archive of St. Petersburg, f. 439, op. 1, d. 1770.

Ob uchitele matematiki kollezhskom assesore Ivane Konovalove [On Ivan Konovalov, collegiate assessor and mathematics teacher]. (1838). Central State Historical Archive of St. Petersburg, f. 114, op. 1, d. 534.

Ob uvol'nenii iz sego uchilischa uchitelya matematiki Kleina [On the dismissal from this school of the mathematics teacher Klein]. (1828). Central State Historical Archive of St. Petersburg, f. 239, op. 1, d. 788.

O dopuschenii kandidata Wilhelma Busse k prepodavaniu na opyte [On the admission of candidate Wilhelm Busse to practice teaching]. (1861). Central State Historical Archive of St. Petersburg, f. 439, op. 1, d. 38.

O naznachenii g. Yankevicha v 3 gimnaziu [On the appointment of Mr. Yankevich to Gymnasium No. 3]. (1856–1862). Central State Historical Archive of St. Petersburg, f. 439, op. 1, d. 450.

O periodicheskikh sobraniyakh uchiteley odnorodnykh predmetov dlya obsuzhdeniya pedagogicheskikh voprosov [On regular meetings of teachers of the same subjects to discuss pedagogical issues]. (1882). Central State Historical Archive of St. Petersburg, f. 276, op. 1, d. 1985.

O predmetakh prepodavaniya i pis'mennykh uprazhneniyakh uchenikov Pervoy gimnazii. [On academic subjects and written exercises for students at Gymnasium No. 1]. (1868). Central State Historical Archive of St. Petersburg, f. 114, op. 1, d. 3178.

O predstavlenii godichnogo otcheta o sostoyanii Larinskoy Gimnazii. [On the annual report about the state of the Larinskaya Gymnasium]. (1838). Central State Historical Archive of St. Petersburg, f. 276, op. 1, d. 58.

O predstavlenii godichnogo otcheta o sostoyanii Larinskoy Gimnazii. [On the annual report about the state of the Larinskaya Gymnasium]. (1837). Central State Historical Archive of St. Petersburg, f. 276, op. 1, d. 31.

O predstoyaschem g. Popechiteliu otchete o sostoyanii Gimnazii. [On the forthcoming report to the Region Supervisor about the state of the gymnasium]. (1840). Central State Historical Archive of St. Petersburg, f. 276, op. 1, d. 250.

O prepodavatele matematiki Vereschagine Iraklii Petroviche [On Vereschagin, Irakliy Petrovich, teacher of mathematics]. (1884). Central State Historical Archive of St. Petersburg, f. 114, op. 1, d. 4138.

O raspredelenii urokov mezhdu prepodavatelyami [On distributing lessons among teachers]. (1881). Central State Historical Archive of St. Petersburg, f. 64, op. 1, d. 943.

O rassmotrenii knigi Lesnevskogo 1. Arifmetika na schetakh, 2. Primenenie algebry [A review of Lesnevskii's book: 1. Arithmetic on an abacus, 2. Applying algebra]. (1852). Central State Historical Archive of St. Petersburg, f. 439, op. 1, d. 3161.

O sluzhbe Busse Frantsa I. [On the service of Franz I. Busse]. (1837–1866). Central State Historical Archive of St. Petersburg, f. 439, op. 1, d. 39.

O sluzhbe direktora 6 gimnazii Knorringa [On the work of Knorring, director of Gymnasium No. 6]. (1862–1864). Central State Historical Archive of St. Petersburg, f. 458, op. 1, d. 17.

O sluzhbe kandidata Peschurova [On the work of the candidate Peschurov]. (1853–1856). Central State Historical Archive of St. Petersburg, f. 439, op. 1, d. 293.

O sluzhbe prepodavatelya matematiki Varaksina I.Z. [On the employment of mathematics teacher I.Z. Varaksin]. (1882). Central State Historical Archive of St. Petersburg, f. 439, op. 1, d. 48.

O sluzhbe pri Sankt-Peterburgskoy gimnazii uchitelya Fyodora Busse [On Fyodor Busse's work as teacher at the St. Petersburg Gymnasium]. (1823–1859). Central State Historical Archive of St. Petersburg, f. 439, op. 1, d. 40.

O vydache spravok o lichnom sostave [On issuing personnel certificates]. (1837). Russian State Naval Archive, f. 432, op. 1, d. 2890.

Perepiska ob uchitelyakh, o prieme uchitelem frantsuzskogo yazyka inostrantsa I. Kochinskogo [Correspondence about teachers; on the appointment of the foreigner I. Kochinsky as teacher of the French language]. (1783–1791). Russian State Naval Archive, f. 432, op. 1, d. 298.

Perepiska ob uchrezhdenii Komiteta po ustroistvu uchebnykh zavedenii [Correspondence on the establishment of the Committee on the organization of educational institutions]. (1826–1828). Russian State Historical Archive, f. 738, op. 1, d. 1.

Po otnosheniiu direktora Tret'ey Sankt-Peterburgskoy gimnazii o soobschenii emu program prepodavaniya matematiki. Tut zhe o novom raspredelenii prepodavaniya matematiki po utverzhdennoy g. Ministrom programme [On communicating the mathematics curriculum to the director of St. Petersburg's Gymnasium No. 3. Also on the new organization of instruction in mathematics in accordance with the program approved by the minister]. (1845–1846). Central State Historical Archive of St. Petersburg, f. 276, op. 1, d. 621.

Po otnosheniiu Grafa Benkendorfa. Po Vysochaishemu Poveleniiu o dostavlenii svedeniya imeet li inostranets Schors attestat na uchitel'skoe zvanie [Count Benckendorff, Official correspondence. A Supreme Order to provide information regarding whether the foreigner Schors is in possession of a teaching qualification diploma]. (1831). Russian State Historical Archive, f. 735, op. 1, d. 376.

Prokhorenkov, A.A. (2010). Rossiiskii uchitel' vtoroy poloviny XIX veka (na primere Pervoy Sankt-Peterburgskoy gimnazii) [Russian teachers of the second half of the nineteenth century (On the example of St. Petersburg's Gymnasium No. 1)]. *Universum. Vestnik Gertsenovskogo universiteta, 1,* 148–156.

Prudnikov Vasily E. (1956). *Russkie pedagogi matematiki XVIII-XIX vekov* [Russian mathematics educators of the XVIII–XIX centuries]. Moscow: Gosudarstvennoe uchebno-pedagogicheskoe izdatel'stvo.

Pyatidesyatiletie S.-Petersburgskoy Larinskoy gimnazii 1836–1886 [Fifty years of St. Petersburg's Larinskaya Gymnasium, 1836–1886]. (1886). St. Petersburg: Tipografiya Stasiulevicha.

Shapin, Steven., & Thackray, Arnold. (1974). Prosopography as a research tool in history of science. *History of Science,* 22, 1–28.

Soloviev, Dmitry N. (1880). *Pyatidesyatiletie S.-Peterburgskoy pervoy gimnazii 1830–1880* [Fifty years of St. Petersburg's Gymnasium No. 1, 1830–1880]. St. Petersburg: Tipografiya II otdeleniya Sobstvennoy E.I.V. Kantselyarii.

Svedeniya o chinovnikakh i prepodavatelyakh Sankt Peterburgskoy Larinskoy gimnazii, sluzhivshikh v ney so vremeni ee osnovaniya v sentyabre 1835 goda [Information about the civil servants and teachers of St. Petersburg's Larinskaya Gymnasium who have worked at the gymnasium since its founding in September 1835]. (1863). Central State Historical Archive of St. Petersburg, f. 276, op. 1, d. 1.

Vereschagin, Irakliy P. (1883). *Sobranie voprosov i zadach priamolineynoy trigonometrii dlya gimnazii i real'nykh uchilish* [Collection of questions and problems in linear trigonometry for gymnasia and real schools]. St. Petersburg: Tipographiya V. Bezobrazova.

Vereschagin, Irakliy P. (1884). *Sbornik arifmeticheskikh zadach dlya srednikh uchebnykh zavedenii muzhskikh i zhenskikh* [Collection of problems in arithmetic for secondary educational institutions for boys and girls]. St. Petersburg: Tipographiya V. Bezobrazova.

Vereschagin, Irakliy P. (1886). *Sbornik algebraicheskikh zadach dlya uchenikov starshikh klassov srednikh uchebnykh zavedenii* [Collection of algebraic problems for upper grade students at secondary educational institutions]. St. Petersburg: Tipographiya V. Bezobrazova.

Zubkov, Igor' V. (2010). *Rossiiskoe uchitel'stvo. Povsednevnaya zhizn' prepodavateley zamskikh shkol, gimnazii i real'nykh uchilish* [Everyday lives of teachers at village schools, gymnasia, and real schools]. 1870–1916. Moscow: Novyi khronograph.

Author Biography

Alexander Karp is a professor of mathematics education at Teachers College, Columbia University. He received his Ph.D. in mathematics education from Herzen Pedagogical University in St. Petersburg, Russia, and also holds a degree from the same university in history and education. Currently, his scholarly interests span several areas, including the history of mathematics education, gifted education, mathematics teacher education, and mathematical problem solving. He served as the managing editor of the *International Journal for the History of Mathematics Education* and is the author of over one hundred publications, including over twenty books.

Arithmetic Patterns in the *Arithmetic* of Petros Argyros

Elena Kiliari, Athanasios Gagatsis and Irene Papadaki

Abstract This paper analyzes the content of Arithmetic (1632), a manuscript of commercial arithmetic written by the Greek-Cypriot Petros Argyros. Arithmetic was probably influenced by the first Greek printed arithmetic *Logariastiki* (Glyzonios 1569), a work with a great deal of similarity to the Italian Abaci. Consequently, the paper devotes attention on the one hand to the Abaci tradition and on the other hand to the early formative stages of Greek and Cypriot mathematics education. The paper's main aim is to draw attention to the analysis of patterns presented in the book. Recreational mathematics was typical for the Abaci tradition and analysis of these problems raises questions about their history and the approaches that were used centuries ago. In the authors' opinion, the very existence of such problems, which presuppose analytic thought and reasoning, compels us to verify the widespread conception of the nature of Abaci mathematics and to define it with greater precision.

Keywords Abaci · Commercial arithmetic · Greek mathematics education
Patterns · Recreational mathematics

1 Introduction

This paper is a continuation of the publications by Kiliari et al. (2015) and Gagatsis et al. (2015) and is focused on the analysis of the recreational problems found in the written by Greek-Cypriot Petros Argyros *Arithmetic.*

The manuscript *Arithmetic* (1632) is a work devoted to practical—commercial mathematics. The word abacus (άβακας in Greek) is written on the first page of the manuscript (see Fig. 1). *Arithmetic* was firstly mentioned by Karas in his work *Sciences in the Ottoman Era* (1992, p. 37). Historical and content analysis of Argyros's arithmetic was presented in our previous article (Gagatsis et al. 2015).

E. Kiliari (✉) · A. Gagatsis · I. Papadaki
University of Cyprus, Nicosia, Cyprus
e-mail: elenakiliari@cytanet.com.cy

F. Furinghetti and A. Karp (eds.), *Researching the History of Mathematics Education*, ICME-13 Monographs,
https://doi.org/10.1007/978-3-319-68294-5_7

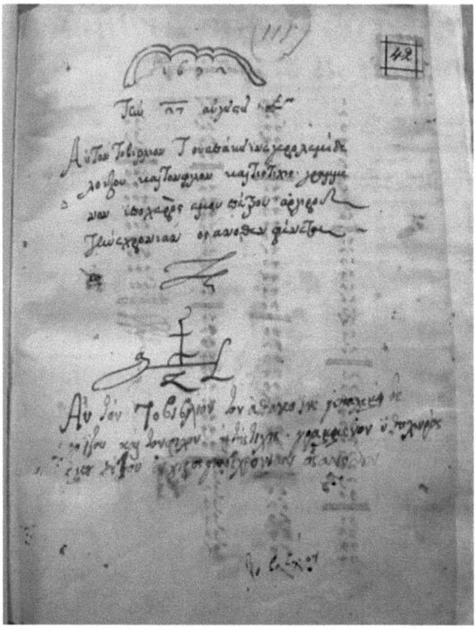

Fig. 1 The first page of *Arithmetic*

As in other similar works of commercial arithmetic, its content is composed of the discussion of the knowledge required for the solution of problems together with problems and their solution methods. An important part of the book is also taken by recreational mathematics, analysis of this part is a goal of this paper. Such an analysis, in the authors' opinion, makes it possible to verify and define with greater precision the widespread conception of the nature of Abaci mathematics.

2 Mathematics Education in the West and the Abaci Tradition

The book of Petros Argyros is deeply rooted in Abaci tradition, therefore, we will start with brief comments on this tradition.

Abaci tradition was generated during the Italian Renaissance in Italian cities-commercial centers where the social and economic conditions favored the development of commercial arithmetic. Works of practical arithmetic are guides to the solution of trade problems. The most representative book of this mathematical category is Fibonacci's *Liber Abbaci* (1202) and often is considered to be the source for later books of practical arithmetic. Researchers though support that *Liber Abbaci* was not the inspiration for the production of commercial books of arithmetic (Hoyrup 2005, 2010). Solving methods in various Abaci are not necessarily related to the methods presented in Fibonacci's work. Moreover, the fact that content presented in *Liber Abbaci* and later in Italian Abaci is similar to Abaci manuals from Iberian Peninsula and Provence regions, indicates that Fibonacci's book did not constitute a model for all Abaci. Historical data shows that the tradition of abacus was created through a broad process of interaction with the non-scientific Arabic tradition, which was originally developed in the Iberian Peninsula. In Castile Spain, three manuscripts of commercial arithmetic have been found, all from the 14th century. The first, entitled *Libro de Arismética* is written in Castilian and contains a collection of practical examples. Two other manuscripts entitled *Manuales de Mercaduria* and *Aritmética Mercantil* are also manuals of commercial arithmetic (Hoyrup 2010).

Nevertheless, there are Abaci inspired by Fibonacci's book like *Compendy de la praticque des nombres* issued in southern France in 1471 that gave a new impetus to the arithmetic tradition of France. The book is based on the methods of solving mathematical problems by Leonardo of Pisa and contributed to the transmission of mathematical thinking of abacus in the rest of Europe (Spiesser 2000). The Italian Abaci tradition in general had influenced the writing of the respective manuals in the rest of Europe. According to Karpinski (1936), the contents and the conditions of the first printed arithmetic practices came from the Italian tradition. From the 14th century, German traders visited Italy, especially Venice, to learn the Italian commercial practice and the currency exchange (Swetz 1987). The *Bamberger Rechenbuch* of Ulrich Wagner, the first German commercial arithmetic book was printed in Nuremberg in 1483, and *Rechenbuch für die Kauffmannschaft* by Johannes Widmann printed in 1489 in Leipzig, were focused on trade issues.

There were two categories of arithmetic books, Abaci and Algorisms. Algorisms were the first to import Hindu-Arabic arithmetic and calculation methods in Europe (Van Egmond 1976), written in Latin. Unlike Abaci, the Algorisms have limited scope and theoretical basis and present the calculation of basic arithmetic without focusing on the methods of solving mathematical problems (mainly commercial). On the contrary in the Abaci taxonomy, the algorithms are part of the necessary equipment to solve the problems, while the problems are the core of the book. As Van Egmond (1976) states, Algorisms and Abaci have fundamentally different philosophy.

According to Van Egmond (1976), the Abaci content is divided into three sections.

(a) The 'preliminary material' contains the four mathematical operations of the Hindu-Arabic numeric system and the rule of three.
(b) The problems: commercial, recreational and geometrical.
(c) The 'Special Subjects' such as the calendar, calculations with square and cube roots, increased difficulty proportion problems, and algebra topics by Al-Khwārizmī (Hannah 2007).

Glyzonios' *Logariastiki*, the first Greek printed book of arithmetic, shows high similarity to Abaci tradition. It's a work of commercial arithmetic presenting mainly commercial problems and partly recreational together with their solution methods, a typical characteristic of Abaci tradition. The book also contains a special subject—the Greek orthodox calendar for the calculation of Easter Sunday.

3 Mathematics Education in Greek Territories During the Ottoman Era

The fall of the Byzantine Empire and subsequently the Ottoman rule had changed things socially and economically and as a result people were gathered around Orthodox faith and tradition. The marginalization of mathematics and science was maintained by the difficult living conditions of the conquered Greeks.

Efforts were made in order to preserve Greek education. These were held by two opposite poles. On the one hand the Byzantine clergy aimed at the continuation of the Greek Orthodox tradition. Their efforts were based in the Turkish occupied areas. On the other hand, Greek diaspora located in the commercial centers of the West attempted to provide the rebirth of ancient Greek culture and ethic. Their actions were placed in the Italian cities and their conquered islands (Kastanis 2004). Education in Greek territories was chaotic and unorganized nearly for the first two centuries of Ottoman rule. No educational system was provided; education was sustained by the scholars, clerics and merchants initiatives to promote the establishment of schools, book publication, mainly in commercial centers (Karas 1991).

Whereas the Hindu-Arabic arithmetic was widely used in the West, in Greek regions it arrives in the mid-16th century. Access to books and manuscripts gave Greek people the opportunity to evolve mathematically. The Byzantine manuscript *Codex Vindobonensis phil. gr. 65* (1436) of the 15th century partly contains practical arithmetic problems, commercial and recreational (the problems from the manuscript were presented firstly by Hunger and Vogel (1963) and later by Chalkou (2003). Apart from this work, there is another manuscript of commercial arithmetic, as reported by Sotirakis (1962), which dates back to the mid-16th century preceding the edition of the first Greek printed book of arithmetic, *Logariastiki* (1569) *by* Glyzounis. As it was discussed above, the content of this manuscript is very similar to the Abaci. The book of Petros, written in 1632, also a practical commercial arithmetic incorporates the Western mathematical model as it has a great resemblance to Abaci tradition.

The development of mathematics in Greece was rather slow due to lack of educational organization and resources. The situation changed in the middle of the 16th century with the operation of primary school where students were taught writing, reading and arithmetic. The teachers were mainly members of the clergy, so the teaching of mathematics was superficial since they did not have the knowledge themselves (Tgigoni 2005). Mathematics teaching in 16th century as Govdelas mentioned in his book *Algebra Elements* (1806), was "neglected" in almost all Common Schools as mathematics was not taught (Karas 1992).

In addition, many people were self-taught by reading books of commercial arithmetic. During the 17th century education recovers in a slow pace through the efforts of merchants and relations with the western world. Commercial cities like Chios, Smyrna and Ioannina had known spiritual recovery. Especially the actions of merchants of Ioannina had resulted in the establishment of institutes of higher education following European standards; the inclusion of mathematics teaching contributed to educational resurrection (Tgigoni 2005).

Briefly describing the further development, one can mention that in the 18th century there was an increase in scientific production, although the high cost of publication considerably reduced publications, because of it several works remained in handwritten form (Karas 1991). Until the outbreak of the French Revolution, Greek Enlightenment focused mainly on the development of Greek philosophical and scientific thought and the cultivation of an education able to arouse the enslaved nation. Within this climate of revival of science and knowledge, the Enlightenment stream started from Europe and was transported in Greece. More mathematics works were published, especially in arithmetic, many of them containing number theory and algebra. Scholars such as Eugene Voulgaris, Athanasios Psalidas (a highly educated merchant), Thomas Dimitriou, Nikiforos Theotokis, had translated and adopted mathematical knowledge from well-known European scholars. During this period we observe adoption of the Western mathematical model and not necessarily production of new scientific knowledge. The birth of the Greek independent nation later in the first half of 19th century brought more positive changes in mathematics education.

Education in the island of Cyprus until 1800 follows the Byzantine tradition. The priests in villages and the monks in monasteries served as teachers. Children were taught reading, writing, and basic arithmetic. There are no evidences concerning mathematics education in particular, until the beginning of 19th century when public schools were first established and mathematics curriculum was designed for the first time (Gagatsis 2011). The Frankish occupation of Cyprus was better than the Byzantine in educational terms, as secondary schools were founded in the outskirts of Nicosia and science and ecclesiastical schools in the city as well, but there are no records of the mathematics curriculum (Peristianis 2000). The 14th century was characterized by commercial development in Cyprus and the cities of Nicosia and Famagusta were developed into urban and commercial centers. During the Venetian period (1489–1571), education was mostly based on private initiatives still the city of Famagusta was known for the respected schools of higher education that taught Greek and Latin (Gagatsis et al. 2015) Later, in the Ottoman era, there were two forms of schools: Private schools—teachers there were local residents of villages with basic knowledge of reading, writing, arithmetic—and community schools, where the teacher's position was held by a qualified person or a 'Greek' school graduate who taught with the assistance of very capable older students (Gagatsis 2011).

4 The Content of *Arithmetic*

The content of *Arithmetic* is described elsewhere (Gagatsis et al. 2015). Briefly we can say that the types of problems found in Argyros's book, a Greek Abaci are:

A. Commercial problems

- "Prices and Products" problems
- "Interest and discount" problems
- "Exchange" problems
- "Barter" problems
- Companies

B. Recreational problems

- Arithmetic sequences

Learning mathematics concepts in Argyros is based on procedural understanding as the reader follows certain steps applying the rule given to solve the problem. *Arithmetic's* structure and appearance, use of plain language and solution methods used reveal its connection with the Italian Abaci. Cyprus was a Venetian territory and the writer uses Italian words (as well as Italian currency) in the content. The majority of Argyros problems are commercial solved by the rule of three. The writer explains the "fundamental principle" of commercial activities in this way:

> Three quantities, two of them alike, the one that varies is multiplied by a third different quantity; the outcome is divided by the invariable quantity. The result is the same with the one you have multiplied earlier (Argyros 1632, p. 23).

This rule is applied step by step in the solution of all trade problems. Obviously, the concept of proportion is not conceptually discussed; emphasis is given on the procedure and not on the functional use of the concept. The pattern problems constitute, however, an exception in the procedural conception of knowledge.

5 Arithmetic Patterns in *Arithmetic*

We can partition recreational problems in Arithmetic in two groups: "typical" recreational problems and arithmetic patterns or sequences. Typically, all problems in commercial books of arithmetic, both commercial and recreational, are followed by the solution method (e.g. Borghi 1484; Glyzounis 1569; *Arte dell' Abbaco* in Swetz 1987; Van Egmond 1976). Argyros provides no solution to the problems which we call arithmetic patterns.

The following problem is a typical recreational problem encountered in commercial arithmetic manuals, where the solution is given after the text.

> Three people split ducats. I do not know the total amount or each one's share. I only say that the second man took half the share of the first. And from what the first and the second took together, the third man took half their share, and his share is 50 ducats and $3\frac{1}{5}$ lire. I want to know how much money they had split. And how much the first one got and how much the second (Argyros 1632, p. 138).

In the given solution Argyros uses the "rule of the false" method. That is, making a guess regarding the first man's share and then finding the correct answer by proportion. He assumes that the first man got 16 ducats and $4\frac{2}{4}$ lire, the second 8 ducats and $2\frac{1}{4}$ lire and the third 12 ducats and $3\frac{1}{2}$ lire. Then he finds the correct amount by the rule of three: if the third man's share is 50 and $3\frac{1}{5}$ corresponds to 12 and $3\frac{1}{2}$, then you can ask to what corresponds the first man's share 16 ducats and $4\frac{2}{4}$ lire (see Fig. 2).

The patterns are also recreational problems but they form a special category. They are presented in a table form. An example of a pattern problem is given below (see Fig. 3). The table presents Cyprus birds and mammals in four columns (Argyros 1632, p. 242). Each animal corresponds to a number which is placed below. Therefore, for the patterns to be observed the number has to be placed next to the right animal.

The translation of the Birds and Mammals table appears in Table 1. Also, the patterns arise after number placement on the right.

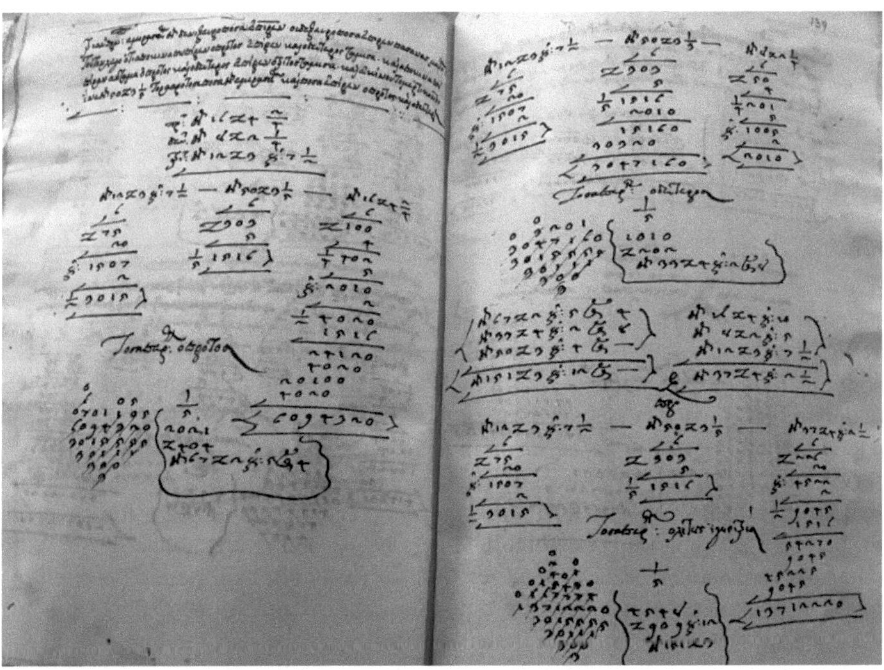

Fig. 2 A typical recreational problem in *Arithmetic*

Fig. 3 Birds and mammals table

The part of Table 1 is followed by the second part (Table 2).

What was the aim of the writer here, which problem did he offer and what solution method did he expect from his readers? As his manual appears to be addressed to a closed educated group of friends we are not aware whether the writer added something to the given in the manuscript. To start with, Argyros makes a serious diversion from the Abaci philosophy as he forces his reader to think "out of the ordinary". He gives no instructions or hints regarding the pattern for the reader to identify the relation between the numbers below and nouns above. Here and in other patterns the use of any algebraic notation was not expected. The book is a practical arithmetic manual of the mid-17th century and no formal algebra was taught, yet problems, such as arithmetic sequences, appear in the book.

The discovery of the pattern depends on the capability of grasping a commonality noticed on the particulars. The finding of the commonality and the prediction

Table 1 Birds, mammals and numbers (first part)

sheep 1	lamb 2	cow 4	wild pigeon 8	buffalo 16
wild 3	wild 3	rabbit 5	pig 9	dove 17
rabbit 5	hen 6	hen 6	goose 10	blackbird 18
sparrow 7	sparrow 7	sparrow 7	owl 11	goat 19
pig 9	goose 10	partridge 12	partridge 12	koudela 20
owl 11	owl 11	turtle dove 13	turtle dove 13	goldfinch 21
turtle dove 13	rooster 14	rooster 14	rooster 14	lark 22
wild hen 15	wild hen 15	wild hen 15	wild hen 15	duck 23
dove 17	blackbird 18	koudela 20	sikalida 24	sikalida 24
goat 19	goat 19	goldfinch 21	pordales 25	pordales 25
goldfinch 21	lark 22	lark 22	aktahinara 26	aktahinara 26
duck 23	duck 23	duck 23	bee-eater 27	bee-eater 27
pordales 25	aktahinara 26	crow 28	crow 28	crow 28
bee-eater 27	bee-eater 27	crane 29	crane 29	crane 29
crane 29	eurasian dove 30	eurasian dove 30	eurasian dove 30	eurasian dove 30
thrush 31	thrush 31	thrush 31	thrush 31	thrush 31

Table 2 Birds, mammals and numbers (second part) (Numbers 24, 25 and 26 represent the names of Cyprus birds and cannot be translated)

1. sheep	8. wild pigeon	14. rooster	20. koudela	26. aktahinara
2. lamb	9. pig	15. wild hen	21. goldfinch	27. bee-eater
3. wild				
4. cow	10. goose	16. buffalo	22. lark	28. crow
5. rabbit	11. owl	17. dove	23. duck	29. crane
6. hen	12. partridge	18. blackbird	24. sikalida	30. eurasian dove
7. sparrow	13. turtle dove	19. goat	25. pordales	31. thrush

of a plausible generalization was assumed by the writer. In this way, the rule of the pattern would appear. Most likely Argyros would expect the reader to observe the relationships among number structures in order to find any term of the pattern and also to identify the change happening between terms. In a sense, although no use of algebraic expression was expected by the writer, some kind of algebraic reasoning was. This reasoning was required for identifying and extending the pattern. The result of it comes up as the general rule of the nth term and sum.

Argyros offers several sequences for each of which the reader is supposed to find the sum. We will limit ourselves to three examples:

(a) 1 3 5 7 9 11 13 15 17 19 21 23 25 27 29 31
(b) **2** 3 **6** 7 **10** 11 **14** 15 **18** 19 **22** 23 **26** 27 **30** 31
(c) **4** 5 6 7 **12** 13 14 15 **20** 21 22 23 **28** 29 30 31

In the first case, we have a sequence of all odd numbers. Using modern language, we can say that the task requires finding the sum of the arithmetic sequence. In sequence (b) we observe two arithmetic sequences—one is 2, 6, 10, … (given in bold fonts) and another is 3, 7, 11…. The last sequence (c) contains four arithmetic sequences.

Modern textbooks offer several approaches to arrive to the sum of the arithmetic sequence in this or that way utilizing the idea that sums of two terms equidistant from the ends of the sequence are constant (1 + 31 = 3 + 29 = 5 + 27…, etc.). We can hypothesize that similar trick was expected from the solver of the problem. Importantly, similar problems were fairly traditional.

6 Discussion and Conclusion

Heeffer (2004, p. 13) cites Rouse Ball who analyzed recreational problems given in the book of Bachet published yet in 1612 and pointed out that "several of Bachet's problems are taken from the writings of Alcuin, Pacioli di Burgo, Tartaglia and Cardan, and possibly some of them are of oriental origin, but I have made no attempt to add such references". Indeed, the origin of some recreational problems hardly can be found—some of them were offered yet in the Antiquity. It is worth mentioning that oriental origin of some of them is quite plausible. For example, a lot of patterns even more sophisticated than those discussed above were given in the famous and influential work of a 16–17 century Middle Eastern scholar Sheikh Bahai *Khulasat al-Hisab or Quintessence of Calculation* (Rofagha 2006).

Although we hardly can restore the reasoning used in the previous time, we can definitely say that there was a reasoning. Moreover, we see the development of some ideas as it is presented in the book of Argyros: arithmetic sequence is followed by the sequence consisting of two arithmetic sequences and then of even more sequences. What for these sophisticated problems were given? Obviously, they are not practical as other problems included in Abaci collections. Still, the ability to wisely calculate something clearly was considered important. Heeffer (2004) remarks that recreational mathematics brings together two sixteenth-century traditions, mercantile arithmetic and natural magic. Indeed, numerical mastery was seen as connected with the magic and wisdom. Anyway, it is possible to conclude that understanding of Abaci tradition as purely aiming at practical purposes is not quite correct—there were some developmental goals (with all reservations regarding the use of this modern terminology). A confirmation of it is that problems aiming to develop some reasoning and wisdom appear even in the manuscript representing transmission of existing tradition and, therefore, containing most typical features.

Acknowledgements The present research is a part of a research project of the foundation A.G. Leventis.

References

Argyros, Petros (1632). *Arithmetic* [Η Αριθμητική].

Borghi, Pietro (1484). *Arithmetic* [Aritmetica]. Venice.

Chalkou, Maria (2003). *The mathematical content of the Codex Vindobonensis phil. Graecus 65 of the 15th century* [Μελέτη του μαθηματικού περιεχομένου του Βιενναίου Ελληνικού Φιλολογικού κώδικα 65] (PhD thesis). Athens: Εθνικό Κέντρο Διδακτορικών Διατριβών.

Gagatsis, Athanasios (2011). *Elements of mathematics education in Cyprus, 1878–1960* [Στοιχεία ιστορίας της μαθηματικής εκπαίδευσης στην Κύπρο, 1878–1960]. Nicosia: Κέντρο Επιστημονικών Ερευνών.

Gagatsis, Athanasios; Kiliari, Elena; & Papadaki, Irene (2015). The practical arithmetic of Petros Argyros in relation with the Abaci and the first Greek printed book of arithmetic Logariastiki. *The International Journal for the History of Mathematics Education, 1*0(2), 83–98.

Glyzonios, Manolis (1569). *Logariastiki* [Βιβλίον πρόχειρον τοις πάσιν περιέχον την τε Πρακτικήν Αριθμητικήν ή μάλλον ειπείν την Λογαριαστικήν]. Venice: Author.

Hannah, John (2007). False position in Leonardo of Pisa's Liber Abbaci. *Historia Mathematica*, 34, 306–332.

Heefer, Albrecht (2004). Récréations mathématiques (1624). A study on its authorship, sources and influence. Ghent: Centre for Logic and Philosophy of Science, Ghent University, available at http://logica.ugent.be/albrecht/thesis/Etten-intro.pdf, accessed on May 18, 2017.

Hoyrup, Jens (2005). Leonardo Fibonacci and Abbaco Culture: a Proposal to Invert the Roles. *Revue d'Histoire des Mathématiques,* 11, 23–56.

Hoyrup, Jens (2010, December). Fibonacci – Protagonist or Witness? Who Taught Catholic Christian Europe about Mediterranean Commercial Arithmetic? Paper presented at the workshop "Borders and Gates or Open Spaces? Knowledge Cultures in the Mediterranean during the 14th and 15th Centuries", Seville, Spain.

Hunger, Herbert, & Vogel, Kurt (1963). *Ein byzantinisches Rechenbuch des 15. Jahrhunderts: 100 Aufgaben aus dem Codex Vindobonensis* Phil. Gr. 65; Text, Übersetzung und Kommentar. Wien: H. Böhlaus Nachf.

Karas, Yiannis (1991). *Sciences in Greece 15th – 16th century* [Οι θετικές επιστήμες στον ελληνικό χώρο 15ος-19ος αιώνας]. Athens: Dedalos.

Karas, Yiannis (1992). *Sciences in Ottoman Era, Manuscripts and Prints, vol. 1 Mathematics* [Οι Επιστήμες στην Τουρκοκρατία, Χειρόγραφα και έντυπα, Τόμος Α' Τα Μαθηματικά]. Athens: Estia Bookshop.

Karpinski, Louis C. (1936). The First Printed Arithmetic of Spain: Francesch Sanct Climent: Suma de la Art de Arismetica Barcelona, 1482. *Osiris,* 1, 411–420.

Kastanis, Nicos (2004). *An introduction in Modern Greek Mathematics Education* [Μια εισαγωγή στη Νεοελληνική Μαθηματική Παιδεία]. Thessaloniki: Aristotle University.

Kiliari, Elena, Gagatsis, Athanasios & Papadaki, Irene (2015). The first Greek printed book: The famous book of arithmetic Λογαριαστική of Manolis Glyzounis. *The International Journal for the History of Mathematics Education, 10(*1), 25–54.

Peristianis, Ieronymos. K. (2000). *History of the Greek Education from the Turkish occupation until the English conquer, 1571–1878* [Ιστορία των ελληνικών γραμμάτων από της τουρκικής κατακτήσεως μέχρι της αγγλικής κατοχής 1571–1878]. Nicosia: Εκδόσεις Επιφανίου.

Rofagha, Nader (2006). *Mathematical and pedagogical contribution of Baha Al-Din Muhammad Ibn Husayn Amil (aka Sheikh Bahai), a sixteenth century Middle Eastern teacher and scholar. Unpublished PhD dissertation.* Columbia University.

Sotirakis, Nicos (1962). *Contribution to the Greek Enlightment research – Mathematics in Ottoman era* [Συμβολή στην έρευνα του Ελληνικού Διαφωτισμού – Τα Μαθηματικά επί Τουρκοκρατίας]. Athens: Εκδόσεις Ελληνική Μαθηματική Εταιρεία.

Spiesser, Maryvonne (2000). Problemes lineaires dans le Compendy de la praticque des nombres de Barthelemy de Romans et Mathieu Prehoude (1471): Une approche nouvelle basee sur des sources proches du Liber abbaci de Leonard de Pise. *Historia Mathematica, 27,* 362–383. Doi:10.1006/hmat.2000.2285.

Swetz, Frank J. (1987). *Capitalism and Arithmetic: The New Math of the 15th Century, Including the Full Text of the Treviso Arithmetic of 1478. (D. E. Smith, Trans.).* Illinois: Open Court.

Tgigoni, Anastasia (2005). *Mathematics of Greek Enlightenment their place and role in education and technical implementation* [Τα Μαθηματικά κατά την περίοδο του Νεοελληνικού Διαφωτισμού, η θέση και ο ρόλος τους στην παιδεία και στην τεχνική εφαρμογή] (Phd Thesis). Panteion University.

Van Egmond, Warren (1976). *The commercial revolution and the beginnings of western mathematics in Renaissance Florence 1300–1500.* Indiana University.

Author Biographies

Elena Kiliari studied primary education at the University of Crete (Bachelor degree 2001) and continued her studies in education at the University of Cyprus (Master in Education 2007). She got her Ph.D. in 2017 at the University of Cyprus with a thesis on "The contribution of *Logariastiki* by Manolis Glyzounis in Greek Mathematics Education (16th–18th centuries)". In addition she is working as a teacher in public elementary schools for the last 16 years. Her particular interests are the history and philosophy of mathematics education stemming from her involvement in philosophy of education studies at the University of Crete.

Athanasios Gagatsis is a Full professor of Mathematics Education at the Department of Education of the University of Cyprus. Athanasios is the Vice Rector of Academic Affairs of the University of Cyprus since December 2010. He was Lecturer and Assistant Professor at the Department of Mathematics of the Aristotle University of Thessaloniki and Visiting Professor and Associate fellow in different European Universities. His research is published in different languages on geometry teaching and learning, on the use of representations in the teaching and learning of different mathematical concepts, on the history of mathematics education and on implicative statistical analysis.

Irene Papadaki studied Greek Literature at the University of Crete (Bachelor degree 1995, Doctorate 2005). As a researcher of the Greek Institute of Venice (1996–1999), a scholar of the Alexander S. Onassis Public Benefit Foundation (2001–2003) and an external collaborator of the Piraeus Bank Group Cultural Foundation (2003–2005), she conducted a ten-year research in Venice on the manuscript and archival sources of post-Byzantine cultural history. Her research interests and publications focus on Literature and Culture in Greek regions of the Eastern Mediterranean and the centers of the Hellenic Diaspora (15th-18th c.), Comparative Analysis of Modern Greek and Italian Literature, and the History of manuscripts and printed books of the Post-Byzantine era. She has taught Modern Greek Literature at the University of Crete and History of the Byzantine and Post-Byzantine World at the Hellenic Open University. Since 2014 she works as Assistant Professor at the Department of Byzantine and Modern Greek Studies of the University of Cyprus.

Frans van Schooten Sr. (1581–1645) Lecture Notes for the First Dutch Course for Engineers, Leiden, 1600–1681

Jenneke Krüger

Abstract This paper is devoted to the important period of, and the important contributors to, Dutch and European mathematics education. In 1600 an Engineering Course, for surveyors and military engineers, was created at Leiden University, the first such a course in the Netherlands. It was a very innovative course: the teaching plan was written by Simon Stevin, who gave instructions concerning the content and teaching methods and the teaching language was Dutch, instead of the customary Latin. Because of this the course was often called Duytsche Mathematique [Dutch Mathematics]. The purpose of this course was to provide modern and high quality instruction to a relatively large group of men who did not know Latin. To attract a sufficient number of students the quality and reputation of the teacher was very important. Professor Frans van Schooten Sr., successor of Professor Ludolf van Ceulen, taught the course from 1611 to 1645. His elaborate lecture notes are preserved in the library of Leiden University. This manuscript gives us insight into the rich content of the program of the Engineering Course and the quality of the teaching preparations. In this paper we discuss the significance of Frans van Schooten Sr. for the Engineering Course, during and after his lifetime, based on relevant characteristics of these lecture notes, with examples in arithmetic and surveying.

Keywords Engineering Course · Leiden · Duytsche Mathematique
Frans van Schooten Sr. · Surveying and fortification · Simon Stevin

J. Krüger (✉)
Freudenthal Institute, Utrecht University, Utrecht, The Netherlands
e-mail: jenneke.kruger@gmail.com

F. Furinghetti and A. Karp (eds.), *Researching the History of Mathematics Education*, ICME-13 Monographs,
https://doi.org/10.1007/978-3-319-68294-5_8

1 Introduction[1]

Around 1620 professor Frans van Schooten Sr. (1581–1645) composed his extensive lecture notes for the Engineering Course for surveyors and military engineers in Leiden. The course was offered by the University of Leiden, however the teaching language was Dutch, instead of Latin. At a time when Latin was the obligatory teaching language at all universities in Europe, a course in the vernacular was sufficiently exceptional to warrant the commonly used name 'Duytsche Mathematique' (Dutch Mathematics). Very soon the name 'Duytsche Mathematique' was used both for the course and its content, so the name had slightly different meanings, depending on the context in which it was used. The course had a well-defined formal aim, formulated by Simon Stevin as part of his teaching plan and written down in the Resolutions of the University governors.

Frans van Schooten taught the course for a longer time than either his predecessors or his successor did, from 1611 to 1645. Nevertheless, his name was not mentioned much after his passing away; he published very little, a trigonometric table for surveyors, which had several reprints and a translation of Euclid which was quickly forgotten. Both his predecessor and his successor remained well-known after their death. His predecessor was Ludolf van Ceulen, a well-known and highly respected reckonmaster and mathematician, who had published some mathematical works, some of which were still in use in the mid-eighteenth century. The successor of Frans van Schooten Sr. was Frans van Schooten Jr., his eldest son, at that time a mathematician of some fame, the center of a circle of Dutch students and mathematicians, amongst whom Christiaan Huygens. Frans Jr. published several mathematical works, in Latin and in Dutch, for mathematicians and practitioners; he was active in making Descartes' work known, through translating Géométrie [Geometry] in Latin and through his private students. He was considered an expert in practical mathematics by mathematics users well into the eighteenth century. When the name Frans van Schooten is mentioned in literature, it is nearly always the name of his son. So was Frans van Schooten Sr. just an insignificant teacher, who dutifully implemented the teaching plan by Stevin, and continued the lectures in the same way as the respected professor van Ceulen, more of the same, until his brilliant son took over?

The question which provides a framework for this paper is

What significance had Frans van Schooten Sr. for the reputation of Duytsche Mathematique, the first mathematics course for surveyors and engineers in the Netherlands?

The answer to this question will help us to understand better an important period in the development of Dutch and European mathematics education. Fortunately, the lecture notes, written by Frans van Schooten around 1620, are preserved as

[1]This paper is based on research for my doctoral dissertation, chapter two (Krüger 2014).

manuscript BPL 1013[2] in the library of Leiden University. They are very extensive, beautifully illustrated and provide us with a unique insight in the planning of lessons for Duytsche Mathematique in the early 17th century. They represent van Schooten's interpretation of the teaching plan by Stevin and so they tell us something about the content and also the style of teaching as planned by Frans van Schooten Sr. As these lecture notes were used by his successor as well, the discussion of the significance of Frans van Schooten Sr. as teacher of the Duytsche Mathematique will to a large extent be based on these lecture notes, however other sources will be taken into account as well. For practical reasons, only the parts on arithmetic and geometry will be used, the rather large part on fortification will be discussed at another occasion.

2 The Dutch Republic Around 1600

Since 1568 the Dutch were fighting an independence war with the king of Spain, a war which was mainly fought around the many small and larger towns in the Netherlands, both the Northern and the Southern Netherlands (at present Belgium). The regions which considered themselves independent of the Spanish king formed a federation of provinces, the Dutch Republic, with the States General, situated in The Hague as a government. Leiden University was established in 1575 by Prince Willem van Oranje [William of Orange], the leader of the rebellion against Spain. Prince Willem was assassinated in 1584; the following year his son, Prince Maurits (1567–1626), became stadtholder of Holland and Zeeland, at a rather young age. From 1590 Maurits was commander of the army of the Republic; he and his cousin Willem Lodewijk (re)captured a significant number of fortified towns. They also managed to hold on to most of these towns, through modernization of the fortifications. For this style of warfare well-trained surveyors and army engineers, who knew modern mathematical methods, using geometry and trigonometry, were indispensable.

Since the last decades of the 16th century, the conquests by the Spanish army of many towns caused a stream of migrants moving from the southern Netherlands to the north, often well-educated people and relatively wealthy. In the northern Netherlands in the 1590s the economic situation improved considerably: trade and navigation increased, and the population of towns, especially in the western part of the country, expanded considerably during the early years of the seventeenth century. This also led to a high demand for qualified surveyors and engineers, as land had to be measured and divided and the expanded towns must be fortified.

Italian architects and engineers were very influential in the Netherlands: from the mid-sixteenth century, Italian advisors were responsible for improving the defensive walls of towns. Their ideas also became widely known through publications of

[2]Further in this paper 'manuscript BPL 1013' will be shortened to 'BPL 1013'.

Italian mathematicians and of others, like Daniel Speckle and Simon Stevin. In the *Stercktenbouwing* [Fortification] (1594) Stevin used the Italian system, in which the preferred ground plan of a fortification was a regular polygon and the bastions were pentagonal.[3] Any surveyor and engineer who worked with the army had to have knowledge of geometry and also trigonometry, as a lot of the work was performed on inaccessible lands.[4] Thus there was a demand for surveyors and engineers, who were able to design and build modern fortifications, but who also could map the land and had other surveying skills.[5] The use of reliable trigonometric tables could improve the results of surveying techniques considerably; by the end of the 16th century surveyors had started to use plane trigonometry, for which they needed instruments to measure angles. The extensive calculations that were necessary could be simplified using decimal fractions, but these were not yet used by mathematical practitioners (Struik 1995). Thus in order to meet the demands of the developing nation there was a great need for good mathematical education.

In the Dutch Republic there was no centralized educational system. Schools were the responsibility of town councils; often there was a primary school, in which children could learn reading, writing and some arithmetic, depending on the skills of the teacher. However parents had to pay the schoolmaster a small sum for the teaching of each subject, arithmetic was the most expansive of these three. In addition there were private schools and private teachers of mathematics, many of whom had arrived from other countries or regions. There was no standard curriculum, nor any form of quality control (Krüger 2010). In 1600 there were two universities in the northern Netherlands, at Leiden and Franeker.

At Leiden University mathematics was taught from 1581 by Rudolphus Snellius (1546–1613), who valued both pure and mixed mathematics. This distinction was common in the 17th century: 'mathematica pura' consisted of geometry and arithmetic, the abstract mathematics, without a relation to matter; 'mathematica mixta' consisted of those topics which had also a qualitative aspect, subsisting in matter. Some examples of mixed mathematics are navigation, astronomy, surveying, optics, mechanics, perspectives and geography[6] (Mulder 1990; Alberts et al. 1999). Both pure and mixed mathematics could be studied for its own sake or could be used for practical purposes, by mathematical practitioners. The University of Franeker (Friesland) was established in 1585; from 1598 Adriaan Metius (1571–1635), a student of Snellius, taught pure and mixed mathematics.[7]

---

[3]See van den Heuvel (2006) on fortifications as knowledge system.

[4]See Bennett and Johnston (1996) on the geometrical aspects of war.

[5]See van Maanen (1987) for the importance of immigration for Dutch society, van Maanen (1987) and Metin (2007) on the influence of Italian style mathematics, Struik (1981) and Taverne (1978) for the relation between town expansions and fortification.

[6]In the course of the 18th century the expression 'applied mathematics' appeared in some countries, it became common in the 19th century. The present meaning of applied mathematics differs from mixed mathematics.

[7]Examples of topics taught by Metius: navigation, surveying, fortification, astronomy (Dijkstra 2012).

Maurits had studied at Leiden University; in 1583 both he and Simon Stevin (1548–1620) were enrolled as students.[8] Stevin and Maurits probably both went to the mathematics courses by Snellius. Maurits was interested in mathematics and its applications and was well aware of its importance for modern warfare. Around 1593 he engaged the services of Simon Stevin as his personal tutor and as designer of the army's quarters. In 1600 the course for surveyors and military engineers, the Duytsche Mathematique, was established at Leiden University on instigation of prince Maurits and Simon Stevin. The aim was to train engineers for the Dutch army thoroughly and quickly. That meant a solid mathematics background with the focus on practical mathematics.

3 Publications on Dutch Mathematics and Mathematics Education in the 17th Century

Mainly the recent and most important publications on the situation in the northern Netherlands are mentioned here. For more detailed information, older publications and articles see Krüger (2014).

Van Maanen (1987) provides an inventory and a systematic description of 17th century mathematical manuscripts in the library of Leiden University. His evidence made it clear that Frans van Schooten Sr. was indeed the author of BPL 1013, which until then was sometimes disputed. Boekholt and de Booy (1987) published on the history of the school in the Netherlands (including the teaching of arithmetic), from the Middle Ages until the 20th century. De Booy (1980) wrote in more detail on primary education in the towns of Utrecht from 1580 until the 19th century. Kool (1999) published research on Dutch arithmetic books from the 15th and 16th century. Those books could of course also be used for self-instruction.

Several authors have described the Duytsche Mathematique and the teaching plan by Stev*in* (Bierens de Haan 1878; Molhuysen 1913; Muller and Zandvliet 1987; Winter 1988). However, not much attention has been paid by these authors to the implementation of the teaching plan, the teachers or the characteristics of the teaching content.

Dopper (2014) published research on Frans van Schooten Jr. who, as the successor of his father, taught Duytsche Mathematique from 1645 until 1660. This publication also contains a systematic description, based on the method used by van Maanen (see van Maanen 1987), of the manuscripts in the Van Schooten collection, in the library of Groningen University. Moreover Dopper gives an overview of the teaching topics and their treatment by Frans van Schooten Jr.

[8]du Rieu (1875).

Two more publications on mathematics teaching at the Dutch universities are relevant for this paper. De Wreede (2007) published research on Willebrord Snellius (1580–1626) , professor of mathematics at Leiden University from 1613 to 1626. Dijkstra (2012) describes the position of mathematics at the University of Franeker during the 17th century.

Van Gulik-Gulikers (2005) published research on mathematical practitioners in which she describes surveying instruments and practice of surveying in the 17th century. Sitters (2007) published his research on Sybrandt Hansz. Cardinaels, a teacher of mathematics and arithmetic, who published a popular collection of geometrical problems, *Hondert geometrische question met hare solutien* (Hundred geometrical questions with their solutions), around 1614.

4 The Duytsche Mathematique

In the Netherlands until well into the 19th century Latin was officially the language of the university courses. Thus, even if Snellius in Leiden and Metius in Franeker might occasionally offer lessons in Dutch, Latin was their official teaching language and the language of examinations. Consequently the Duytsche Mathematique could not be recognized as an academic course, the students who only came for Duytsche Mathematique were called ‘auditores’. However, the University appointed the professor, paid his salary, provided teaching locations and teaching materials, in short the University took responsibility for this mathematics course in vernacular. The choice for Leiden University as the place for this course was obvious. Not only did prince Maurits and Stevin both study at Leiden University; one of the curators, Jan Cornets de Groot, was a friend of Stevin. Another curator, Jan van Banchem, came from a family of surveyors and must have seen the value of a high level course for surveyors and engineers.

The Duytsche Mathematique was exceptional at a time when learning mathematics for professional purposes was mainly a matter of private instruction. The course was also unique because there was a formal plan for teaching, written by the mathematician and engineer Simon Stevin and usually named the *Instructie* [Instruction] (Krüger 2010, 2015). It was added as an addendum to the minutes of the meeting of the curators with prince Maurits and Stevin on 9 January 1600 (Molhuysen 1913). Stevin, who himself was a good mathematician, engineer and teacher, gave the aim of the course, an outline of the content, the order of the topics, what teaching activities he deemed necessary, i.e. practical work following theory, the teaching language etc. The *Instruction* formed the framework for the course; moreover it is the first example of such a formal plan for mathematics teaching in Dutch language.

4.1 The "Instruction"

The first paragraph of the *Instruction*,[9] which follows on a short introduction, states the intended aim and the global content of the course.

His Excellency has deemed it useful for the sake of the country and for those who will enroll to become an engineer, that there will be a plan for the teaching thereof which will take place at the Academy in Leiden, as follows.[10]

1. The general opinion is that the auditors should be trained as quickly as possible to serve the country as engineers. To this goal they shall learn arithmetic or the counting and surveying, but of each of these only so much as is necessary for a practising ordinary engineer. Those who have come thus far are allowed to study more in depth if they wish to do so.

So the aim of the Engineering Course was to train the 'auditors' as fast as possible, in order that they could serve the country as engineers. In the Regional Archives in Leiden there is a manuscript, a copy of the *Instruction*, provided by the curators for the printer Jan Paedts Jacobsz. (Paedts 1600, RAL LB 36949). The first paragraph differs slightly from the same paragraph in the book of Molhuysen (1913). The most important difference is the expansion of the professions mentioned. Instead of "for those who will enroll to become an engineer,…" it has "for those […] who will become an engineer and other Mathematical Arts."[11]

The addition of "other Mathematical Arts" suggests that the curators preferred a broader aim than the training of engineers only. Wine gauger, schoolmaster and building master were examples of other "Mathematical Arts".

Stevin was an advocate of the teaching of theory prior to the practical work, of the use of Dutch language instead of Latin, both in teaching and in mathematics and of the use of decimal notation by mathematical practitioners, as he wrote in his often translated *De Thiende* [The Tenth] (Stevin 1585) . All these characteristics are found in the *Instruction*, which consists of 13 short paragraphs, on content and other aspects of the course.

The second and third paragraphs of the *Instruction* are on arithmetic.

> In the counting will be taught the four operations on whole numbers, on broken numbers and on decimal numbers, also the rule of three in each type of number.
>
> When they have become skilled in this way of counting, they know sufficient arithmetic for ordinary engineering.

One could not be sure that the students would be sufficiently skilled in basic arithmetic, so Stevin included this in his teaching plan. He emphasized the teaching of decimal fractions (decimal numbers). At the same time he limited the extent of

---

[9]See for Dutch text Molhuysen (1913): https://adcs.home.xs4all.nl/stevin/varia/ingenieurs.html#resolutie.

[10]All translations from the original are by the author of this paper.

[11]See for Dutch text in RAL LB 36949: http://fransvanschooten.nl/fvs_ordre.htm.

arithmetical techniques which should be taught. The use of decimal fractions, in the notation used by Stevin, was not at all common at the time. In manuscripts and publications, on surveying and fortification, one finds this notation rarely or not at all, as will be discussed in Sect. 7.

On geometry the *Instruction* contains the following passages (paragraph 3, 4, 7 and 11).

They will then start with surveying on paper in the way of surveyors, that is not to construct any lines through known lines, but only to find the area of plane figures through measured lines, drawn to scale, using arithmetic in decimals. [...] they shall only learn with rectilinear figures, thereafter with curved figures in the way of surveyors, how to measure a plane by means of various subdivisions, such as triangles or other figures to check the different calculations. [...] Once they have sufficient skills in measuring on paper and understand on a small scale what has to be done in reality, they shall start with the actual surveying in the field, they will be shown that instead of using ruler, compass and square on paper, one uses other tools in the field, with the same purpose. [...] After that they will learn to draw on paper the perimeter of the land they have measured, and again the reverse, from a drawing on paper, in the open field lay out the plan using pickets. [...] As they have learnt so much, that they can thus usefully serve the country, those who wish to do so are allowed, as has been mentioned before, to continue to study at Leiden in winter, in order to become more accomplished engineers.

On teaching activities and teaching language Stevin remarks (paragraph 12 and 13).

The lessons in counting and geometry will take half an hour in general; the other half hour will be filled with answering questions from individual students and explanations about what they haven't understood from the lectures. As those who work in engineering, rarely talk in Latin with each other, but instead in each country the vernacular is used, so the lessons will not be given in Latin, French or similar languages, but only in Dutch.

There was no mention of an examination procedure in the *Instruction*.

4.2 Teachers and Students

To ensure that the new course would attract students an innovative teaching plan was not sufficient. Much depended on the quality of the teacher. What was needed was someone who was an expert in the required mathematics and the practice of surveying techniques, who was willing to teach a group of uneducated students in Dutch and who could teach well. Maurits and Stevin chose Ludolf van Ceulen (1540–1610), a well-known and respected 'reckonmaster', a private teacher in mathematics, who had published some mathematical treatises and who also taught fencing (Krüger 2014; Wepster 2010). Van Ceulen had not studied at a university, he did not speak or write Latin, but he had a very good knowledge of mathematics and he had a good reputation as a teacher. He participated in government committees, for example to find a method to determine the longitude at sea or to compose a table for taxes of inhabitants of Leiden. A second professor was appointed together with him, a surveyor and cartographer, also living in Leiden, Symon Fransz. van Merwen (1548–1610). As van Ceulen was already 60 years old when he was appointed, it is likely that van Merwen was supposed to take care of the lessons in the practice of surveying techniques. Soon after the start of the course van Ceulen and van Merwen informed the University governors that the students had repeatedly asked to receive a testimony on their surveying capabilities. In August 1600 the University governors decided that there would be an optional final exam, which would serve as a testimony in the admittance procedure for surveyor by the provincial authorities. A final exam resulting in a testimony for military engineer was not realized.

The students had very different educational backgrounds and they were of very different ages, roughly from 16 to 45 years old (Krüger 2014). Some were regular students, enrolled at the University; most were already craftsmen, such as carpenter, stonemason, surveyor, draughtsman or teacher. The reputation of van Ceulen as a reckonmaster and teacher was sufficiently good to attract students during the first ten years of the existence of Duytsche Mathematique.

Not much is known about the teaching by van Ceulen and van Merwen. In 1615 a book by van Ceulen was published posthumously, *De arithmetische en geometrische fondamenten* [Fundaments of arithmetic and geometry]. This book has six chapters, on arithmetic, geometric constructions, geometric transformations, geometrical applications, advanced applications and regular polygons, with inscribed and circumscribed circles, a specialty of van Ceulen. It seems likely that these topics were to some extent part of his lectures in Duytsche Mathematique. However, there is much more on species of numbers and on circles than was necessary for Duytsche Mathematique; there is not much on surveying and there is no use of decimal numbers as mentioned by Stevin. The Regional Archives in Leiden possess a handwritten booklet by Symon van Merwen, the colleague of van Ceulen. The title is *De vijff spetie inde tiende getalen* [The five species in decimal numbers]. There is no date but van Merwen refers to himself as professor of mathematics in Leiden; he also mentions professor van Ceulen, so it must be written between 1600 and 1610. It is a small booklet, containing 148 written pages. The first 12 of those are on the use of decimal numbers, with a notation which is reminiscent of the one used by Stevin. However, the other 136 pages are about different types of numbers, very much like the first chapter in the book by van Ceulen, with many examples concerning currencies. Decimal numbers remained for van Merwen a separate topic, for which he had little use. This booklet possibly shows mainly topics which were taught in private lessons.

More is known of the work of van Ceulen's and van Merwen's successor, Frans van Schooten Sr.

5 Frans van Schooten

Frans van Schooten Sr. was the son of migrants from the southern Netherlands. His father, a baker, was citizen of Leiden since 1584. Frans, the oldest of four brothers, studied Duytsche Mathematique with van Ceulen and van Merwen and took more lessons in mathematics with van Ceulen; he was officially admitted as a surveyor in 1608. For some years he was an assistant in van Ceulens small private school for mathematics. During 1610, after van Merwen died, van Schooten assisted professor van Ceulen with the lessons in Duytsche Mathematique. After Ludolf van Ceulen also passed away in December 1610, van Schooten continued the lessons on request of the students, in a temporary position and without an official appointment by the University. He taught both the theory and the practice of surveying, the practice probably on the field which had been purchased by the University for that purpose. From 1611 until 1614 at least three formal requests to the curators of the University were made by Frans van Schooten and his students, to have van Schooten appointed as successor to professor van Ceulen. However, there was a serious competitor, a candidate who got the support of prince Maurits. Samuel Marolois (1572–before 1627) was born in the Netherlands shortly after his parents emigrated from France. He was a military engineer and technical advisor; he taught mathematics and produced books on geometry, perspective, architecture and fortification. These books were popular, as they contained many engravings of good quality. However, at the start of 1615 Frans van Schooten was appointed Professor Duytsche Mathematique. He continued his teaching of theory and practice in this official capacity until December 1645.

Frans van Schooten Sr. left some teaching texts on mathematics. The most impressive of those is BPL 1013, the extensive lecture notes for the Duytsche Mathematique. Van Schooten probably wrote the manuscript around 1620; it was also used by his successors, Frans van Schooten Jr. and Petrus van Schooten, so there was a long-lasting influence of this teaching text. Next to teaching Duytsche Mathematique Frans van Schooten worked as a surveyor and cartographer for the army during the summer months, he was an advisor on fortification and examiner for the surveyor admission committee in Holland (Dopper 2014; Krüger 2014).

In 1627 he published a trigonometric table, *Tabulae Sinuum Tangentium Secantium ad Radium 10 000000. Met 't gebruyck der selve in Rechtlinische Triangulen* [Table of sine, tangent, secant based on radius 10000000. With the use of the same in plane triangles]. The book had several reprints, possibly also because of the clear explanation and the systematic treatment of examples, with many drawings, in rectangular, acute and obtuse triangles. In the book van Schooten explained that the accuracy of the tables is dependent on the accuracy of the instruments for measuring angles in the field.

> As working on the land it is with the instruments available not possible to measure angles more accurately than in minutes, I have calculated the tables in as many rectangular triangles as the number of minutes which can be represented on a quarter of a circle (Tabulae, p. 2).

So, the tables have an accuracy of minutes.

As far as we know his students appreciated him, Duytsche Mathematique had a good reputation, also in neighboring countries; the University governors evidently thought well of him, even if he did not have a university education.

6 BPL 1013

BPL 1013 is a fairly large manuscript, 425 × 280 mm. The writing on the pages is in brown ink, on each page is some text and pen drawings, of geometrical figures (Fig. 1), landscapes with geometrical figures (Figs. 2, 3 and 4) or geometrical figures enhanced with a landscape (Fig. 7). Occasionally calculations are added, written with a different pen, sometimes one sees a correction of an error in a calculation. The text is structured following the teaching plan by Stevin, the style is what one would expect of a teacher who prepares carefully his lessons, with some explanations and short notes as scaffolding for his teaching. The carefully executed illustrations may well have served as visualization for his students, to support explanations and exercises. A lot of the content consists of mathematical exercises, mostly with solutions. The last part of the manuscript, roughly one third, is on fortification and contains some watercolor drawings, pen drawings and calculations. The manuscript was in very bad condition due to ink-corrosion. Recently it has been restored and preserved; it is now digitized.[12]

On f.113^r, just before the treatment of fortification, is written

> Begonnen den 25 November Anno 1622 door Frans van Schooten professor der Fortificatien en Dependerenden Scientien in de Universiteit tot Leyden" [Started 25 November 1622 by Frans van Schooten, professor of fortifications and related sciences at the University of Leiden]

Van Schooten must have written the preceding content in the years before 1622.

As already indicated, the topics in the lecture notes and their order are according to Stevin's *Instruction*, however, Frans van Schooten left out the start of arithmetic, i.e. the four operations with whole numbers and rational numbers and the rule of three (Table 1). These operations were probably known to the students; van Schooten may have taught them separately to those who lacked knowledge in arithmetic. He included some topics which had nothing to do with surveying, such as the construction of a spiral or a geometric rose and wine gauging. In the part on surveying he added notes on a variety of geometrical methods, which were not commonly used.

The content of the lecture notes may be described as in Table 1.

Some differentiation is noticeable in the content. The numbers 1–3 seem to cover a basic course, 4, 5 and 6 are more advanced geometry and cover everything the best surveyors had to know; 5 has a more mathematical theoretical character and 7

[12]https://www.bibliotheek.universiteitleiden.nl/bijzondere-collecties, search for BPL 1013.

may be considered a further specialization, that of military engineer, which at that time included building fortifications. The differentiation noticeable in the content suggests that van Schooten catered for different demands of students.

6.1 Interesting Features of BPL 1013

The order of topics is roughly the same as in the *Instruction*; the treatment of topics is very well structured. Within topics there is a built-up from simple to more complex situations, with occasionally a reminder of what was previously taught.

Remarkable is the consistent use of decimal notation, starting on f.4^r and always linked to units of measurement, with the rod as unit of length. This is a very early example of the use of this notation throughout a manuscript. I know of no other publications or manuscripts from the early 17th century, in which decimal notation is used throughout the text, with such apparent ease; at most this notation is used occasionally or demonstrated as a special type of number.

Another interesting feature is the systematic use of instruments to measure, combined with the use of trigonometric tables and the demonstration of how to create a triangulation network, starting from a line of known length (f.65^v).

The manuscript is strong on visualization; van Schooten used illustrations throughout the manuscript, with every topic and exercise, very neatly executed; some are of a high artistic quality (f.66^r–f.89^r).

He paid much attention to the minimalizing of errors in the results. In his publication on trigonometric tables (Schooten van 1627) he pointed out the importance of choosing a suitable radius for measurements, depending on the magnitude of the length to be measured and citing Stevin on this topic. As mentioned above, he also drew attention to the limitations of the measuring instruments available.

Van Schooten systematically showed more than one method to solve a problem. This served as a way to check outcomes of measurements and calculations, as Stevin had prescribed. But it also gave the future practitioner a choice of methods, depending on personal preference and, in surveying, the tools available.

6.2 Arithmetic (See Table 1, Row 1)

This part differs slightly from Stevin's plan as there is no treatment of the four main operations; presumably the students already mastered basic arithmetic. The manuscript starts with calculations of square and cube roots from whole numbers and fractions, followed by the introduction of decimal notation. For example to introduce calculation of a cube root (f.2^v), van Schooten uses a drawing of a cube with a volume of 110 592 cubic feet. Calculation of the cube root gives the length of the edge, 48 feet. This is followed by a table of roots 1–9 with their cubes, and

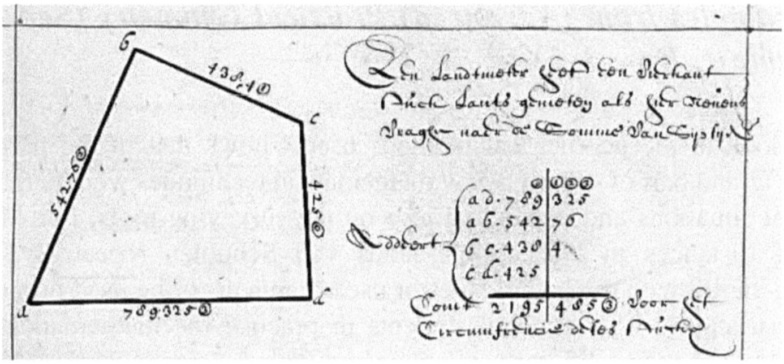

Fig. 1 Exercise to practice calculations in decimal numbers, f.4^{v}

calculation of the edge of a cube with volume 30 517 578 125 cubic rods, which turns out to be 3 125 rod. On the last page of this topic one finds another example of taking the cube of a very large number and two small tables of cubes of some fractions, based on calculation of the third power of the fraction (f.3^{v}). All examples are illustrated with a drawing of a cube, with the numbers written in it. Looking at the handwriting and some other characteristics the drawings probably were added some time after the calculations were written. One reason may be that the calculations were written by van Schooten while he was studying with or assisting Ludolf van Ceulen and that he added the drawings as a didactic tool when he was teaching independently as van Ceulens successor.

Introduction of the unknown decimal notation as promoted by Stevin takes place on f.4^{r}, linked to the surveyor's rod, divided in ten equal parts. At the time, division in twelve equal parts was still more common, though division in ten parts occasionally was used (van Gulik-Gulikers 2005). The notation is the same as used by Stevin in *The Tenth* (*De Thiende* 1585), the indication of value above the digits (Fig. 1).[13] However, van Schooten preferred the more convenient indication of the number of decimals at the end (Figs. 1, 2, 3 and 4). The first example is a length of 346 rods, 8 feet, 7 inches, 5 grain, which in modern notation would become 346,875.[14] Most examples are on measuring in plane geometry, such as in Fig. 1. From here on, van Schooten used decimal notation only in all calculations.

The geometry (see Table 1, row 2) starts with 32 definitions, each one illustrated. The part on constructions and transformations seems to be rather similar to the posthumous publication by van Ceulen (1615). The topics were of importance for the practice of surveying and fortification, with some additions such as the construction of a spiral and of a geometric rose. There are only mathematical contexts.

[13]All illustrations are from ms BPL 1013, University Library, Leiden.

[14]This means that a rod (roede) was divided in 10 feet (voeten). A foot was divided in 10 inches (duimen). An inch was divided in 10 grains (greinen).

6.3 Examples from (Advanced) Practical Geometry (See Table 1, Row 3–5)

A closer look at the geometrical methods used, shows that in the part on basic surveying (3 and part of 4) only a few mathematical techniques were taught, applied in different situations and depending also on the surveying tools. For example, in measuring distances in inaccessible lands van Schooten repeatedly combined measuring angles with the law of sines or use of tangents or he used proportionality in similar triangles. This enabled students to practice the mathematical methods they could use in many different situations and with different surveying tools, such as a Dutch circle or an astrolabe for the measuring of angles or a quadrant. See for examples Figs. 2, 3 and 4, in which the distance to a tower or between two towers is to be determined, the context in each case is measurement from the opposite bank of a river or an island in the river. So, some or all of the towers are on inaccessible land for the surveyor.

In part of 4 and in 5 (Table 1) many more techniques are present, many of which most surveyors might never need. In Advanced surveying techniques (see Table 1, row 4) there are many examples of measuring heights, not just the habitual church tower or statue, as found in the Netherlands. Indeed, there are 16 examples of measuring heights and depths of buildings on top of or half way a mountain, a situation in which an engineer for the army would find himself when working in neighboring countries. Some of these drawings are of a very high quality; they look the work of a professional engraver or painter (van Maanen 1987). The

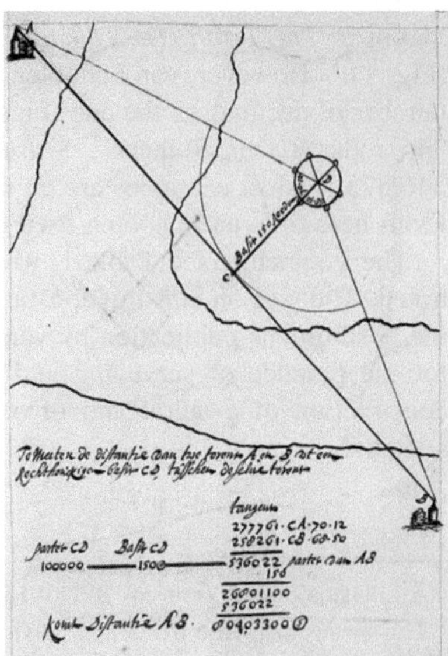

Fig. 2 Measuring angles, use of tangent, f.58^r

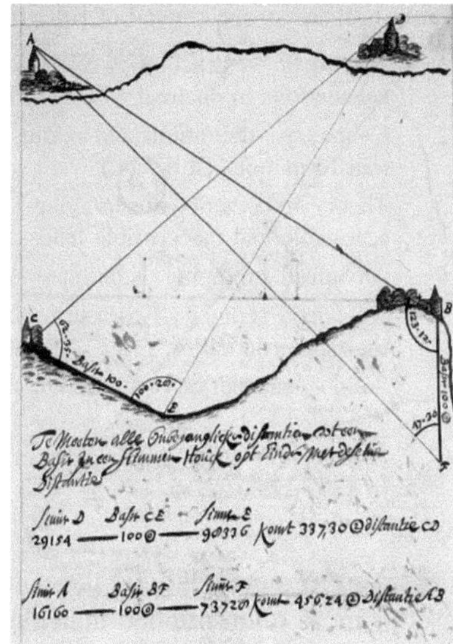

Fig. 3 Measuring angles, use of law of sines, f.57^{v}

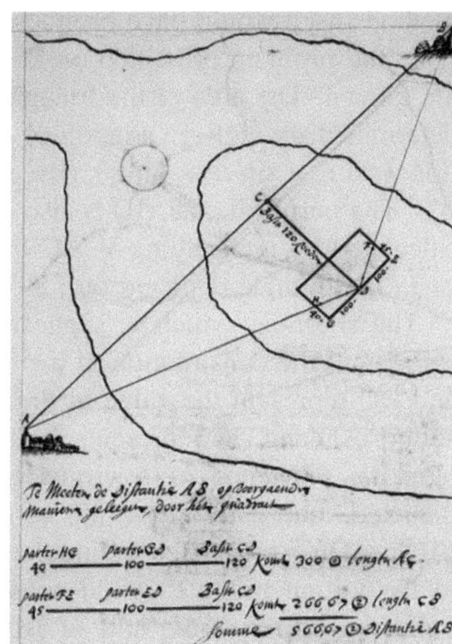

Fig. 4 Use of similar triangles, f.58^{v}

mathematical methods are of a higher complexity than in previous parts of the manuscripts, in the sense that one needs combinations of previous used techniques.

Some mathematical procedures, known at the time, do rarely or not at all occur in the manuscript, e.g. the law of cosines. Examples of less common mathematical methods in Advanced geometry (Table 1, row 5) are on ff.88^{v}–89^{v} (Figs. 5, 6 and 7); in each case the length of the sides of the triangle has to be determined. It is

Table 1 Index of the content of BPL 1013

1	Arithmetic—extraction of roots, decimal notation of fractions, calculation of area, calculations in decimal notation (ff.1^r–8^r)
2	Geometry—definitions and axioms, a selection of propositions from Euclid, constructions, transformations (ff.8^v–44^r)
3	Theory and practice of surveying—preparation, use of instruments, measuring distances in accessible and inaccessible lands, calculations (ff.44^v–64^r)
4	Advanced surveying techniques—making maps, measuring heights, depths (ff.64^v–83^r)
5	Advanced geometry—measuring lands and calculations, with and without use of trigonometric tables, using a variety of geometric techniques (ff.84^r–116^v)
6	Solids—calculations on different shapes and materials, calculating content, wine gauging (ff.117^r–132^r)
7	Fortification—definitions, plans of fortifications, bastions, calculations of the works (ff.134^r–225^v)

important to remember that these are notes to use during the lessons, so the method used is more often implied than elaborately written down. The description below is an interpretation based on drawings and calculations written on the folia discussed. As the Duytsche Mathematique was a course for mathematical practitioners the methods used would have been less formal than in lectures on 'mathematica pura'.

In the problem presented on f.88^v (Fig. 5) two angles and the sum of the sides are known. The area of the triangle is also requested. The lengths of the sides are determined through a variation of the law of sines. In slightly more modern notation: (sin A + sin B + sin C): (BC + AC + AB) = sin A: BC = sin B: AC = sin C: AB. The perpendicular BD is also determined by using proportionality. The area is calculated by multiplying ½ AC by BD.

In the triangle depicted on f.89^r (Fig. 6) the area and two angles are known.

The solution, which is suggested in and partially written under the drawing, starts with the construction of a circle with center B and radius BD. The sides AD and DC represent the value of the tangent of respectively ∠ABD and ∠CBD, the sum of AD and DC represents the 'parts' of AC. See Fig. 8. Van Schooten assumes here the existence of a virtual triangle, which is proportional to the triangle depicted, but with the value of the tangent as the lengths of the sides. The trigonometric table used to find these values is based on a radius of 100,000, so BD in the virtual triangle would consist of 100,000 units or 'parten [parts]' as van Schooten writes. Take half of BD as 50,000 ('parts') and calculate the area in 'parts' as ½ BD × AC. Proportionality in the form of the rule of three gives the real value of BD. Repeated use of the rule of three combined with the cosecant of ∠A and ∠B provides the real length of the sides.

In the triangle with the beautiful illustration on f.89^v (Fig. 7) the area is known, as well as the proportion of the three sides (13:14:15). In this case the angles have to be determined as well. Van Maanen (1997) provides a good example of treatment of this interesting problem in a modern secondary school.

Fig. 5 f.88^{v}

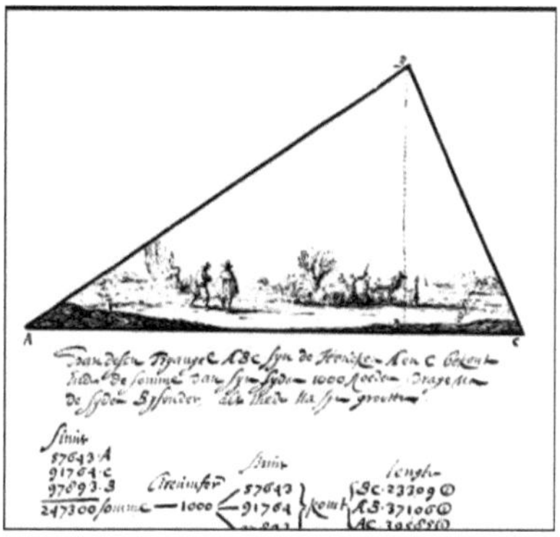

Fig. 6 f.89^{r}

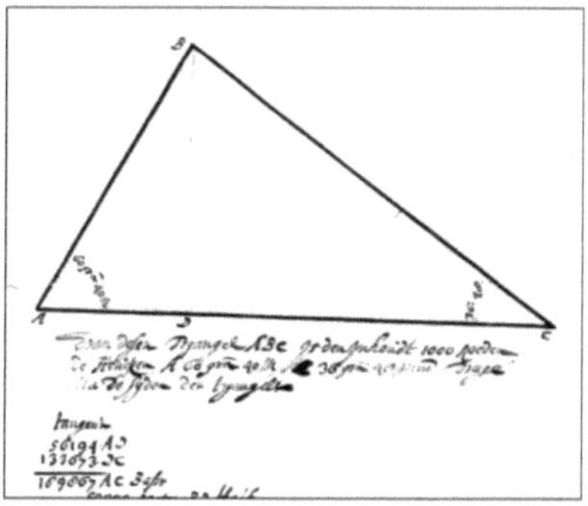

Fig. 7 f.89^{v}

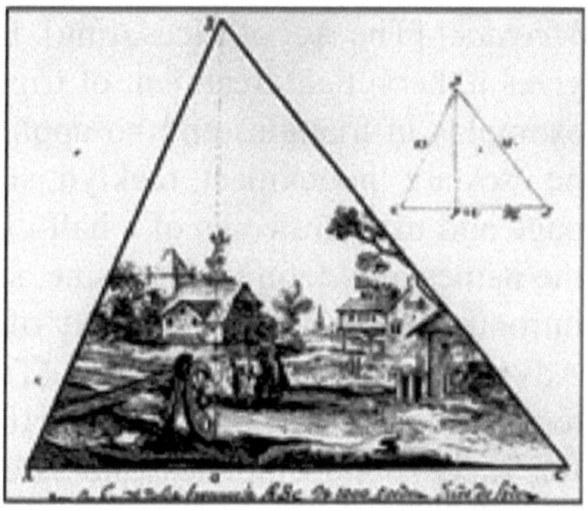

7 Comparison with Other Texts

Around 1620 there were some books on surveying and fortification available in Dutch language; they were of course relatively expensive, but the University library possessed copies. How does BPL 1013 compare to these books, which were available at the time when Frans van Schooten Sr. started lecturing? The most

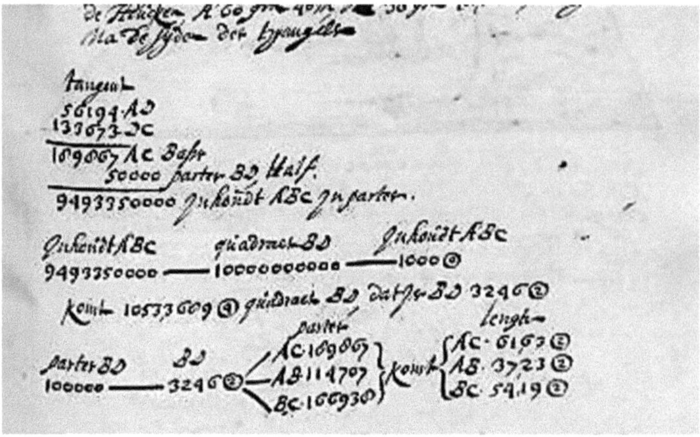

Fig. 8 The calculations of the solution to the problem on f.89[r]

important for comparison are publications[15] by Stevin (1585–1608), Sems and Dou (1600), Marolois (1614) and van Ceulen (1615).

Publications by Stevin which were relevant to Duytsche Mathematique are *De Thiende, 't Weereltschrift* and *De Meetdaet.*

In *De Thiende* [The Tenth], published in 1585, Stevin introduced decimal notation and promoted its use by mathematical practitioners, such as surveyors. The notation used in *De Thiende* had didactical value, it made very clear in which way surveyor's measurements could very simply be translated into decimal notation. In his calculations van Schooten used a simpler notation, also used by Stevin in his *De Meetdaet* [The Act of Measuring]. In *'t Weereltschrift* [Describing the world] Stevin gives a theoretical treatment of trigonometry, starting with many definitions, many examples in triangles and no applications in surveying (Stevin 1605a). The terms he uses are 'houckmaet, raeklyn, snylyn'; van Schooten has an introduction of one page and uses drawings of a half-circle, divided into degrees and minutes and with the names and definitions of sine, secant and tangent added. All exercises after this introduction are in the context of surveying (Figs. 2, 3 and 4) until part 5 on Advanced geometric techniques (Table 1; Figs. 5, 6, 7 and 8). So the approach and terminology in BPL 1013 is practical, starting with the theory which is necessary for surveying techniques and later on more in-depth theoretical examples. *De Meetdaet* (Stevin 1605b) is on geometrical constructions, transformations and surveying techniques. The topics in part 2–4 of BPL 1013 (Table 1) are very similar to what one finds in *De Meetdaet*, but the applications in BPL 1013 are more systematic in the treatment of different instruments, increase in complexity and in the continuous use of surveying contexts through illustrations. Though some topics

[15]Year of first edition between brackets, Stevin published a number of works which were relevant to Duytsche Mathematique.

and their treatment are comparable, it is clear that van Schooten was inspired by, but did not imitate Stevin.

Johan Sems and Jan Pietersz. Dou were surveyors. They published *Practijck des lantmetens* [Practice of surveying] together with *Van het ghebruyck der geometrische instrumenten* [The use of geometrical instruments] as one volume in 1600, when the lectures Duytsche Mathematique started. The topics are roughly the same as in BPL 1013, but they occur in a different order and the treatment is less well structured. The authors use a division in rods, feet and inches in twelve and in ten parts. Other differences with BPL 1013 are: no use of decimal notation for fractions, with as a consequence extensive calculations with fractions and the different measurements, not much of Euclid, mentioning of sine and related terms in the definitions and no mentioning of secant or tangent. There is nothing on fortification. This volume could be used by students, if they had some money to spend, for the surveying part of the course. The lack of decimal notation made the calculations more complicated, the trigonometry was less up-to-date than in BPL 1013. The text by van Schooten is also more concise; that may be due to the difference between teaching notes as preparation for a course by a mathematical expert and a publication meant for self-instruction.

Opera Mathematica [Mathematical Works] by Marolois contained chapters on geometry, perspective, architecture and fortification. His books were illustrated with many engravings of high quality; there were several reprints during the 17th century, usually with improvements. Albert Girard added the use of trigonometric tables; Frans van Schooten Sr. made improvements for a French edition of *Fortificatie* [Fortification], in 1628. Marolois paid much attention to perspective, not a topic in Duytsche Mathematique. Marolois did not use decimal fractions systematically and he preferred the more common division of measuring units in 12 parts. Frans van Schooten Sr. referred to Marolois in his lectures on fortification, with regards to calculation of angles of a regular polygon as basis (BPL 1013 f.140^{v}), but the treatment of fortification in BPL 1013 is different from Marolois.

De arithmetische en geometrische fondamenten by van Schooten's former teacher, Ludolf van Ceulen, has been discussed in 4.2. The chapters on geometrical constructions and transformations are very similar to BPL 1013, ff.8^{v}–44^{v}, again bearing in mind the difference in style between a book and preparatory notes for teaching. However, the rest of the book is very different from van Schooten's manuscript, though van Ceulen is mentioned a few times as an authority.

The analyzed manuscripts from the 17th century on similar subjects, in the libraries of Leiden and Utrecht, show a less complete treatment of the topics, none uses decimal notation consistently, sometimes not at all. A few examples from Leiden[16] are BPL 2084 (first half of 17th century) and BPL 1970 (1658). Both originate from Zeeland, the southwestern part of the Netherlands and treat similar topics as in BPL 1013, but much more limited in scope. They are instructional texts, BPL 1970 is a copy from another text, the scribe mentions several experts amongst

[16]See van Maanen (1987) for more information on the manuscripts in Leiden.

whom are Ludolf van Ceulen, Frans van Schooten and the surveyors Sems and Dou. Another example is BPL 1351 (1655), notes from a student of Kechelius, a private mathematics teacher in Leiden. In this manuscript one finds more or less the same subjects as in BPL 1013, without use of decimal notation and sometimes written in Latin. In Utrecht UBU 1363 is a manuscript on geometry, in two parts, written around 1700. The scribes are Anemaet (1690) and Hub. Anemaet (1726–1727). The manuscript deals with geometry and surveying, with limited treatment of the topics. It probably represents student's notes.

8 Discussion and Conclusion

The Duytsche Mathematique had to attract sufficient number of students to make it worthwhile for the University to continue paying the salary of a professor. As the students, or 'auditores', who came solely for Duytsche Mathematique, usually were not registered as such, there is no direct way to find information about the numbers. Those who were registered as university students sometimes also followed the lectures Duytsche Mathematique, as becomes clear from various sources; some examples are Martinus Wimmers (1607), Tielmannus Wyntgens and Steffen van Blitterswijck (1609), Jean Gillot (1630), Levyn Baers (1631), Willem van Schoonhoven (1633), Caspar Poorten (1639) and Johannes Steen (1644) (Krüger 2014; Witkam 1969). An important source of information is formed by the archives on admission of surveyors in Holland until 1646; there are also some letters of contemporaries in which Duytsche Mathematique is mentioned. Quite a few of the candidates for admission as surveyor mentioned that they had studied Duytsche Mathematique; this served as a testimony.

During the first ten years the reputation of professor Ludolf van Ceulen as a mathematician and as a teacher must have been part of the attraction. Between 1611 and 1614 a group of students lobbied for the appointment of Frans van Schooten as successor of van Ceulen and van Merwen. Such a request was sometimes signed by more than 30 people. In the request of 1612 they wrote that

> … the lessons have been continued by Frans van Schooten during one year and ten months, both in the lecture hall and in the fields… (UBL AC1 Archief van Curatoren, 1574–1815, inv. 42)

So, Frans van Schooten taught both the theory and the practice of surveying. Amongst the signatures were those of at least three well-known surveyors from Leiden who evidently thought well of his lessons. The governors of the University paid for his services at the end of each year; in 1612 he also did some surveying for the University; at the end of that year the governors asked him to continue teaching in 1613. At the end of 1613 he was remunerated for the costs of four wooden instruments which he had ordered for the teaching. After van Ceulen's passing away, the number of students clearly remained sufficiently high to appoint a new professor, who actually did the work of the previous two professors. Frans van

Schooten got appointed, with support from the students. During the period 1615–1645 one finds in the various archives many names of surveyors, who indicate that they studied Duytsche Mathematique, with professor van Schooten. There is less information about engineers, however there are engineers who studied Duytsche Mathematique, for example Axel Arup, Geraert van Belcum, Joris Gerstecoorn and Jean Gillot. Some of the engineers were employed by the king of Sweden. Comments from contemporaries also was in general favourable. In 1646, after his father passed away, Frans Jr. stated in a letter to Christiaan Huygens that the Duytsche Mathematique had produced engineers, surveyors, wine gaugers, reckonmasters (teachers of mathematics) and other professionals; students also came from abroad.

We may conclude that Frans van Schooten attracted students sufficiently to maintain the course.

What made his lectures so valuable? We only have his lecture notes, the manuscript BPL 1013, there are no reports on his teaching known. I discuss the content and some characteristics which are important in teaching.

If the content of BPL 1013 is compared with the publications on surveying and fortification, which were available around 1620, it becomes clear that none of the books treated all the topics which occur in BPL 1013. Also, van Schooten was the only one, apart from Stevin, who used decimal notation consequently and in a seemingly natural way. For those who mastered decimal notation, calculations were less cumbersome than with ordinary fractions. The manuscripts from that time which have been analysed also show a less comprehensive treatment and far less consistent use of decimal notation, or not at all. Van Schooten also showed and let the students practice with several instruments and up-to-date mathematical techniques, with attention to limitations of measurements accuracy and the question of accuracy of calculations.

The characteristics of the manuscript which are important in teaching:

- careful planning, resulting in a well-structured, coherent and consistent curriculum, in which relevant mathematical techniques for surveyors and engineers were taught and practised;
- extensive use of visualization; throughout the manuscript, illustrations are used with every exercise and with every definition, axiom or proposition;
- a gradual increase in complexity within topics;
- many exercises, a great many examples from surveying, but also from other practices, such as building;
- distinction between the core of the curriculum and more advanced topics;
- differentiation in breadth (more topics) and in depth (advanced geometry).

In BPL 1013 neither logarithms nor algebra are mentioned. Logarithmic tables were new and not yet available in a Dutch edition.[17] Algebra was not mentioned in

[17]Adriaen Vlacq and Ezechiel de Decker published the first Dutch work on logarithms in 1626/1627.

the teaching plan by Stevin; at the time when BPL 1013 was composed it was still coss algebra, for which neither surveyors nor engineers had much use. There is a manuscript on coss algebra by Frans van Schooten Sr. (Dopper 2014), he probably taught it privately, but it was not part of Duytsche Mathematique. Topics such as taking square roots and wine gauging, both in BPL 1013, were not mentioned by Stevin in his teaching plan. However, calculating roots was a necessary arithmetic skill for practitioners. Wine gauging was a very useful skill as most towns employed a professional wine gauger. This is an example of aims, which were more diverse than the training of military engineers, of Duytsche Mathematique.

BPL 1013 shows a very extensive collection of teaching notes by someone who evidently was an accomplished mathematician, surveyor, and an excellent teacher. As Frans van Schooten Sr. taught the Duytsche Mathematique for a long time, 35 years and must have had many students, he was influential in reaching the aims of Duytsche Mathematique during those years.

However, he did more. Through his well-designed and carefully written and illustrated lecture notes he left a legacy for his son who was his successor. Frans van Schooten Jr. (1615–1660) made use of these notes of his father during many years and so the influence of Frans van Schooten Sr. lasted for years after he passed away.

We may conclude that in a quiet way Frans van Schooten Sr. during and after his lifetime was very important for the reputation of the Duytsche Mathematique and for the achievement of the aims of this course.

References

Manuscripts

Regional Archives Leiden:

Merwen, Symon Fransz van (before 1610). *De vijff spetie(n) inde tiende getalen.* NL-Ldn-RAL, AB 8399

Library of Leiden University:

AC1 Archief van Curatoren, 1574 – 1815, inv. 42

Schooten, Frans van Sr. (1620). BPL 1013

An. (1658). BPL 1970

An. (1612 – 1650). BPL 2084.

Kechelius, Samuel Carolus (1655). BPL 1351

Library of Utrecht University:

Anemaet, Dingeman & Anemaet, Hubertus (1690, 1726). UBU 1363

Publications

Alberts, Gerard; Atzema, Eisso; & Maanen, Jan van (1999). Mathematics in the Netherlands. A brief survey with an emphasis on the relation to physics, 1560–1960. In K. van Berkel, A. van Helden & L. Palm (Eds.), *A history of science in the Netherlands* (pp. 367–404). Leiden: Brill.

Bierens de Haan, David (1878). Handschrift van Franciscus van Schooten. [Manuscript by Frans van Schooten] In D. Bierens de Haan, *Bouwstoffen voor de geschiedenis der wis- en natuurkundige wetenschappen in de Nederlanden* (pp. 250–280). Amsterdam: Koninklijke Akademie van Wetenschappen.

Bennett, Jim & Johnston, Stephen (1996). *The geometry of war: 1500–1750.* Oxford: Museum of the History of Science.

Boekholt, Petrus & Booy, Engelina de (1987). *Geschiedenis van de school in Nederland.* [History of schools in the Netherlands] Assen: van Gorcum.

Booy, Engelina de (1980). *Kweekhoven der wijsheid. Basis en vervolgonderwijs in de steden van de provincie Utrecht, van 1580 tot begin 19e eeuw.* [Nurseries of wisdom. Primary and further education in the towns of the province of Utrecht, from 1580 until early 19th century]. Zutphen: Walburg Pers.

Ceulen, van Ludolf (1615). *De arithmetische en geometrische fondamenten.* [Foundation of arithmetic and geometry] Leiden: Joost van Colster and J. Marcus.

Dijkstra, Arjen (2012*). Between academics and idiots. A cultural history of mathematics in the Dutch province of Friesland (1600–1700).* (Doctoral dissertation). University of Twente.

Dopper, Jantien (2014). *A life of learning in Leiden. The mathematician Frans van Schooten (1615–1660).* (Doctoral dissertation). University of Utrecht.

Gulik-Gulikers, van Iris (2005). *Meetkunde opnieuw uitgevonden.* [Reinvention of geometry] (Doctoral dissertation). University of Groningen.

Heuvel, Charles van den (2006). De vesting als mathematisch en cultureel kennissysteem. [Fortification as a mathematical and cultural system of knowledge]. *Scientiarum Historia, 32,* 99–117.

Kool, Marjolein (1999). *Die conste vanden getale.* [The art of numbers] Hilversum: Verloren.

Krüger, Jenneke (2010). Lessons from the early seventeenth century for mathematics curriculum design. *BSHM Bulletin 25,* 144–161. http://dx.doi.org/10.1080/17498430903584136.

Krüger, Jenneke (2014). *Actoren en factoren achter het wiskundecurriculum sinds 1600.* [Actors and factors behind the mathematics curriculum since 1600] (Doctoral dissertation). University of Utrecht. http://dspace.library.uu.nl/handle/1874/301858.

Krüger, Jenneke (2015). A first mathematics curriculum: Stevin's Instruction for engineers (1600). *International Journal for the History of Mathematics Education, 10*(1), 79–88.

Maanen, van Jan (1987). *Facets of seventeenth century mathematics in the Netherlands.* (Doctoral dissertation). University of Utrecht.

Maanen, Jan van (1997). New maths may profit from old methods. *For the Learning of Mathematics, 17*(2), 39–46.

Marolois, Samuel (1628). *Opera mathematica.* Amsterdam: Jan Janssen.

Metin, Frédéric (2007). *Adam Fritachs new fortification.* Prague: workshop ESU5.

Molhuysen, Philipp (1913). *Bronnen tot de geschiedenis der Leidsche universiteit,* [Sources for the history of the University of Leiden] *Vol. 1.*'s Gravenhage: M. Nijhoff. https://adcs.home.xs4all.nl/stevin/varia/ingenieurs.html#resolutie.

Mulder, Henry (1990). Pure, mixed and applied mathematics. The changing perception of mathematics through history. *Nieuw Archief voor Wiskunde, IV* (8*),* 27–42.

Muller, E. & Zandvliet, Kees (Eds.) (1987). *Admissies als landmeter in Nederland voor 1811.* [Admissions as surveyor in the Netherlands before 1811] Alphen aan den Rijn: Canaletto.

Paedts, Jan Jacobsz (1600). *Instructie door S. Stevin.* NL Ldn-RAL. www.fransvanschooten.nl/fvs_ordre.htm.

Rieu, G. du (1875*). Album studiosorum academiae lugduno batavae.* Den Haag: M. Nijhoff.

Schooten, Frans van (1627). *Tabulae SINUUM Tangentium Secantium ad Radium 10 000000.* Amsterdam: Willem J. Blaeu (in de Sonnewijzer).
Sems, Johan & Dou, Jan Pietersz. (1600). *Practijck des Lantmetens. Van het gebruyck der geometrische instrumenten.* [Practice of Surveying. About the use of geometrical instruments]. Leyden: Jan Bouwensz.
Sitters, Matthijs (2007). *Sijbrandt Hansz Cardinael 1578–1647, rekenmeester en wiskundige: zijn leven en zijn werk.* [Sijbrandt Hansz. Cardinael 1578–1647, reckon master and mathematician, his life and his work] (Doctoral dissertation). University of Groningen.
Stevin, Simon (1585). *De Thiende.* [The Tenth] Leiden: Christoffel Plantijn.
Stevin, Simon (1594). *Stercktenbouwing.* [Fortification] Leiden: Francoys van Ravelenghie.
Stevin, Simon (ca. 1605a). 't Weereltschrift. [Describing the world]. In *Wisconstighe gedachtenissen.* Leiden: Jan Bouwensz.
Stevin, Simon (ca. 1605b). Meetdaet. [The act of measuring]. In *Wisconstighe gedachtenissen.* Leiden: Jan Bouwensz. http://hdl.handle.net/1874/40317.
Struik, Dirk (1981). *The land of Stevin and Huygens.* Dordrecht: Reidel.
Struik, Dirk (1995). *Geschiedenis van de wiskunde.* [History of mathematics]. Utrecht: Het Spectrum.
Taverne, Ed (1978). *In't land van belofte: in de nieue stadt.* [In the promised land, in the new city]. Maarssen: Gary Schwartz.
Wepster, Steven (2010). Ludolph van Ceulen in Hollandse kringen. [Ludolf van Ceulen in Dutch circles] *Nieuw Archief voor Wiskunde, 5*(11), 63–69.
Winter, Pieter van (1988). *Hoger beroepsonderwijs avant la lettre.* [Higher vocational instruction before the term was coined]. Amsterdam: Noord-Hollandsche Uitgevers Maatschappij.
Witkam, Henricus (1969). Jean Gillot, een Leids ingenieur (2). [Jean Gillot, an engineer from Leiden (2)] *Leids Jaarboekje, 61*, 39–70. Leiden: Sijthoff.
Wreede, Liesbeth de (2007). *Willebrord Snellius (1580–1626). A Humanist reshaping the mathematical sciences.* (Doctoral dissertation). Utrecht University.

Author Biography

Jenneke Krüger works as a self-employed curriculum researcher and author, and is connected to the Freudenthal Institute for Science and Mathematics Education (Utrecht University, Netherlands). She obtained a master's degree in biology from Utrecht University, a teaching degree in mathematics from the State University at Groningen, after studying mathematics at universities in Adelaide, London and Groningen and a Ph.D. from Utrecht University. Her Ph.D. research was on history of mathematics education; the influences on Dutch mathematics curricula for secondary education, starting from the 17th century. The title of the Ph.D. thesis is *Actoren en factoren achter het wiskundecurriculum sinds 1600* (Actors and factors behind the mathematics curriculum since 1600). Her main interests are education and history of education, especially in natural sciences, computing science and mathematics. Concerning history of mathematics education, she specializes in the level of education following primary education.

Arithmetic in the Spanish Army at the End of the 19th Century: The Textbooks by Salinas and Benítez

Antonio M. Oller-Marcén and Vicente Meavilla-Seguí

Abstract The last third of the 19th century was a prolific period with respect to the production of mathematics textbooks by Spanish authors, many of which were army officers who wrote for military students. Among them, Manuel Benítez y Parodi and Ignacio Salinas y Angulo collaborated on several books on arithmetic and algebra that were reprinted several times. In this work, we focus on their books on arithmetic. In particular, following a qualitative approach, we carry out a text-driven analysis of their arithmetic textbooks that focus on the mathematical content as well as on the pedagogical aspects of the texts. This analysis will give us an overview of the arithmetical requirements for the different levels of the Spanish army (from enlisted to officers) at the end of the 19th century. For example, we conclude that aspirants to corporal received a training equivalent to elementary primary education, while army officers had an arithmetical training equivalent to that of architects or engineers.

Keywords Arithmetic · Textbook · 19th century · Army · Spain

1 Introduction

The history of Spain during the 19th century was rather turbulent (Esdaile and Charles 2000). The country was involved in several different wars (including a civil war) and the Spanish Empire finally collapsed, different types of governments (monarchy, republic, provisional governments) were in power and up to six different constitutions were promulgated during this century. This political and social instability had an impact on education and, in particular, on mathematics education. There were 25 mathematical curricula in less than 75 years and many different

A.M. Oller-Marcén (✉)
Centro Universitario de La Defensa de Zaragoza, Saragossa, Spain
e-mail: oller@unizar.es

V. Meavilla-Seguí
Departamento de Matemáticas, Universidad de Zaragoza, Saragossa, Spain

F. Furinghetti and A. Karp (eds.), *Researching the History of Mathematics Education*, ICME-13 Monographs,
https://doi.org/10.1007/978-3-319-68294-5_9

educational regulations appeared (Ausejo and Matos 2014). The most important was the so-called Moyano Law of 1857 (Puelles 1985). This law laid the groundwork for the Spanish education system for more than a century. It established three stages: primary education (elementary and upper), secondary education, and university education. It established the secularization of education[1] and that elementary primary education was mandatory (for ages 6–9) and free for those who could not afford it. The law also established a high degree of administrative centralization (curricular and linguistic uniformity throughout the country, for instance). It determined the requisites to become a teacher and the required organization of school centers. Finally, it included a certain degree of liberty in the creation of school centers. For a detailed account on this law, we refer the interested reader to Montero's article published in 2009.

At the end of the 19th century, military education was also one of the main concerns of the government and there were several different attempts at military reform that, consequently, had an impact on military education (Ruiz 2004). In 1887, for instance, there were 13 different centers of military education: the General Military Academy, specialization academies of cavalry, artillery, engineering, military administration and general staff (attended by cadets after a common period at the General Academy), the School of Equitation, and six preparatory schools for entrance to the General Military Academy (Baltar 2009). Later, in 1893, the General Military Academy was closed and it was replaced by the specific academies for each arm (cavalry, artillery, infantry, engineering and military administration). The preparatory schools were also closed and some new institutions were created (Laguna 1998). These attempts at reform also had an impact on curricula: for example, the Academy of Military Engineers had seven different curricula between 1803 (when it was established) and 1893 (Velamazán and Ausejo 1989).

It must be pointed out that entrance to one of the military academies was not the only way into the army. In addition to becoming a cadet at one of them (which eventually would lead to becoming an army officer), it was also possible to enter the army as a private. Once a private, it was possible to be promoted to sergeant and, after a period of training, enter one of the academies to, eventually, become an officer. Direct entrance to one of the academies required certain economical expenses; hence, people coming from less favored social classes usually followed the second path, which only rarely led to becoming an army officer (Fernández 1978).

The conditions for entering the academies varied over time. The required age ranged between 14 and 16 years. By the end of the 19th century, the applicant had to pass oral examinations on different subjects (with arithmetic being the only one related to mathematics) with a level equivalent to the contemporary secondary education. In the case of privates, no previous education was required (other than the mandatory elementary primary education). This explains the need (see Sect. 7) for textbooks including basic knowledge of subjects such as arithmetic, grammar, or calligraphy.

[1]Although the Catholic Church still had privileges.

In spite of the previously described instability, the last third of the 19th century in Spain was, regarding mathematics, "a period of consolidation when the distance that separated the study of this science in Spain from other countries was reduced" (García 1982, p. 126). In this regard, we must mention the role of the so-called *sowers* (Ausejo 2008). With this term, attributed to Gino Loria (Ríos 1988), we refer to four mathematicians: José de Echegaray, Zoel García de Galdeano, Eduardo Torroja, and Ventura Reyes Prósper, who were committed to the importation into and dissemination through Spain of "modern" mathematics (Oller-Marcén 2017). García de Galdeano, in the first issue of *El Progreso Matemático*, stated that

> todo este movimiento general se sigue en España [...] pero esto que se circunscribe a un reducido número de personas [...] no trasciende a otro público más numeroso [...] La propaganda de las ideas es el remedio más eficaz para la desaparición de este mal [this general movement [modern mathematics] is followed in Spain [...] but it circumscribes to a reduced number of people [...] it does not transcend to a more numerous public [...]. The dissemination of ideas is the best remedy for this illness[2]] (*El Progreso Matemático*, year I. Zaragoza, January 20, 1892. No. 1, p. 2).

Nevertheless, according to Sánchez (1992, p. 58) "el ámbito en el que se movieron los físicos y matemáticos españoles de aquella centuria fue, con muy pocas excepciones, el de la enseñanza" [the area where Spanish physicists and mathematicians of that century worked was, with very few exceptions, that of teaching]. Within this context, mathematics textbooks experienced a popularization without precedents (Gómez 2011). Thus, we find a huge number of mathematics textbooks published all throughout this century, both original and translated (mainly from French authors).

During the 18th century, mathematics was an important element in the training of cadets and officers in the Spanish Army (Navarro 2011). For instance, it was the opinion of the unknown author of a manuscript devoted to the teaching of cadets during the 18th century in Spain that (Anonymous n.d., p. 5): "esos indispensables principios no pueden penetrarse sin el auxilio de las Matemáticas" (these essential principles [of the art of war] cannot be apprehended without the help of mathematics). Although this is just an example of a single personal opinion, we can easily imagine that this was a commonplace sentiment at that time.

Even if their importance decreased with respect to the preceding century, military academies were still important places regarding the development of mathematics and its teaching at the end of the 19th century (Peralta 2000). Thus, quite a number of textbooks written by and for the military were published (Velamazán 1990). In fact, Maz (2005) points out the existence of several army and navy officers among the main authors of mathematics textbooks of that time. We may mention, for instance, Miguel Ortega y Sala (colonel of engineers), Antonio Terry y Rivas (admiral of the navy), Juan Montero Gabutti (military engineer), Cirilo

---

[2]Here and throughout the paper, quotations and titles of books and articles have been translated by the authors.

Aleixandre (military engineer), Manuel Benítez y Parodi (major-general), and Ignacio Salinas y Angulo (major-general).

Different aspects regarding the teaching of mathematics in the Spanish army during the 18th and 19th centuries have been studied already from different perspectives. From the institutional point of view, Riera and Riera (2008) describe the remarkable Academy of Mathematics of Barcelona, while Navarro (2011) focuses on the Military School of Ávila. Regarding curricular aspects, Velamazán and Ausejo (1989) describe and analyze the evolution of the different curricula in the Academy of Engineers throughout the 19th century. Finally, there are works that focus on personal contributions such as Navarro and Velamazán (2006) that analyzes the scientific contributions of José de Odriozola, or Velamazán (2015), that studies the mathematical background and contributions of Antonio Sangenís.

In this work, we will focus on two of the textbook authors mentioned above: Manuel Benítez and Ignacio Salinas. The main reason is that they collaborated on multiple editions of various mathematics textbooks that were reprinted several times during the 19th and 20th centuries (with a gap of more than 70 years between the first and the last known editions). This illustrates the success of their textbooks, which became quite popular. In particular, they were used extensively as official textbooks in different military academies and also in the Naval Academy (Comas 2015). Moreover, some of them were adapted and published for a broader public. In fact, some of their books (see Sect. 4) became important textbooks for secondary education students and for prospective engineers and architects (Vea 1995).

For the reasons stated above, the textbooks written by Salinas and Benítez are worth analyzing. In particular, we are interested in those arithmetic textbooks that were specifically addressed to the military. Thus, the main objectives of this paper are:

1. To give an overview of the life and works of Manuel Benítez and Ignacio Salinas.
2. To analyze the books about arithmetic written by these authors.
3. To present a general outlook on the arithmetic training in the Spanish Army at the end of the 19th century.

2 Methodological Issues

In order to achieve the previous objectives, we follow a qualitative approach, performing exploratory and descriptive research that uses documents as a resource while focusing on their content (McCulloch 2004; Prior 2016).

The sources of information varied according to the different objectives. To obtain biographical information of the authors, we have reviewed journals like *La Correspondencia Militar* or *La Ilustración Española y Americana* that contained information about military and civil society news and, most importantly, *La Gaceta de Madrid* that, at that time, published all the laws and decrees promulgated by the

Government. These journals are digitized and can be accessed through the database of the National Library of Spain (hemerotecadigital.bne.es). Regarding the textbooks, some of them can also be found in digital format in the National Library (bibliotecadigitalhispanica.bne.es) while others belong to the personal collection of one of the authors.

Scott (1990) establishes four criteria to assess the documents used as sources of information in this type of research. They are authenticity, credibility, representativeness, and meaning. Authenticity and credibility are granted by the use of original sources. In the case of textbooks, we are not concerned about representativeness because we are analyzing all the arithmetic textbooks written by the considered authors. Finally, meaning (i.e., the clarity of the information) is also granted by the nature of the analyzed documents, which are journals and textbooks.

Textbook analysis is an important topic in mathematics education, admitting several approaches and different perspectives (Fan et al. 2013). From a historical point of view, Schubring (1987) points out that textbook analysis allows us to study how different mathematical concepts evolved and were transmitted during a particular period of time. Maz and Rico (2015) use textbook analysis focusing not only on concepts but also on pedagogical, curricular and social aspects.

In this work, we have followed the phases of content analysis as described by Mayring (2014, p. 54). In particular, we carry out a text-driven analysis of the arithmetic textbooks by Salinas and Benítez (Krippendorff 2014, pp. 340–342). According to objectives 2 and 3 above, the main focus of our research is on the mathematical content presented by the authors as well as on the pedagogical aspects of the texts.

3 Bio-Bibliographical Sketch

In this section, we will briefly present some biographical information about the generals Benítez and Salinas. We will also give an overview of their writings that includes books of non-mathematical content.

Manuel Benítez y Parodi was born in Seville on August 21, 1845. He had a quite successful life (Méndez 1922). From the military point of view, he reached the rank of major-general and received several decorations and medals. He was also named the Civil Governor of Seville and later of the Canary Islands.

Regarding his scientific activity, he was made a member of the Royal Academy of Exact, Physical and Natural Sciences on May 3, 1905 (although he died before taking up his post) and vice-president of the Spanish Mathematical Society, which was created partly at his proposal. In addition to the textbooks that we will present later, he also was the author of a memoir (which remained unpublished) with the title *Exposición razonada y metódica de los desarrollos en serie de las funciones matemáticas* [Reasoned and methodical exposition of the development of mathematical functions into series], for which he received an award in 1893 (Octavio de Toledo 1912).

Fig. 1 Portrait of Manuel Benítez y Parodi (*La Ilustración española y americana*, year LV. Madrid, December 8, 1911. No. XLV, p. 335)

As a teacher, he was described by Octavio de Toledo as (1912, p. 195) "inimitable; tenía la envidiable habilidad de hacer interesante y atractiva la materia que exponía" (inimitable, he had the enviable skill to make interesting and attractive the topic that he was presenting). He was also interested in the teaching of mathematics and in the dissemination of mathematical ideas. In fact, he was one of the official collaborators of *El Progreso Matemático*, the first Spanish journal solely devoted to mathematics (Hormigón 1981, 1993). Benítez died on November 29, 1911 (Fig. 1).

In contrast to the many biographical details available about Benítez, very little is known about Ignacio Salinas y Angulo. Most of this information can be found in his service record, quoted in his promotion to major-general (*Gaceta de Madrid*, nº 70, March 1908). He was born on February 1, 1843 and he received a series of medals and decorations. He was the sole author of three books of juridical content (Salinas 1879, 1883, 1899). Of special historical interest is his *Defensa del General Jáudenes* [Defense of the General Jáudenes] (Salinas 1899). This book is a transcription of the discourse that General Salinas pronounced in the name of General Jáudenes in the court-martial that was judging his acts during the surrender of Manila[3] and contains a very detailed description of the precarious situation of the Spanish Army in the Philippines at that time.

Regarding his scientific activity, not much is known. He seems to have done some topographic works and, according to Lecoq (2014) he was present at the

[3]General Fermín Jáudenes y Álvarez was the Governor of the Philippines for less than one month in 1898, during the Spanish-American war, which ended up with the loss of Cuba, Puerto Rico, Philippines, and the few remains of the former Spanish Empire. The battle of Manila was considered dishonorable and has sometimes been referred to as the "Mock Battle of Manila".

Fig. 2 Signatures of Salinas (left) and Benítez (right) as found at the front page of *Pizarras por papeletas para el examen de Aritmética en la Academia General Militar*

border demarcation between Spain and Portugal that took place on July 18, 1894 (Salinas and Oliver 1867). Nevertheless, no further scientific activity is known for him, apart from the mathematics textbooks that he wrote together with Benítez (Fig. 2).

General Salinas resigned his position of Chief of the Northern Captaincy General Staff on January 20, 1904 due to "the poor state of his health" but was promoted to major-general on March 8, 1908 (*Gaceta de Madrid*, nº 21, January 1904). He died on March 14, 1910 (*La Correspondencia Militar*, nº 9841, March 1910).

The Salinas-Benítez partnership wrote the text of juridical content *Ordenanzas del Ejército armonizadas con la legislación vigente* [Army regulations harmonized with current legislation] (Salinas and Benítez 1914). They also wrote the following textbooks with arithmetical content:

1. *Aritmética* [Arithmetic] (Salinas and Benítez 1898a).
2. *Aritmética: obra redactada con estricta sujeción al programa oficial para los exámenes de ingreso en la Escuela Preparatoria de Ingenieros y Arquitectos* [Arithmetic: work written with strict subjection to the official program of the entrance examinations to the school of engineers and architects] (Salinas and Benítez 1887).
3. *Cálculos de la aritmética: de texto para el ingreso en la Academia General Militar y apuntes relativos a los mismos* [Calculations of the arithmetic: textbook for the entrance to the General Military Academy and notes related to them] (Salinas and Benítez 1892a).
4. *Pizarras por papeletas para el examen de Aritmética en la Academia General Militar* [Slates by sheets for the arithmetic exam at the General Military Academy] (Salinas and Benítez 1892b).
5. *Manual para las clases de tropa* [Handbook for teaching the troops] (Salinas and Benítez 1894, 1913).

In addition, they also wrote a book on algebra in two parts (Salinas and Benítez 1892c, 1898b). Nevertheless, as we pointed out, we are interested here in arithmetic textbooks addressed to the military. Hence, we will focus only on points 1, 3, 4 and 5 above.

For the moment, the circumstances under which these two authors met and worked together over the years remain unknown. Salinas was two years older than

Benítez, but they entered the Special School of the Army General Staff one year apart (Salinas in 1860 at the age of 17 and Benítez in 1861 at the age of 16). From the moment they left the school, their professional paths did not seem to cross again (according to their service records). Benítez held more "political" positions and was more involved in teaching and administrative duties than in campaigns. By contrast, Salinas seemed to be a man of more action. He spent five years in the Philippines and his service record shows that he participated in some combat in the Philippines as well as in some skirmishes caused by strikes back in Spain.

4 *Aritmética* (Arithmetic)

Aritmética by Salinas and Benítez was designated as the official textbook for all the Military Academies by a Royal Order on June 28, 1884 after a contest performed on April 30. It was awarded a medal in the 1888 Barcelona Universal Exposition. This text was extensively used and reprinted. The last edition dates from 1955, more than 70 years after its first edition.

The edition of 1898 that we have been able to examine contains 299 pages and is divided into six books. Books 1–4 constitute the first part which is devoted to abstract numbers. The second part is made up of books 5 and 6 and it introduces numbers as quantities, the decimal system, and the classical applications of proportionality. In Table 1, we present the contents of this book.

As we see, Salinas and Benítez deal with topics that were present in most arithmetic textbooks at that time. Salinas and Benítez explicitly cite the names of Juan Cortázar (1809–1873), Bernardino Sánchez Vidal (1829–1878), and Joseph-Alfred Serret (1819–1885). Consequently, it seems plausible that they consulted the works of these authors. However, we must point out the absence of logarithms, progressions, or different numbering systems that in some contemporary texts were sometimes given, at least, in appendixes.

The authors alternate between an explanatory discourse, which motivates and presents examples, and explicit statements and rigorous proof of theoretical results. At the end of each section, they present a great number of unsolved exercises. In many cases, these problems are presented in a military context, even if they are just "realistic problems;" i.e., mere simulations of reality (Díaz and Poblete 2001, p. 37). As an example of these type of problems, let us present the following problem taken from Book 1, Chap. 2 (p. 24): "Un general, al salir para una expedición con 180,000 hombres, dejó 600 para guarder una plaza; al propio tiempo recibió un refuerzo de 800 hombres, de los cuales se quedaron 450 en los hospitals; pidió 3500 hombre más, pero no pudieron mandarle más de 2370, y, de estos, tuvo que dejar 1750 en diferentes puestos: ¿con cuántos hombre llegó a su destino?" [A general leaves on an expedition with 180,000 men, leaving 600 to keep a place safe. At the same time, he received a reinforcement of 800 men, of which 450 stayed in the hospital. He further asked for 3500 more men, receiving only 2730, of which 1750 had to be left behind in different employments. How many men did he arrive

Table 1 Contents of the *Aritmética* by Salinas and Benítez

Part 1	Book 1. Integer numbers	Chapter 1. Preliminary notions
		Chapter 2. Fundamental operations
		Chapter 3. Divisibility of numbers
		Chapter 4. Greatest common divisor
		Chapter 5. Least common multiple
		Chapter 6. Prime numbers
		Chapter 7. Applications of prime numbers
	Book 2. Fractions	Chapter 1. Properties of ordinary fractions
		Chapter 2. Operations with fractional numbers
		Chapter 3. Complex fractions and fractional equalities
		Chapter 4. Continuous fractions
		Chapter 5. Decimal fractions
		Chapter 6. Reduction of fractions
	Book 3. Powers and roots	Chapter 1. Powers
		Chapter 2. Square roots
		Chapter 3. Cubic roots
	Book 4. Incommensurable and approximate numbers	Chapter 1. Incommensurable numbers
		Chapter 2. Numerical approximations
Part 2	Book 5. Concrete numbers	Chapter 1. Decimal metric system and its relation with the old system
		Chapter 2. Operations with concrete numbers
	Book 6. Proportional magnitudes	Chapter 1. Ratio and proportion
		Chapter 2. Questions of commercial arithmetic

with?]. We can also find verbal problems whose data are taken from other disciplines. These "multidisciplinary problems" were rather common in Spanish textbooks during the second half of the 19th century (Meavilla and Oller 2014). In this respect, some problems are identical (even the numerical data) to problems found in a previous work by Antonio Terry y Rivas (Meavilla and Oller 2015). Surprisingly, this author was not cited. A detailed comparison of both texts shows that up to seven problems were the same. In Table 2, we present two different examples.

It is worth mentioning that the authors include some interesting etymologic and historic comments in the form of footnotes. For example, we have the following comment in Book 1, Chap. 1 (p. 2): "Palabra [algoritmia] utilizada por Hoené Wronksi, matemático polaco de principios del siglo XIX, para designer la ciencia de los números y adoptada por los matemáticos alemanes" (Word [algoritmia] used by Hoené Wronski, a Polish mathematician from the beginning of the 19th century, to designate the science of numbers and adopted by German mathematicians). These comments prefigure, in some sense, the later interest in introducing history in the teaching of mathematics (Tzanakis and Arcavi 2000).

Table 2 An example of two identical problems

Salinas and Benítez (1898a, p. 35)	Terry (1880, p. 14)
Cada vez que un hombre respira introduce 665 centímetros cúbicos de aire en sus pulmones; suponiendo que respire 18 veces por minuto, ¿qué cantidad de aire introducirá en sus pulmones durante una hora? [Every time that a man breathes, he introduces 665 cubic centimeters of air in his lungs; assuming that he breathes 18 times per minute, what quantity of air will he introduce in his lungs during an hour?]	Cada vez que un hombre respira introduce 665 centímetros cúbicos de aire en sus pulmones; respira poco más o menos 18 veces por minuto: ¿qué cantidad de aire introduce en sus pulmones durante una hora? [Every time that a man breathes, he introduces 665 cubic centimeters of air in his lungs; he breathes approximately 18 times per minute: what quantity of air does he introduce in his lungs during an hour?]
Salinas and Benítez (1898a, p. 257)	Terry (1880, p. 112)
Descartes nació el 3 de abril de 1596 y murió el 11 de febrero de 1650; Pascal nació el 19 de junio de 1623 y murió el 19 de agosto de 1662; Newton nació el 15 de diciembre de 1642 y murió el 18 de mayo de 1727; ¿qué edad alcanzó cada uno de estos insignes matemáticos? [Descartes was born on April 3, 1596 and died on February 11, 1650; Pascal was born on June 19, 1623 and died on August 19, 1662; Newton was born on December 15, 1642 and died on May 18, 1727; what age did each of these distinguished mathematicians reach?]	Descartes nació en 3 de Abril de 1596 y murió en 11 de febrero de 1650. Pascal nació en 19 de Junio de 1623 y murió en 19 de Agosto de 1662. Newton nació en 15 de Diciembre de 1642 y murió en 18 de Marzo de 1727. Se pregunta: ¿cuánto tiempo vivió cada uno de estos insignes matemáticos? [Descartes was born on April 3, 1596 and died on February 11, 1650. Pascal was born on June 19, 1623 and died on August 19, 1662. Newton was born on December 15, 1642 and died on May 18, 1727. It is asked: for how long did each of these distinguished mathematicians live?]

To emphasize the importance and positive reception of this book, it is worth noting that it was soon adapted in order to reach a broader public. In fact, only three years after *Aritmética* was published, in 1887, they published *Aritmética: obra redactada con estricta sujeción al programa oficial para los exámenes de ingreso en la Escuela Preparatoria de Ingenieros y Arquitectos*. This text was just a reprint of *Aritmética* with few variations. The order of the contents was slightly modified and some of the contents (continuous fractions; compound interest and annuities; time and angular magnitudes in the section devoted to the decimal metric system) were replaced by some other ones (binary and duodecimal numerical systems or some numerical methods) that might have been of special interest to engineers. Consequently, the structure of this book mirrors that of *Aritmética*. Only some parts were removed because they had disappeared from the official program.

5 Reactions to *Aritmética*

As we already pointed out, *Aritmética* was extensively reprinted. In fact, three editions were published before the end of the 19th century and at least 14 more editions appeared during the first half of the 20th century. Of course, this makes it evident that the book was also extensively used both inside and outside the Army.

As further evidence of the importance of this textbook in a military context, it is worth mentioning the publication of several works trying to clarify some theoretical issues, presenting more detailed proofs of some results, or providing solutions to the proposed exercises.

These texts were written by teachers at the military academies that used the Salinas and Benítez book as a textbook for their classes. Hence, they were familiar with the difficulties that the students could face and, consequently, they tried to prevent them. Moreover, these teachers were military men and not mathematicians, so it is possible that they also faced difficulties themselves when using the book. Whatever the case, the existence of these works makes clear that *Aritmética* by Salinas and Benítez was the basis of the arithmetical training of Spanish officers for over 70 years.

The first of these reactions to *Aritmética* was published very early, in 1885, only one year after the publication of the original work, and it was written by Juan Montemayor and Pedro Bazán (Fig. 3), two captains who were teachers at the General Military Academy. Its title was *Apuntes para el estudio de la Aritmética de D. Ignacio Salinas y D. Manuel Benítez y resolución de los ejercicios prácticos* [Notes for the study of Arithmetic by Mr. Ignacio Salinas and Mr. Manuel Benítez and solutions to the practical exercises] (Montemayor and Bazán 1885). The title of this book explains the intention of the authors, but they made it even clearer in a small introduction where they stated that (p. 3) "nos proponemos facilitar el studio de la Aritmética" (we intend to facilitate the study of *Arithmetic*). To do so, they gave clarifications to some theoretical questions, presented the proofs of some theorems and corollaries, and provided the solution to all the exercises that were left to the reader in the original work. They also proposed some new problems to their readers.

Before the end of the 19th century yet another reaction to *Arithmética* was going to be published by the military engineer Francisco de Lara y Alonso. It was the *Pizarras con aclaraciones para facilitar el estudio de la Aritmética de los señores Salinas y Benítez* [Slates with clarifications to facilitate the study of Arithmetic by Messrs. Salinas and Benítez], whose second edition was released in 1900 (Lara 1900). With this book, the author pursued the goal of (p. 3) "facilitar la preparación de los que aspirer a seguir la Carrera de las armas" (facilitating the training of those pursuing a career in the Army). In fact, he provided the solutions to most of the exercises proposed in the original text as well as the (p. 3) "demostraciones de teoremas que nuestra práctica en la enseñanza nos han hecho ver que ofrecían dificultades a los alumnos" (proofs of the theorems that, according to our teaching

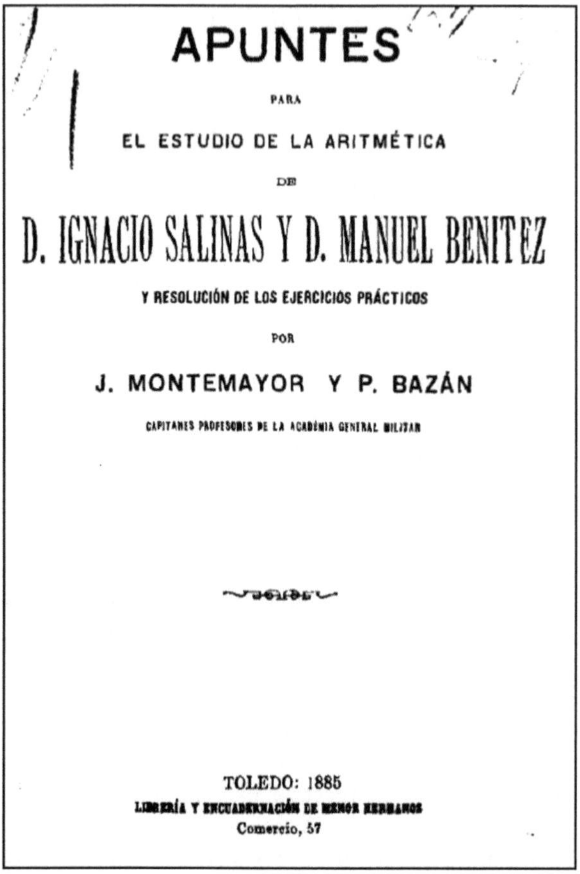

APUNTES

PARA

EL ESTUDIO DE LA ARITMÉTICA

DE

D. IGNACIO SALINAS Y D. MANUEL BENITEZ

Y RESOLUCIÓN DE LOS EJERCICIOS PRÁCTICOS

POR

J. MONTEMAYOR Y P. BAZÁN

CAPITANES PROFESORES DE LA ACADEMIA GENERAL MILITAR

TOLEDO: 1885

LIBRERÍA Y ENCUADERNACIÓN DE MENOR HERMANOS

Comercio, 57

Fig. 3 Front page of the book of Montemayor and Bazán (1885)

experience, are more troublesome for our pupils). This text was republished at least twice, in 1912 and 1918.

The existence and proliferation of this kind of texts, trying to complete (in some sense) and to make easier to understand the original work, seem to have annoyed the authors. In 1892, Salinas and Benítez themselves published the *Cálculos de la aritmética de texto para el ingreso en la Academia general Militar y apuntes relativos a los mismos*. In the prologue, they justify the publication, claim their sole authority to make notes and comments on the original work and try to show that their work had no gaps to be filled (p. iii):

> Desde que publicamos la aritmética […] han ido apareciendo sucesivamente diversos folletos que, con la denominación genérica de apuntes, se ofrecen a los profesores de preparación y a los alumnos con el propósito de explicar o ampliar el referido texto. Estos trabajos, inspirados sin duda en el laudable empeño de facilitar la enseñanza, entendemos nosotros que no están en armonía con el espíritu que informa la obra […] y llevan la confusión al alumno y aun a los profesores, quienes pudieran creer que los mencionados apuntes tienen nuestra autorización, nunca otorgada ni tampoco pedida, contraviniendo así,

sus autores, a los preceptos de la ley de propiedad literaria. Estas consideraciones nos han inducido a la publicación del presente trabajo con el fin exclusivo de evidenciar que el texto contiene todo cuanto el programa exige, sin necesidad de otras explicaciones [...] las notas correspondientes a las distintas partes del cálculo, no son en verdad para suplir deficiencias de la obra, sino para confirmarla y contestar a objeciones y dudas así como para diferir a los deseos de algunos profesores, cuya ilustración y aptitudes para la enseñanza nos complacemos en reconocer

[Since we published Arithmetic [...] various booklets have been published, with the generic name of notes and have offered to coach teachers and students in order to explain or expand the referred text. These works, certainly inspired by a laudable effort to facilitate teaching, are however not in harmony with the spirit that informs the work, [...] and lead students and even teachers to confusion, believing that the aforementioned notes have our authorization, which was never requested nor granted thus contravening its authors the precepts of the law of copyright. These considerations have led us to the publication of this work, with the sole purpose of demonstrating that the text contains everything that the program requires, without further explanations [...] notes are not really to fill gaps in the work, but to confirm and answer objections and doubts [...], and to defer to the wishes of some teachers, whose enlightenment and teaching skills we are pleased to recognize].

However, the authors did not seem to succeed, because still more examples of texts based in the work by Salinas and Benítez can be found on the libraries of different military education institutions in Spain. For instance, in the Artillery Academy at Segovia we can find three more books:

- *Apuntes de aritmética adaptados al texto de Salinas y Benitez* [Notes on arithmetic adapted to the text of Salinas and Benítez] (Salgado 1922), written by Captain Manuel Salgado Biempica.
- *Apuntes y aclaraciones de aritmética al texto "Salinas y Benítez"* [Notes and clarifications to the text "Salinas and Benítez"] (Equis 1910), written by Captain Equis (sic).
- *El indispensable para la preparación militar: apuntes de aritmética: cuaderno el más completo y aclaratorio de las teorías de mayor dificultad contenidas en la aritmética de Salinas y Benítez, de texto para el ingreso en las Academias Militares* [Essential for military preparation: notes on arithmetic: the most complete and explanatory notebook on the most difficult theories contained in Arithmetic by Salinas and Benítez, textbook for the entry into the Military Academies] (Barrera 1916), written by Eduardo Barrera Bau.

It is noteworthy that one of these books (Equis 1910) was published anonymously, perhaps to avoid copyright problems. In any case, the titles of these works show again the difficulties faced by the students and the need of clarifications in order to understand the original text.

6 The *Pizarras Por Papeletas Para El Examen de Aritmética En La Academia General Militar* (Slates by Sheets for the Arithmetic Exam at the General Military Academy)

In 1892, soon after the publication of *Cálculos de la Aritmética* (see the previous section), Salinas y Benítez also published the textbook *Pizarras por papeletas para el examen de Aritmética en la Academia General Militar.*

The main goal of this textbook was to help those students that wanted to enter the General Military Academy. Although this Academy was closed one year later, it is reasonable to assume that this book was used in the specific academies of each arm that replaced it.

The different topics of the arithmetic exam in the General Military Academy were distributed into 38 sheets. During the exam, each student had to pick up one (at random) and then he had to present that topic in front of the examining board. Thus, the book contains the 38 sheets, corresponding to the different topics covered by the public examination syllabus that granted access to the Academy. The content of each sheet is presented, in words of the authors (p. iii), "en la misma forma en que los alumnos han de escribirlos en la pizarra" (in the same way as the students must write them in the blackboard).

Consequently, the book is just a collection of schemes including the main aspects that should be covered by a candidate in the event of having to present that topic during his public examination. In Fig. 4, for instance, we see the sheet corresponding to the topic "Prime numbers." It is clear that the text would not be useful in order to learn anything about this topic. In fact, the authors provided the numbers of the sections of *Aritmética* that contained the relevant information (in this case, sections 96, 97 and 98). Therefore, we can conclude that this book was conceived as a complement to *Aritmética.*

7 The *Manual Para Las Clases de Tropa (Manual for Teaching the Troops)*

On April 11, 1892 a contest to choose the official textbook for the regimental Academies of the Army was held. The contest was won by the *Manual para las clases de tropa* written by Salinas and Benítez. It contains the mathematical knowledge required for privates, corporals and sergeants, and also grammar, history, laws, moral education, etc. It was structured into two books, the first containing the contexts for privates and corporals and the second those for sergeants. The report of the board that chose the *Manual para las clases de tropa* pointed out that the division into two books was an important economic improvement for the

PAPELETA 9.ª

Números primos.—Principios fundamentales y determinación de estos números.—Definiciones.—Primeras proposiciones.—Formación de una tabla de números primos.—§ *del 96 al 98.*

Operaciones con los números incomensurables.—Medida de la magnitud incomensurable.—Concepto de las operaciones con números incomensurables.—Generalización de las reglas de cálculo.—§ *del 206 al 208.*

***

$$p,\ a;\quad N, N', N'', N''' \ldots\ldots;\quad N > N' > N'' > N''' > \ldots\ldots > 1.$$

$$p;\quad 1.2.3.5.7.11 \ldots\ldots p = \pi(p);\quad \pi(p)+1;\quad [\pi(p)+1] - \pi(p) = 1.$$

$n+n=2.n$
$2.n+n=3.n$
$3.n+n=4.n$
. . . .

$2, 3, 5 \ldots\ldots p;\ q, q^2$

$N < q^2 \ldots\ldots$; $N' > p$; $N' \lessgtr q$; $N : N' = N''$; $N = N'.N''$ … $N'' < q$.

1	**2**	**3**	4	**5**	6	**7**	8	9	10
11	12	**13**	14	15	16	**17**	18	**19**	20
21	22	**23**	24	25	26	27	28	**29**	30
.	.	.	.	.	.	.	.	.	.

Fig. 4 Sheet about prime numbers

students. According to the authors, in the case of aspirants to corporal, their book (with a price of 2.5 pesetas) replaced a set of books costing from 25 to 30 pesetas.[4]

The first book was aimed at privates and corporals and it was divided into two parts. The first part was devoted to the Academy of privates aspiring to the rank of corporal and the second one to the Academy of corporals. With respect to privates aspiring to the rank of corporal, the authors state (p. 66): "las demostraciones rigurosas se han sustituido a veces por razonamientos fáciles de comprender […] la expresión explícita y terminante de las reglas prácticas de las operaciones permite al alumno el no tener que deducirlas por sí mismo, haciendo un trabajo de abstracción al que no está acostumbrado, y al que no puede aspirarse" (rigorous proofs have been sometimes replaced by easy-to-understand arguments […] the explicit expression of practical rules prevents students from deducing them alone by avoiding a degree of abstraction to which he is not accustomed and that we cannot aspire to […]). The mathematical content of this part of the book mainly involves the four rules for integers and the nomenclature of the decimal metric system. Decimal numbers are explained without having introduced fractions. In the second

[4]In 1888, according to De la Torre (2015, p. 14), a corporal of the lowest degree earned a monthly salary of less than 25 pesetas.

part of this first book, devoted to corporals, the authors introduce fractions, their composition, and the use of the decimal metric system.

The second book was devoted to the Academy of sergeants. In this book, the authors slightly raise the mathematical level of the text but they do not seem to be confident about the possibility of doing the same in practice. In this respect, for instance, the authors present proofs in a smaller typeface so that they can be omitted. This book was also divided into two parts according to the two levels of the training of sergeants. The first part was devoted to arithmetic and dealt with the concepts of ratio and proportion as well as with their main usual applications (rule of three, rule of interest, proportional distributions, etc.). The second part involved geometry, both plane and spatial.

We can conclude that the didactical treatment of the mathematical contents presented in the *Manual para las clases de tropa* was aimed at clarity and systematization rather than at rigor. It was a very practical text and a large collection of proposed problems (mostly extracted from *Aritmética*) was included at the end of each section.

8 A Final Comment

For decades, the arithmetical training of the Spanish Army rested on the official textbooks written by Salinas and Benítez. Finally, we can establish a rough comparison between the arithmetical requirements for the different levels of the Spanish army at the end of the 19th century with those of other professionals or social groups.

If we compare the arithmetical training of soldiers, capes and sergeants with the official contents presented in the Moyano Law, we can conclude that sergeants had more or less the same arithmetical training as a teacher of upper primary education (Meavilla and Oller 2016). Privates aspiring to the rank of corporal received a training equivalent to elementary primary education, while corporals could be roughly compared to upper primary education students. The arithmetic knowledge required of candidates for officers at the Military Academies was similar to secondary education arithmetic contents at that time (Vea 1995). Finally, from the arithmetical point of view, Spanish officers had a training that was equivalent to that of engineers and architects at the time.

We can conclude that arithmetic was an important topic in the intellectual training of Spanish army at the end of the 19th century. Nevertheless, with the only possible exception of officers, the focus was mainly on practical applications rather than on conceptual issues. In the case of officers, it would be interesting to perform a detailed comparison between the textbooks written by Salinas and Benítez and those books written by the teachers that actually had to teach the discipline in the Academies.

Acknowledgements This work has been partially supported by Gobierno de Aragón and the European Social Fund via the research group "S119-Investigación en Educación Matemática."

References

Anonymous (n.d.). *Aritmética para la instrucción de los cadetes del Regimiento de Reales Guardias Españoles, sacada de diferentes autores y coordinada por un oficial del mismo cuerpo* [Arithmetic for the instruction of the cadets of the Regiment of Spanish Royal Guards excerpted from different authors and coordinated by an officer of the same corps]. Manuscript MSS/10874 of the Spanish National Library.

Ausejo, Elena (2008). La Asociación Española para el Progreso de las Ciencias en el Centenario de su creación [The Spanish Association for the Progress of Science in the centenial of its creation]. *Revista Complutense de Educación, 19*(2), 295–310.

Ausejo, Elena, & Matos, Jose M. (2014). Mathematics education in Spain and Portugal. In A. Karp & G. Schubring (Eds.), *Handbook on the history of Mathematics Education* (pp. 283–302). New York: Springer.

Baltar, Francisco (2009). Los colegios preparatorios militares [The preparatory military schools]. In G. Vicente y Guerrero (Ed.), *Actas del I Congreso sobre Historia de la Enseñanza Media en Aragón* (pp. 199–234). Zaragoza: Institución Fernando el Católico.

Barrera, Eduardo (1916). *El indispensable para la preparación militar: apuntes de aritmética: cuaderno el más completo y aclaratorio de las teorías de mayor dificultad contenidas en la aritmética de Salinas y Benítez, de texto para el ingreso en las Academias Militares* [Essential for military preparation: notes on arithmetic: the most complete and explanatory notebook on the most difficult theories contained in Arithmetic by Salinas and Benítez, textbook for the entry into the Military Academies]. Alicante: Imprenta de Rafael Costa.

Comas, Joaquín (2015). *La enseñanza de las matemáticas en la Armada Española en el siglo XIX* [The teaching of mathematics in the Spanish Navy during 19th century]. (Doctoral dissertation, Universidad de Zaragoza, Spain). Retrieved July 3, 2015, from https://zaguan.unizar.es/record/32763.

De la Torre, Rafael (2015). *La Academia Especial de Sargentos De Zamora (1885–1890). Una Oportunidad Perdida* [The Special Academy of sergeants from Zamora (1885–1890). A lost opportunity]. Lleida: Ministerio de Defensa.

Diaz, María Verónica, & Poblete, Álvaro (2001). Contextualizando tipos de problemas matemáticos en el aula [Contextualizing types of mathematical problems in the classroom]. *Números, 45*, 33–41.

Equis, C. (1910). *Apuntes y aclaraciones de aritmética al texto "Salinas y Benítez"* [Notes and clarifications to the text "Salinas and Benítez"]. Madrid: Establecimiento Tipográfico de Layunta y Compañía. Madrid: Establecimiento Tipográfico de Layunta y Compañía.

Esdaile, Charles J. (2000). *Spain in the liberal age: From Constitution to Civil War, 1808–1939*. Oxford: Blackwell.

Fan, Lianghuo; Zhu, Yan; & Miao, Zhenzhen (2013). Textbook research in mathematics education: development status and directions. *ZDM The International Journal on Mathematics Education, 45*, 633–646.

Fernández, Fernando (1978). *El Ejército Español en el siglo XIX* [The Spanish army during 19th century]. Madrid: Siglo XXI.

García, Ernesto (1982). La matemática en la España del Siglo XIX [Mathematics in Spain during 19th century]. In M. Hormigón (Coord.), *Actas II Congreso de la Sociedad Española de Historia de las Ciencias* (Vol. 2, pp. 115–130). Jaca, Spain: SECHYT.

Gómez, Bernardo (2011). Marco preliminar para contextualizar la investigación en historia y educación matemática [Preliminary framework for contextualizing research on history and mathematics education]. *Epsilon, 28*(1), 9–22.

Hormigón, Mariano (1981). El Progreso Matemático (1891–1900). Un estudio sobre la primera revista matemática española [El Progreso Matemático (1891–1900). A study about the first Spanish mathematical journal]. *Llull, 4*, 87–115.

Hormigón, Mariano (1993). García de Galdeano and El Progreso Matemático [García de Galdeano and El Progreso Matemático]. In E. Ausejo, & M. Hormigón (Eds.), *Messengers of Mathematics: European Mathematical Journals (1800–1946)* (pp. 95–116). Madrid, Spain: Siglo XXI.

Krippendorff, Klaus (2014). *Content Analysis. An Introduction to Its Methodology*. London: SAGE.

Laguna, Francisco (1998). La formación del militar profesional en la España renovada [Professional military training in the renewed Spain]. *Cuadernos de Estrategia, 97*, 159–180.

Lara, F. (1900). *Pizarras con aclaraciones para facilitar el estudio de la Aritmética de los señores Salinas y Benítez* [Slates with clarifications to facilitate the study of Arithmetic by Messrs. Salinas and Benítez]. Toledo: Imprenta, Librería y Encuadernación de Rafael Gómez-Menor.

Lecoq, Nuno Joaquim (2014). A Herdade da Contenda. Uma Paisagem Única como Paisagem de Referência [The Contenda Estate. A Unique Scenery as a Reference Landscape]. (Doctoral dissertation, Universidade de Évora, Portugal). Retrieved June 24, 2015, from https://dspace.uevora.pt/rdpc.

Mayring, Philipp (2014). *Qualitative Content Analysis. Theoretical Foundation, Basic Procedures and Software Solution*. Klagenfurt: Beltz.

Maz, Alexander (2005). *Los Números Negativos en España en los siglos XVIII y XIX* [Negative numbers in Spain during 18th and 19th centuries]. (Doctoral dissertation, Universidad de Granada, España). Retrieved June 15, 2015, from http://hdl.handle.net/10481/556.n

Maz, Alexander, & Rico, Luis (2015). Principios didácticos en textos españoles de Matemáticas de los siglos XVIII y XIX [Dydactical principles in Spanish mathematical texts from 18th and 19th century]. *Revista Latinoamericana de Investigación en Matemática Educativa, 18*(1), 49–76.

McCulloch, G. (2004). *Documentary research in education, history, and the social sciences*. New York: Routledge/Falmer.

Meavilla, Vicente, & Oller, Antonio M. (2014). Multidisciplinariedad en algunas aritméticas españolas del siglo XIX [Multidisciplinarity in some 19th century Spanish arithmetic texts]. *Revista UNION, 37*, 121–137.

Meavilla, Vicente, & Oller, Antonio M. (2015). Los textos matemáticos de Antonio Terry y Rivas [The mathematical texts of Antonio Terry y Rivas]. *Números, 90*, 89–103.

Meavilla, Vicente, & Oller, Antonio M. (2016). La formación de maestros y maestras elementales en España a finales del siglo XIX. El caso de la Geometría [Elementary male and female teacher training in Spain at the end of the 19th century. The case of Geometry]. *Contextos Educativos, 1*, 79–96.

Méndez Bejarano, Mario (1922). *Diccionario de escritores, maestros y oradores naturales de Sevilla y su actual provincia (Tomo I)* [Dictionary of writers, teachers and speakers from Sevilla and its province (Volume I)]. Sevilla: Tipografía Gironés.

Montemayor, J., & Bazán, P. (1885). *Apuntes para el estudio de la Aritmética de D. Ignacio Salinas y D. Manuel Benítez y resolución de los ejercicios prácticos* [Notes for the study of Arithmetic by Mr. Ignacio Salinas and Mr. Manuel Benítez and solutions to the practical exercises]. Toledo: Librería y Encuadernación de Menor Hermanos.

Montero, Antonio (2009). Una ley centenaria: la Ley de Instrucción Pública (Ley Moyano, 1857) [A centennial law: the Law of Public Instruction (Moyano Law, 1857)]. *Cabás, 1*.

Navarro, Juan (2011). Las Matemáticas en la escuela militar de Ávila (1774) [Mathematics at the Military school of Ávila (1774)]. *La Gaceta de la RSME, 14*(2), 309–332.

Navarro, Juan, & Velamazán, María Ángeles (2006). El militar José Odriozola y su contribución a la ciencia en España en el siglo XIX [The military man José Odriozola and his contribution to science in 19th century Spain]. In J.A. Pérez, J.C. Martín, E. Wulff, J.F. Casanueva, F. Herrera & F.J. González (Coord.). *Actas del IX Congreso de la Sociedad Española de Historia de las Ciencias y de las Técnicas* (pp. 925–937). Cádiz: SEHCYT.

Oller-Marcén, Antonio M. (2017). The problem section of *El Progreso Matemático*. In K. Bjarnadóttir, F. Furinghetti, M. Menghini, J. Prytz, & G. Schubring, G. (Eds.), *"Dig where you stand" 4. Proceedings of the fourth International Conference on the History of Mathematics Education* (pp. 235–246). Rome, Italy: Nuova cultura.

Octavio de Toledo, Luis (1912). Sección Biográfica [Biographical section]. *Revista de la Sociedad Matemática Española*, 6, 193–196.

Peralta, Javier (2000). *La matemática española y la crisis de finales del siglo XIX* [Spanish mathematics and the crisis at the end of 19th century]. Madrid: Nívola.

Prior, Lindsay (2016). Using documents in social research. In D. Silverman (Ed.). *Qualitative research* (pp. 171–185). London: SAGE.

Puelles, Manuel de (1985). *Historia de la Educación en España II. De las Cortes de Cádiz a la Revolución de 1868* [History of education in Spain II. From the Cádiz courts to the revolution of 1868]. Madrid: Ministerio de Educación y Ciencia.

Riera, Juan. & Riera, Luis (2008). La Academia de Matemáticas de Barcelona: ilustración e inquisición a finales del siglo XVIII [The Academy of Mathematics of Barcelona: illustration and inquisition at the end of 18th century]. *Actes d'Història de la Ciència i de la Tècnica, 1*(2), 153–160.

Ríos, Sixto (1988). Julio Rey Pastor (1888–1962). *Gaceta Matemática, s. 2, 1*(2), 129-135.

Ruiz, Jesús María (2004). La enseñanza militar en el alto mando: Historia, organización y metodología [Military training in the high command: History, organization and methodology]. *Educación XX1, 9*, 199–220.

Salgado, Manuel (1922). *Apuntes de aritmética adaptados al texto de Salinas y Benitez* [Notes on arithmetic adapted to the text of Salinas and Benítez]. Burgos: Imprenta y librería editorial Hijos de Santiago Rodríguez.

Salinas, Ignacio (1879). *Legislación militar aplicada al ejército de Filipinas* [Military legislation applied to the Philippines army]. Manila: Establecimiento Tipográfico de Plana y Cª.

Salinas, Ignacio (1883). *Dirección de los ejércitos: Exposición de las funciones del estado mayor en paz y en guerra* [Army management: exposition of the roles of the General Staff during peace and war]. Madrid: Imprenta y Litografía del Depósito de la Guerra.

Salinas, Ignacio (1899). *Defensa del General Jáudenes, hecha por el general de brigada D. Ignacio Salinas y Angulo. Leída el 29 de septiembre de 1899* [Defense of the General Jaúdenes, made by the bridadier general D. Ignacio Salinas y Angulo. Rea don September 29, 1899]. Madrid: Imprenta y Litografía del Depósito de la Guerra.

Salinas, Ignacio, & Benítez, Manuel (1887). *Aritmética: obra redactada con estricta sujeción al programa oficial para los exámenes de ingreso en la Escuela Preparatoria de Ingenieros y Arquitectos* [Arithmetic: work written with strict subjection to the official program of the entrance examinations to the school of engineers and architects]. Madrid: Imprenta de Ricardo Fé.

Salinas, Ignacio, & Benítez, Manuel (1892a). *Cálculos de la aritmética de texto para el ingreso en la Academia General Militar y apuntes relativos a los mismos* [Calculations of the arithmetic: textbook for the entrance to the General Military Academy and notes related to them]. Madrid: Est. Tipográfico de Ricardo Fé.

Salinas, Ignacio, & Benítez, Manuel (1892b). *Pizarras por papeletas para el examen de Aritmética en la Academia General Militar* [Slates by sheets for the arithmetic exam at the General Military Academy]. Madrid: Est. Tipográfico de Ricardo Fé.

Salinas, Ignacio, & Benítez, Manuel (1892c). *Álgebra (Primera parte)* [Algebra (first part)]. Toledo: Imprenta, Librería y Encuadernación de Menor Hermanos.

Salinas, Ignacio, & Benítez, Manuel (1894). *Manual para las clases de tropa (Tomo II. Academia de sargentos)* [Handbook for teaching the tropos (Volumen II. Academy of sergeants)]. Madrid: Tipografía de la Viuda e Hijos de Rubiños.

Salinas, Ignacio, & Benítez, Manuel (1898a). *Aritmética* [Arithmetic]. Madrid: Imprenta del Depósito de la Guerra.

Salinas, Ignacio, & Benítez, Manuel (1898b). *Álgebra (Segunda parte)* [Algebra (second part)]. Madrid: Librería de Hernando y Comp[a].

Salinas, Ignacio, & Benítez, Manuel (1913). *Manual para las clases de tropa (Tomo I. Academia de aspirantes a cabo. Academia de cabos)* [Handbook for teaching the tropos (Volume I. Academy of corporals)]. Madrid: Imprenta del Asilo de Huérfanos del S. C. de Jesús.

Salinas, Ignacio, & Benítez, Manuel (1914). *Ordenanzas del Ejército armonizadas con la legislación vigente (6[a] edición, corregida y aumentada)* [Army regulations harmonized with current legislation (6th edition, corrected and expanded)] . Madrid: Imprenta de Alrededor del Mundo.

Salinas, Ignacio, & Oliver, Antonio (1867). *[Itinerario topográfico de la línea de ferrocarril] de Utrera a Morón [de la Frontera]* [Topographical itinerary of the railway from Utrera to Morón de la Frontera]. Centro Geográfico del Ejército — Colección: PCGE — Ubicación: AR — Signatura: Ar.M-T.6-C.33-813.

Sánchez, José Manuel (1992). Las ciencias Físico-Matemáticas en la España del siglo XIX [Physical and mathemtical sciences in 19th century Spain]. In J.M. López Piñero (Ed.). *La Ciencia en la España del siglo XIX* (pp. 51–84). Madrid: Marcial Pons.

Schubring, Gert (1987). On the methodology of analyzing historical textbooks: Lacroix as textbook author. *For the Learning of Mathematics, 7*(3), 41–51.

Scott, John (1990). *A matter of record, documentary sources in social research.* Cambridge: Polity Press.

Terry, Antonio (1880). *Ejercicios y problemas de Aritmética: parte originales y parte escogidos de los principales autores que tratan de la materia (Primera Parte: Enunciados. Tomo I)* [Exercises and problems of Arithmetic: some original and some borrowed from the main auhors who treat this subject (First part: statements. Volume I)]. Madrid: Pedro Abienzo, Impresor del Ministerio de Marina.

Tzanakis, Constatinos, & Arcavi, Abraham (2000). Integrating history of mathematics in the classroom: An analytic survey. In J. Fauvel, & J. Van Maanen (Eds.), *History in mathematics education: the ICMI study* (pp. 201–240). Dordrecht: Kluwer.

Vea, Fernando (1995). *Las Matemáticas en la Enseñanza Secundaria en España en el Siglo XIX* [Mathematics in Spanish secondary schools during 19th century]. Zaragoza: SEHCYT.

Velamazán, María Ángeles (1990). L'Enseignement des Mathématiques dans les Ecoles Militaires en Espagne au XIXème siècle [The teaching of mathematics in Spanish military schools during the 19th century]. *Cuadernos de Historia de la Ciencia, 6*, 23–37.

Velamazán, María Ángeles (2015). Revisión de la formación y la aportación matemática del ingeniero militar Antonio Sangenís (1767–1809) [A review of the training and mathematical contributions of the military engineer Antonio Sangenís (1767–1809)]. *LLULL, 38*, 371–397.

Velamazán, María Ángeles, & Ausejo, Elena (1989). Los planes de estudio de la Academia de Ingenieros del Ejército de España en el siglo XIX [The curricula of the Academy of Military Ingineers in 19th century Spain]. *LLULL, 12*, 415–453.

Author Biographies

Antonio M. Oller-Marcén was born in Zaragoza (Spain) in 1981. He graduated in sciences (mathematics section) in 2004 and he obtained a M.Sc. in algebra in 2006 at the University of Zaragoza. He obtained a Ph.D. degree in mathematics education at the University of Valladolid in 2012. He is an associate professor of mathematics at the Centro Universitario de la Defensa de Zaragoza. He is currently interested, among many other topics, in the history of mathematics in Spain during 19th century. He has also published several papers on different aspects of mathematics education.

Vicente Meavilla-Seguí was born in Mahón (Spain) in 1949. He graduated in sciences (mathematics section) at the University of Zaragoza in 1976 and he obtained a Ph.D. degree in pedagogy at the Autonomous University of Barcelona in 1998. Currently retired, he has taught mathematics at different levels, from high school to university. His main research interest is history of mathematics, focusing on its impact and applications on the teaching of mathematics. He is the author of several papers and books on this subject.

The New Math and School Governance: An Explanation of the Decline of the New Math in Sweden

Johan Prytz

Abstract The main aim of this paper is to explain why certain parts of the Swedish New Math reform failed. The secondary aim is to nuance the idea about the decline of the New Math in Sweden. The analysis is focused on governance, but also ideological, economical and sociological aspects are considered. The main sources are syllabi, official reports and textbooks. Another type of source is a database of historical textbooks in mathematics. Even though the suggested explanation to great degree concerns the Swedish context, it has a more general relevance. Previous research on New Math in other countries is mainly focused on ideological and sociological aspects of the reforms. This paper also highlights the interplay between factors related to textbooks, governance and economics. It is relevant to consider these factors in other countries as well since textbook development was a part of New Math reforms in many countries.

Keywords New Math · Reform · Sweden · Decline · Explanation Governance

1 Introduction

From an international perspective, the New Math movement was an attempt to achieve radical changes of school mathematics. The suggested changes were also comprehensive since they concerned contents as well as teaching methods in several topics and at several grade levels. The outcomes of these attempts have often been described as definite failures (e.g. Kline 1973, p. 102; Damerow and Westbury 1984, p. 22). A Swedish commentator even associates the New Math with madness (Unenge 1999, pp. 63–64). Recently, more nuanced views on the New Math have been expressed, for instance by Kilpatrick (2012, pp. 568–569) in an international overview regarding the New Math.

J. Prytz (✉)
Department of Education, Uppsala University, Box 2136, Uppsala, Sweden
e-mail: johan.prytz@edu.uu.se

F. Furinghetti and A. Karp (eds.), *Researching the History of Mathematics Education*, ICME-13 Monographs,
https://doi.org/10.1007/978-3-319-68294-5_10

This paper concerns the New Math reform in Sweden, which was planned in the 1960s and implemented in the 1970s. The study concerns grade levels 1–9. One aim is to nuance the idea about the decline of the New Math in Sweden. The main aim, however, is to explain why certain parts of the Swedish New Math reform failed. The analysis is focused on governance of school mathematics. Despite the national focus, the intention is to highlight causes of failure that are worthwhile to consider in studies of the New Math in other countries. In brief, the basic idea of my explanation is that the conditions to implement the reform in the 1970s were different from when the reform was planned in the 1960s.

2 Disposition and Sources

Apart from a survey of the literature this paper contains four sections that lead forward to the final section, which contain the suggested explanation.

In Sect. 4, a successive and long-lasting reform process is described. The purpose is to show that the New Math reform in Sweden was a final step in a longer process of centralising the governance of the Swedish school system. In relation to the explanation, the important conclusion is that people involved in the planning of the New Math reform were operating in a context where centralised governance was an established ideal. Another important conclusion is that centralistic governance began to lose importance in the 1970s. The section is mainly based on secondary literature about the Swedish school system, but also syllabi.

Section 5 concerns how the textbook market changed at the same time that the school system was further centralised after 1950. In relation to the explanation, the important conclusion is that the publishing companies in the 1970s were potent actors in the sense that they both had good knowledge about the demand and necessary resources to produce new textbooks. The section is based on official reports regarding the textbook market and a database of historical textbooks in mathematics. The database is described in Prytz (2016).

In Sect. 6 it is further confirmed that the New Math reform pursued centralistic principles of governance. The nature of the reform is also described and the main point is that the reform was more radical and far-reaching than previous reforms. In relation to the explanation, the important conclusion is that the New Math reform was the first attempt to apply centralistic governance in an active way in school mathematics: it had never been tried before. Moreover, it was a reform that relied heavily on textbooks. The sources in this section are mainly official reports regarding the preparations of the reform, syllabi and a database of historical textbooks in mathematics.

In order to explain a decline, it is pivotal to identify some sort of failure. That is the aim of Sect. 7. Three different curriculum levels are considered: intended, implemented and achieved. For each level, failures and success of the New Math reform are identified. The main sources are syllabi, textbooks and reports about tests, national exams and evaluations.

In the final Sect. 8, the main conclusions of the previous sections are synthesised in an explanation of why certain parts of the Swedish New Math reform failed. This explanation is compared with previous research on the New Math in other countries and other types of possible explanations.

3 Survey of the Literature

There is only little research about the New Math reform in Sweden. Kilborn (1977) carried out detailed analyses of the work of authoring the syllabus of 1969, which introduced the New Math in grade levels 1–9. However, he does not greatly consider the long-lasting trials of textbook materials nor the implementation process. Despite that, he concludes that the functionality of syllabus, on a number of points, had not been tested (Kilborn 1977, p. 1). Unenge (1978) briefly considers the implementation process and concludes that the material of the New Math caused difficulties (Unenge 1978, p. 57).

A commonality of the studies by Kilborn (1977) and Unenge (1978) is that they identify the syllabus and the textbooks per se as problematic, which imply that they were the cause of a failure. This idea is criticised by Prytz and Karlberg (2016) on the basis of the reports from the trials of the New Math material in the 1960s. From the reports it is evident that teachers who used the New Math material could achieve results as good as teachers who used traditional textbooks. This indicates that the problems that occurred in the implementation process were not caused by the New Math material per se (Prytz and Karlberg 2016, p. 89). For that reason, it is important to consider other aspects of the reform process.

In fact, studies of New Math reforms in other countries do consider other aspects. In the following, I discuss two such studies. They are relevant to my study since their perspective highlights probable causes of why the New Math reform, in some respect, failed in Sweden.

The study by Phillips (2015) on the New Math in the USA is focused on ideology. By identifying a general shift in ideology, Phillips (2015) also offers an explanation for the decline of the New Math. He purports that in the 1950s and early 1960s, the New Math was seen as a viable solution to educational as well as more general societal needs. The New Math was supposed to foster a rational (as in scientific) mode of thinking (Phillips 2015, pp. 44–46). At the same time, the New Math should bring high school students closer to university mathematics, which in turn should benefit the education of engineers and scientists, which in turn should enhance scientific as well as industrial progress (Phillips 2015, p. 22). According to Phillips (2015), this way of reasoning about education and society rested on a positive appreciation of and respect for science. In the 1970s, however, a more skeptical attitude versus science had gained influence in the public debate. Moreover, new problems in education, related to, for instance, social justice and environmental issues, was considered more pressing than some decades before. In the eyes of many politicians, administrators and the public, the New Math was not the solution to these problems. Thus, it became increasingly difficult to defend the New Math as an essential part in mathematics teaching in the 1970s (Phillips 2015, pp. 131–144). Since Hultén (2008, pp. 196–205) observes a similar general ideological shift in Sweden in the 1970s, it is tempting to believe that the Swedish New Math reform suffered a similar fate.

Another, clearly sociological, approach to reforms and changes in mathematics education is presented by Gispert and Schubring (2011). They study changes in mathematics education in France and Germany, 1900–1980. They consider two so-called key moments when mathematics education went through major reforms: the early 20th century and the 1960s and 70s. They analyse groups of people, so-called 'spheres', involved in mathematics education; in particular how these spheres acted, who belonged to the spheres and how the spheres were related to each other during reforms. Three spheres in the reform processes are identified: (1) experts—the holders of the knowledge at stake; (2) actors in policy and economy; and (3) professionals with concrete experience in the field, in this case the teachers.

Gispert and Schubring (2011) show that the composition of the spheres and their mutual relations changed between the key moments. For example, in both countries in the beginning of the century the expert sphere primarily comprised university mathematicians. In the 1950s and 60s, the situation had become quite different; apart from university mathematicians, the sphere of experts was also comprised of school teachers, psychologists and didacticians. Gispert and Schubring (2011, pp. 100–103) explains changes in mathematics education by such sociological changes. As will become evident in this paper, similar sociological changes also happened in Sweden in connection with the reform of the New Math.

However, these two perspectives alone cannot explain the failures of the Swedish New Math reform. By focusing my analysis on governance of school mathematics, new probable causes are identified. I argue that some of these causes were probably not typical for Sweden. Thus, it should also be relevant to consider this perspective in other countries.

4 Centralization of the Swedish School System, 1900–1970

By the beginning of the 20th century, the major school type in Sweden was *Folkskolan* (in literal translation, the *People school*) (1–6). In short, it gave a basic education for the broad masses. In 1936, *Folkskolan* was extended to include a seventh grade. However, this regulation was not implemented throughout the country until 1950. At the same time, mainly in the urban parts, *Folkskolan* was extended to eight or nine grade levels and received more advanced courses.

Folkskolan had only one programme and was open for both boys and girls and they followed more or less the same syllabus. However, there were deviations and one of them concerned mathematics; before 1919, it was not necessary for girls to study geometry.

A second school type was *Läroverket*, which by the beginning of the twentieth century covered grade levels 4–12. A literal translation of the term *Läroverk* is the *Learning office*; in England the corresponding school type was grammar school. The uniqueness of *Läroverket* was that it gave admission to further studies at university. However, far from all of its students went on to university or even graduated.

In 1905, the lower and upper parts of *Läroverket* were divided into two more distinct parts: *Realskolan* (4–9, later changed to 5–9) and *Gymnasiet* (9–12). The literal translation of *Realskolan* is *Realistic school* and just the name indicated the orientation of its course programme; it contained no classical languages and it was supposed to prepare for further vocational educations and working life as well as university. Moreover, it was the only programme that gave access to *Gymnasiet*, which in turn gave access to university. Regarding *Gymnasiet*, similar terms for school types preparing for university we find in for instance Germany (*Gymnasium*) and Italy (*Ginnasio*). Students bound for *Gymnasiet* usually left *Realskolan* after grade 8. The ninth year of *Realskolan* ended with a more comprehensive exam process for a formal degree. In 1936 a new type of programme within *Realskolan* was created. It was supposed to be more practical and oriented towards working life.

Realskolan was, from its opening year, open for both girls and boys, that was the normal case, but there were some all-boys schools. *Gymnasiet*, on the other hand, was only for boys until 1927. However, even before that, girls could have upper secondary education in *Flickskolan* (*Girls school*). *Flickskolan* comprised grade levels 5–11.

A difference between *Folkskolan* and *Läroverket* was that the former was a municipal institution and the latter mainly a state institution, even though a few schools were private. But, beginning in 1909, municipal versions of *Realskolan* were started. However before that, there were municipal *Högre* (*Higher*) *Folkskolor*.

In addition to these school types, there were various forms of vocational education on the secondary level in, for instance, technology, trade, nursing and housekeeping.

The number of school types was reduced from the early 1960s onwards. In 1962, *Grundskolan* (1–9) was introduced and, over a 10-year period, it replaced *Folkskolan*, *Realskolan* and a portion of *Flickskolan*. *Grundskolan* (in literal translation, the *Basic school*) was a compulsory nine-grade school. *Gymnasiet* was replaced by the *Gymnasieskolan*. In brief, it comprised the programmes of *Gymnasiet* and *Flickskolan*, but several vocational programmes were also added.

The classes in *Grundskolan* were in principal undifferentiated in the sense that the classes were not formed on the basis of ability or plans for future education. This was a major difference in comparison to the old school system. In the old system, students aiming for *Gymnasiet* and university left *Folkskolan* for *Realskolan*. During the first six grades of *Grundskolan*, the students followed the same syllabus. In grade 7, the students choose between courses. This differentiation can be seen as a continuation of the old parallel system with *Folkskolan* and *Realskolan*. The students that aimed for further theoretical studies could chose more advanced courses in mathematics, English and a second foreign language.

The introduction of *Grundskolan* in 1962 was preceded by a long-lasting and successive change in governance. Governance became more centralised as the state successively took greater control over the school system, judicially, financially as well as ideologically (Larsson and Westberg 2011, pp. 331–333).

Financial centralisation consisted of the state giving more and more targeted school funding to the municipalities. The targeted funding also concerned textbooks in *Folkskolan*. From 1946, the municipalities were supposed to provide free textbooks to the students and the state provided much of the money for it; at that time, the state covered 80–90% of the municipalities' costs for teaching material. However, as early as the mid-1930s, state funding for teacher salaries required that the municipalities provided free textbooks (SB 1961, p. 176).

An early example of ideological centralisation is the more comprehensive national policy documents, which included the mathematics syllabus. This type of policy documents appeared already in the late 1910s (Larsson and Westberg 2011, p. 331). Another example is the national textbook review board, which was established in 1938 (Johnsson Harrie 2009, p. 12). All textbooks used in school had to be approved by this review board. Compliance with the syllabus was one criterion throughout the existence of the review board (Johnsson Harrie 2009, pp. 115–116).

Regarding the syllabi in mathematics and governance in the period of 1900–1960, it is important to notice that the syllabi were not used as a tool to achieve major changes. If we consider the mathematical content, there were few changes. The novelty in the new *Folkskolan* syllabus of 1919 was that percentage and equations became subtopics; the former introduced in grade 6 and the latter in grade 7. In the syllabus of 1955, the main novelty was that grades 8 and 9 were added. These courses resembled the corresponding ones of *Realskolan*, but applications had a more prominent place. Another difference was that equations were introduced in grade 6 rather than 7 (NP 1900, pp. 12–13, 29; UP 1920, pp. 58–60; SÖ 1955, pp. 123–124). In the *Realskolan* syllabi of 1905, 1928 and 1933, the formulations about the content were more or less the same. Another similarity was their brevity. They were more or less a list of topics and subtopics (Bergqvist-Nordfelt 1910, pp. 204–205; Bergqvist and Wallin 1928, pp. 226–227; Wallin and Grimlund 1939, pp. 316–317).

The sections on teaching methods were longer, except for the *Folkskolan* syllabi of 1900, and they were successively extended. The latter can of course be seen as an example of more centralised governance. On the other hand, the basic ideas of the methodological guidelines were similar, both between the two school types and over time: not too much, not too fast, not too complicated, keep the interest of the students up, precision, training and repetition. For each subtopic and grade, these guidelines were further specified, mainly in terms of what to teach. Beyond general formulations about the need for visual aids, there were no or few guidelines regarding how explanations and illustrations should be designed (NP 1900, pp. 32–34; Bergqvist-Nordfelt 1910, pp. 205–214; UP 1920, pp. 67–69; SÖ 1935, pp. 129–142, 1955, pp. 124–129).

The introduction of one compulsory school type—*Grundskolan*—marks a peak in the centralisation of governance, even though the municipalities were still a part of the system. Financially, the state provided the funds and settled the amounts for various ends, for instance teacher salaries and school houses. The municipalities, on the other hand, ran the schools, not at least the building of school houses, and employed the teachers (Larsson and Westberg 2011, pp. 331–333).

The ideological centralisation increased as well, if we consider the comprehensiveness of the policy documents connected to *Grundskolan*. This included the syllabus in mathematics, which received an unprecedented level of detail in the prescriptions regarding what to teach and how to teach it (Prytz 2015, pp. 311–312).

But the ideological centralisation also got new elements in connection to the preparation and introduction of *Grundskolan* in the 1950s and 1960s. Different types of large-scale development projects were launched by the state in order to improve the teaching—the contents as well as the teaching methods (Lindensjö and Lundgren 2000, pp. 60–64). Several of these projects concerned school mathematics.

The first mathematics syllabus of *Grundskolan* was preceded by a large survey about the need for mathematics in society; people representing different parts of society took part in the survey, for instance industry, state administration and education (Lindensjö and Lundgren 2000, p. 64).

Another type of development project concerned both content and teaching methods; the New Math project in Sweden is a good example of this development. Together with the other Nordic countries, which took part in the same project, it involved thousands of students during the period of 1961–1966. A cornerstone of the project was textbook development, which included trials and experimental testing with experimental and control groups. More details about this project are accounted for in a coming section.

About the same time as the New Math project, there was a project about mathematics teaching and individualisation (Lindensjö and Lundgren 2000, p. 68). It mainly concerned grade levels 7–9. Individualisation was a key concept in the reform of *Grundskolan*. Rather than having differentiation as a basic pedagogical tool, the teaching in heterogeneous classes was supposed to be individualised in order to fit as many students as possible. Also in this project, textbooks were of central importance. The basic idea was to develop so-called programmed textbooks, which meant that students with different abilities were meant to follow different tracks. The students were also supposed to work on their own, at their own pace. And the teachers were not supposed to have whole class teaching. These books were based on the New Math. However, much of these plans were never realised. This project is analysed in an ongoing study.

A common feature of these development projects is that they involved science in a new way. Rather than just having the scientific disciplines as a source of what to teach, scientific methods such as large scale surveys and experimental testing were applied to determine both the content and the methods of teaching. Psychology and education, two fairly new disciplines, played an important role. This new involvement of social sciences indicates that textbook producers as well as the old scientific disciplines, for instance mathematics, lost some of its influence over the school subjects.

Centralistic governance of the Swedish school system had its characteristics after the Second World War: it was a kind of 'input management'. For instance, the state specified what the municipalities should use the funds for; the teacher received clear and detailed instruction about what and how to teach via national syllabi. On the other hand, the system was not designed to react to results, for instance student performance. 'Output management' was non-existent according to Oftedal Telhaug et al. (2006, pp. 255–256).

The centralisation of the school system in the period of 1900–1970 was not an isolated phenomenon in Sweden. It was an integrated and fundamental part in the construction of a centralised welfare state, which was supposed to care for more and more of its citizens' needs. In the endeavour to achieve this goal, the politicians viewed science as an important tool (Oftedal Telhaug et al. 2006, pp. 249–250, 255).

This is also a period when Sweden went from being a fairly poor country to being a very rich country—and great economic resources were allocated to the building of the welfare state, especially after the Second World War. Here we should note that Sweden did not take part in the war and did not need to rebuild cities, industry, infrastructure, etc.

However, centralisation of the Swedish school system peaked in the 1960s and the early 1970s; thereafter the system was decentralised. This is a standard narrative in overview works on the Swedish school system (Marklund 1983, pp. 395–398; Lindensjö and Lundgren 2000, pp. 93–96; Larsson and Westberg 2011, pp. 333–335). According to this narrative, ideas about decentralisation were discussed in several state investigations in the 1970s; ideas that were realised in the 1980s. Nonetheless, in 1974, textbook review became optional for all subjects except for the subjects in the humanities and social sciences (Johnsson Harrie 2009, p. 12). Thus, more influence over mathematics education was handed over to textbook producers when the New Math reform was being implemented.

5 Changes in the Textbook Market

A fundamental change in the textbook market from 1950 to 1970 was the schools' spending on textbooks; during this period, it increased ten times (LU 1971, p. 53). The increasing number of students was far from the only cause. Other identified causes were a higher price per book, higher technical quality, free textbooks and more course programmes that required specially designed textbooks (LU 1971, pp. 53, 179, 200). Regarding the price per book, it did not increase ten times; from

1960 to 1971, it doubled (LU 1971, p. 53). For the publishing companies, this increased spending on textbooks probably meant good business. A report about the textbook market from 1971 stated that, in principal, every new textbook yielded profits (LU 1971, p. 190).

This increased spending on textbooks between 1950 and 1970 coincides with changes in the production of new textbooks in mathematics. In *Folkskolan* and *Realskolan*, the influxes of brand new textbooks increased a bit in the late 1940s and in the 1950s, in comparison to the previous decade, see Diagrams 1 and 2. However, the influx of brand new textbooks increased greatly in conjunction with the introduction of *Grundskolan* in 1962. We can even speak of a shift in proportions, see Diagrams 1, 2 and 3.

An important factor in this context is of course the number of students. If we consider all students in grades 1–9 in all school types, the numbers decreased from the beginning of the century to about 1940. After that the numbers increased; more rapidly after about 1950. A probable cause of this phenomenon was the expansion of *Folkskolan* to 7, 8 or even 9 grade levels, but also more students entering lower secondary educations. If we consider just lower secondary educations, there was a steady and, in relative numbers, great increase throughout the period of 1900–1960 (SCB 1974, 1977). Both these trends levelled out about the same time that *Grundskolan* was introduced in the 1960s (SCB 1973, 1980).

The greater influx of brand new textbooks in mathematics in the period of 1960–1980 suggests greater competition and increased marketing, not at least since the

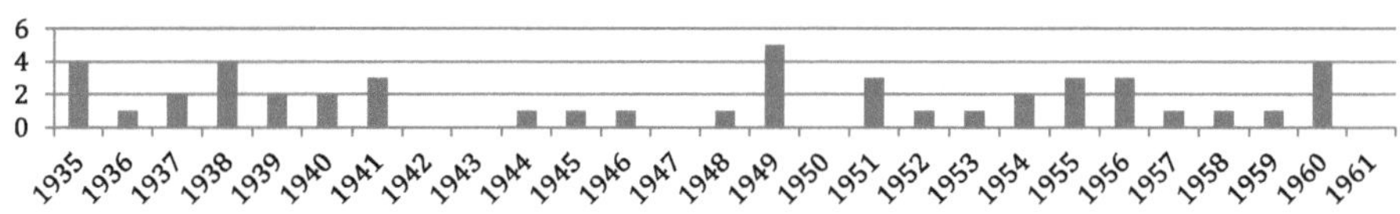

Diagram 1 Numbers of new series in Arithmetic for *Folkskolan*, grade levels 1–2, 3–6, 7–9

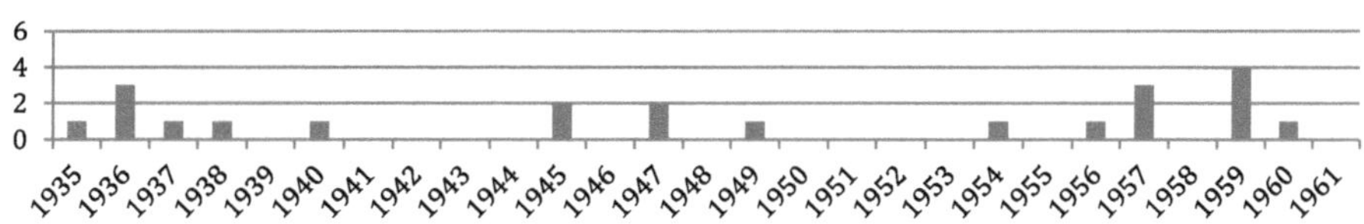

Diagram 2 Numbers of new textbooks for *Realskolan*, all topics and grades 5–9

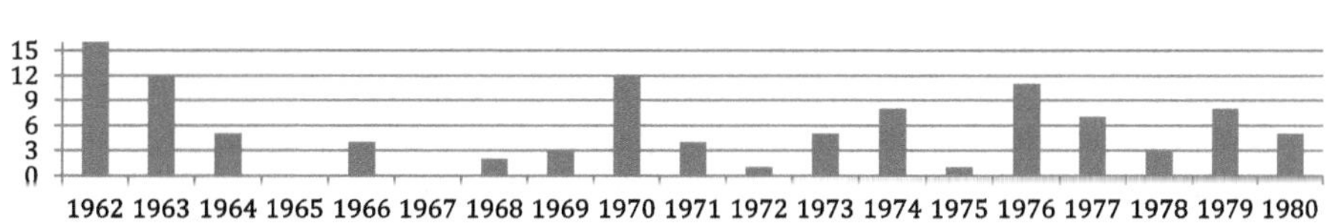

Diagram 3 Numbers of new series in Mathematics for *Grundskolan*, grades 1–3, 4–6, 7–9

increase in the number of students had levelled out. This suggestion is confirmed by official reports about the textbook market that covers the second half of the 1960s and the 1970s. They state that the textbook market, in general, was price-insensitive. Instead the companies competed through product development and marketing. In the 1970s, marketing was the cost that increased the most for the companies (Prytz 2017). Since mathematics was one of the major topics, it is not probable that mathematics was an exception in these matters, especially since the official reports did not say so.

The marketing to the teachers was done in various ways: advertisement, brochures, book samples and through schoolbook consultants. The consultants often had a background as teachers and were trained to inform teachers about teaching materials. Another task was to gather information from the teachers about their use of textbooks (Prytz 2017).

Another change in the textbook market in the 1960s was mergers of publishing companies and the creation of corporations of companies (Prytz 2017). A result of the mergers seems to have been that more resources could be directed toward marketing; its share of the total costs of textbook production increased from 7.6 to 11.9% in the major companies and from 7.8 to 8.3% in the smaller companies.

Thus, as the state spent more resources on the school system, the publishing companies invested more in textbook development and marketing. In the 1970s, the companies even had usurped a dominant position in disseminating information about textbooks (Prytz 2017). Moreover, in the 1970s, the publishing companies had good two-way communications with the teachers, in particular via so-called schoolbook consultants. The important conclusion for my explanation is that the publishing companies in the 1970s were potent actors in the sense that they both had good knowledge about demand and resources to produce new textbooks.

6 Preparations for the New Math Reform, 1960–1968

The centralisation of mathematics education took off in conjunction with the introduction of *Grundskolan* in 1962. One part was the syllabus, which became more comprehensive. For instance, the *Folkskolan* syllabus of 1919 consisted of approximately 1800 words. The first *Grundskolan* syllabus consisted of 10,500 words (Prytz 2015, pp. 311–312).

But centralisation of school mathematics also meant a new type of decision making about the content of the courses; a decision making that involved scientific methods. The new syllabus was preceded by a large-scale survey regarding what the new syllabus should comprise. The survey included people from university, administration and industry, in addition to ordinary citizens.

One of the results that probably had impact on the syllabus concerned geometry, especially the purely theoretical part with proofs and the axiomatic-deductive method. The need for the latter outside the school system was considered small (SB 1960, pp. 482–483). But, its relevance within the school system was also questioned. Until then, the main argument for having this type of geometry as a part of the syllabus concerned formal reasoning and transfer. The basic idea was that the logical reasoning the student learned in geometry should be beneficial to them in other school subjects and contexts. The final report concluded that there was no scientific support for that claim (SB 1960, pp. 485–487). In the new syllabus, geometry became a smaller topic, especially in grades 7–9. Moreover, the new syllabus explicitly stated that the geometry course should not have an "axiomatic structure" (SÖ 1962, p. 188).

But in general, the first syllabus of *Grundskolan* did not bring many other radical changes. In brief, the content of the courses in grades 1–6 corresponded quite well with the *Folkskolan* courses, grades 1–6. The advanced courses in grades 7–9 corresponded quite well with the *Realskolan* courses for the same grades. And finally, the less advanced courses in grades 7–9 corresponded to *Folkskolan* courses for the same grades (Wallin and Grimlund 1939, pp. 316–317; SÖ 1955, pp. 123–124, 1962, pp. 164–166).

With regard to the methodological aspects, the new syllabus of 1962 confirmed principles rather than challenged them. These principles, but also the more specific directives, concerned student activity and the use of visualisations and concrete materials (SÖ 1962, pp. 44–52). If we consider debates in teacher journals, but also methodological literature, in the first half of the twentieth century, these principles were central (Prytz 2007, pp. 99–106). The key concept then was *åskådning* (spatial intuition, similar meaning to the German term *anscchauung*) and it involved both student activity and the use of visualisations and concrete materials. Further, in debates on geometry teaching in *Realskolan* and the axiomatic method, these ideas were considered important, even though some debaters were skeptical. For further details, see Prytz (2007, pp. 102–103).

This means that even if mathematics education was centralised in the sense that the national syllabus much more comprehensively prescribed what to teach and how to teach it, the syllabus was not used to achieve extensive changes. The important point is that the New Math reform was the first attempt to achieve wide-ranging changes in mathematics education through centralistic governance.

The Swedish New Math reform was prepared in cooperation with Norway, Denmark and Finland. Iceland—the fifth Nordic country—did not take part in the cooperation. In 1960, a committee was formed to organise the work: *Nordiska kommittén för modernisering av matemtikundervisningen* (*The Nordic committee for the modernization of mathematics education*). The committee comprised four representatives from each country plus a chairman, the Swede Lennart Sandgren (1926–2009) and a secretary, Håstad (1931–), who was also a Swede. In this case, the secretary functioned as an executive officer. Both Sandgren and Håstad had a background in mathematics. Sandgren had a Ph.D. degree and Håstad the lower Ph. L. degree. They also worked briefly as school teachers. Later on, both held high positions within the central school administration in the 1960s and early 1970s; especially Sandgren, who was *Undervisningråd* (senior counsellor) at *Skolöverstyrelsen* (National Board of Education) and later state secretary at the Ministry of Education.

The objective of the committee was to develop preliminary syllabi which could be tested in schools. This required, the committee concluded, that experimental textbook material was developed and tested. The committee also made surveys about the need for mathematics in different sectors of society (NK 1967, p. 7). This means that an important part of the reform was to replace an old type of textbooks with a new type of textbooks. Thus, the state took a much more active role in the development of textbooks, rather than just wielding control of the textbooks through a textbook review board.

To meet its objectives, the committee engaged about 30 people, where the majority was engaged in designing syllabi and textbook material. The authors had different backgrounds: some mathematicians, some school teachers, some headmasters and some teacher educators (NK 1967, p. 221). However, only one of the Swedes had authored textbooks for grades 1–9 before 1962. In that respect, textbook development was not based on experience of authoring textbooks for the Swedish market. Apart from the people who authored the textbook material, five experts on psychology and education were engaged in the project (NK 1967, p. 221). For grades 4–6, they did not develop new material on their own. Instead they translated textbook material from the School Mathematics Study Group in the USA.

The textbook material was tested in about 1310 classes (1–12) in the Nordic countries during the period 1961–1966. Of these, 90 classes belonged to grades 1–6 and 450 classes belonged to grades 7–9 (Prytz and Karlberg 2016, p. 73). Throughout the trials, the teachers sent in reports about their experiences with the new material. By the end of the development project, the new material was further tested by a method with experimental and control groups. A number of Swedish classes that had followed the New Math syllabus and used the new material for two or three years were compared with classes than had followed the regular syllabus and used traditional textbooks. In grades 1–3, the experimental group comprised 123 students in grades 4–6 they were 120 and in grades 7–9 they were 142 (NK 1967, p. 95). The comparison was done through a written test (national exams) by the end of the period. All in all, the experimental and control groups achieved more or less the same average scores. There was one exception though, the students in *Realskolan* achieved less well with the New Math material (Prytz and Karlberg 2016, pp. 82–89). Here we should note that *Realskolan* had a positive selection of students with regard to talent and/or motivation. This indicated that the New Math, at least the experimental texts, suited the more talented students less well.

According to the final report regarding the development project, published in 1967, the purpose of the comparative testing was not to determine whether the New Math was better or worse than traditional teaching. Instead, the purpose was to investigate in what direction the reform should proceed. The report also stated that it was impossible to make a reliable evaluative comparison between the New Math and traditional teaching. One argument was that the syllabi were so different that there was no test by which the two could be assessed fairly. A second argument was that there was too little information about the teachers and the students to make a proper comparison of the two methods of teaching (NK 1967, pp. 94–96, 215).

If we consider the results presented in the final report, I would say that the people behind the New Math project had reason to be quite confident that a new syllabus and new types of textbooks should work. Not only due to the final comparative testing, but also due to the reports from the teachers. According to the final report, there were both positive and negative experiences, but nobody suggested that the project should be abandoned. After having read through a great number of these teacher reports, that picture is confirmed (RA 1).

The next step in the planning of the New Math reform was to produce the actual syllabus for the 1969 curriculum reform. This group was led by Håstad, who by this time was responsible for mathematics at the National Board of Education. Thus, the connection with the development project then in progress was good. Then after about a year, the remarkable happened, Håstad and another member—Magne (1918–), the psychology expert—left the group, due to conflicts within the group and other people at the National Board of Education. The work was down for a few months. In the end, Sten Sandström, in co-operation with Ingvar Olsson, stepped in and completed the syllabus. Olsson was a teacher educator (7–9) and later he took the position handling mathematics education at the National Board of Education when Håstad left. Surprisingly, Sandström was an expert on chemistry education at the National Board of Education (Kilborn 1977, pp. 19–23). According to Kilborn (1977, pp. 23–25), Sandström and Olsson rewrote several parts. On the basis of interviews conducted by Kilborn, it appears as if Sandström and Olsson reduced the use of concepts and symbols from set theory. However, this did not entail that the new syllabus deviated from the New Math or was not a New Math syllabus.

The syllabus of 1969 was different than the previous syllabus in many respects. With regard to the content of the courses, an overall difference was that topics that had before only been addressed in grades 7–9, were moved to earlier grades. Equations should appear in grade 1, statistics in grade 2 and functions in grade 6, but the coordinate system should be introduced as early as grade 3 as part of geometry. Geometry, which had been introduced in grade 4, should appear as early as grade 1 (Prytz 2015, p. 316). Some concepts or topics were brand new: mappings in grade 1, the position system with other bases than 10 in grade 3, vectors in grade 7 and trigonometry in grade 9. And then, of course, there was set theory. However, it was not a topic on its own. Concepts from set theory were instead included in all other topics (SÖ 1969b, pp. 4–26).

The basic rationale of having set theory in the syllabus was methodological. The idea was that concepts and notations from set theory should be used to introduce and explain other mathematical concepts, not at least together with concrete materials in the earlier grades. Another idea was that the involvement of set theory in various topics should make the students realize the connection between the topics. These ideas were clearly expressed in the syllabus (SÖ 1969a, pp. 138–139). This idea of having set theory as common foundation for all school mathematics in order to facilitate better learning was one of the key elements of the New Math movement (see Kilpatrick 2012, pp. 563–564; Phillips 2015, pp. 87–88, 94; Bjarnadóttir 2014, p. 451). This claim was based on Piaget and Bruner's work in educational psychology; both maintained that there were similarities between mental structures and mathematical structures that could be beneficial in teaching (Phillips 2015, pp. 87–88).

However, it was not only set theory that was a methodological innovation. A second one was the number line. In the syllabus, the number line was integrated in the other topics in the same way as set theory concepts were. But, this was not addressed in the syllabus (SÖ 1969b, pp. 4–26).

Thus, the reform of the New Math concerned methodology as much as it concerned content. As I see it, set theory and the number line constituted a framework for how explanations, visualisations and other concrete materials should be designed. That type of framework did not exist in the previous syllabi.

A final step in the preparations was to educate all teachers in the New Math. This was done through a distance course programme called *Delta*. The focus of the programme was the mathematical content and not so much concrete teaching methods. In total, 42,900 teachers for grades 1–6 and 4700 teachers for grades 7–9 were registered in the programme in the period of 1969–1971. Observe that teachers in grades 1–6 taught all subjects, while teachers in grades 7–9 normally taught two subjects. 40,100 of the teachers finished the programme (KU 1975, pp. 56–57). The statistics regarding the total numbers of teachers for this period is unclear. But, if we consider the statistics for 1977, there were 41,939 teachers in grades 1–6. In grades 7–9, there were about 20,000 teachers, but far from all were mathematics teachers (SCB 1980, p. 336). This means that a great majority of the teachers received in service education in the New Math.

In summary, the Swedish New Math reform was an attempt to radically change mathematics education both with regard to the content and the methodology. Important to notice is that the Swedish New Math reform also meant active centralistic governance, which had not been tried before. Another innovation was that it not only involved the syllabus. A key element in the reform was textbook development, which was done on the basis of controlled testing rather than established experience. Thus, a key element in the reform was to get the teachers to change textbooks. In that perspective, the mandatory national textbook review must have been critical.

7 Implementations of New Math—Failures and Successes, 1969–1980

In this subchapter about the implementation of the New Math, the outcome of the implementation process is put into the foreground. The analysis concerns three curriculum levels: intended, implemented and achieved. For each level, failures and success of the New Math reform is analysed. With regard to the intended curriculum, the analysis is focused on the actions of the central school authorities and their conformity with the original reform plans. The analysis of the implemented curriculum is focused on the teachers' and textbook producers' actions and their conformity with the original reform plans. The analysis of the achieved level is focused on students' achievements in various exams.

7.1 Intended Curriculum

In a comparison of the syllabi of 1969 and 1980 it is easy to agree with commentators who deem the New Math reform a failure. In the syllabus of 1980, the central elements of the New Math were removed. For instance, formulations about having set theory as a common foundation for all the topics, but also for illustrations and other concrete materials, was nowhere to be found. This applies not only to the syllabus, but also to the commentary material. Even all formulations about the number line disappeared. Also the topics trigonometry and vectors disappeared (SÖ 1969a, pp. 137–141, 1969b, pp. 4–26, 1980, pp. 98–107, 1982, pp. 4–59). Thus, in relation to the visions of the New Math, expressed by its proponents in the 1960s, the short presence in the syllabus must be considered a failure. Still, there are traces of success. By the syllabus of 1969, the subtopics algebra, statistics, geometry and functions were to be introduced as early as grades 1–3. A change that made mathematics a much more complex school subject in grades 1–6 in comparison to all previous syllabi, where the dominating subtopic was arithmetic with a little bit of geometry in grades 5–6. This change remained in later syllabi (Prytz 2015, pp. 315–316).

However, there are earlier signs of failure. As early as the beginning of the 1970s, there seems to have been low confidence in the New Math within the central school administration. In an interview with Sven-Erik Gode (1939–), the predecessor of Håstad and Olsson as responsible for mathematics at the National Board of Education in the period of 1972–1977, a low level of confidence in the New Math is apparent. Here we should note that Gode did not have a central position in the preparations of the reform in the 1960s. From the very beginning at his new position he felt compelled to handle deficiencies of the New Math reform. A survey to the teachers indicated that the syllabus was too thorough. The results from the first national exams (standardprov) related to the 1969 syllabus, made in 1973, was, in part, not well conceived. Comparisons showed that students following the 1962 syllabus had better skills in arithmetic than students following the 1969 syllabus. Gode's view was that the critique of the New Math was reasonable. His policy then became prioritising certain parts of the syllabus, which appears to have been numerical counting and more simple geometry and algebra. Moreover, the teachers were encouraged to choose the teaching methods that they saw fit to achieve certain goals (Nämnaren 1983, pp. 10–11).

The decision to encourage teachers to more freely decide how to teach is a clear example of a deviation not only from the New Math reform but also a centralistic mode of governance. Recall that the New Math, to a great extent, concerned teaching methods.

This shift in policy was also visible in a syllabus document named *Basfärdigheter i matematik* (*Basic skills in mathematics*) issued in 1973 by the National Board of Education. It complemented the main syllabus of 1969. Its purpose was to handle problems related to low-achieving students and the preparatory studies began as early as 1971 (SÖ 1973, p. 2). Low-achievers were assumed to comprise about 15% of the students (SÖ 1973, p. 5). The new complementary syllabus specified knowledge and skills that should be included in all regular teaching and that should be prioritized in the teaching of low-achieving students (SÖ 1973, p. 5). Thus, it did not prescribe a new separate course for low-achievers; *Basic skills* was something that concerned all students.

In short, *Basic skills* comprised two parts: the first part contained an analysis of the syllabus and which parts should be considered basic skills; the second part contained guidelines regarding how the teaching should be planned and carried out.

The text contained several references to set theory, but that was mainly in the first part where the mathematical content of the syllabus of 1969 was analysed (SÖ 1973, pp. 5–37). The second part contained sections about diagnoses, planning, individualisation and organisation, but also a section on teaching methods. In the section on teaching methods, the New Math received little attention.

Sets were mentioned in this section on teaching methods, but it was mainly done in connection with ten-sets. It then concerned how students with concrete materials could form subsets of ten elements within sets comprising more than ten elements (SÖ 1973, pp. 58–59). Much more attention was given to ten-base material, which was used to illustrate various algorithms (SÖ 1973, pp. 59–68). On the other hand, sets of ten and ten-base-material must be considered a very small part of the New Math as it was expressed in the syllabus of 1969.

Other important elements in the syllabus of 1969 related to the New Math were not mentioned at all. One such element was the number line (SÖ 1969b, pp. 4–6). Despite the fact that the number line concerns the very basics of elementary school mathematics—numbers and arithmetical operations—it was not included.

Another feature of the syllabus of 1969 that disappeared in *Basic skills* was the use of set theory to join different topics. In the syllabus of 1969, basic concepts from set theory were used in the descriptions of the content of arithmetic, geometry, algebra and functions, but also in the descriptions of which illustrations that ought to be used (SÖ 1969b, pp. 6–12, 24, 26). Algebra was not even mentioned in *Basic skills* since it was considered too difficult for low-achieving students, especially in grades 7–9 (SÖ 1973, p. 43). However, geometry and functions was something that low-achievers were supposed to learn (SÖ 1973, pp. 13–35). But, the idea to join the subtopics by means of set theory was not applied.

An important detail is that the syllabus of 1969 did not apply to all students until the study year 1972/73. The reason for this was that the syllabus of 1969 took effect in the study year 1970/71 and then only in grades 1, 4 and 7. The old syllabus was replaced successively until the study year 1972/73 (SÖ 1969a, p. 5). This means that important steps away from the syllabus of 1969 and the New Math were taken the same year the reform affected all students.

An important aspect of the intended curriculum is national exams; they give indications to teachers and textbooks producers of which content is relevant. In the period 1970–1980, national exams (*standardprov*) were given in grades 3, 6 and 9. The first exam adapted to the syllabus of 1969 was given in 1973. From the very beginning, concepts and notation from set theory and other new concepts such as vectors and mappings were rare. Up to about 1975, the items that included such concepts comprised, in general, 3–5.5% of all items in the exams. This can be considered a reflection of the policy not to underscore the importance of the New Math. On the other hand, it may also reflect the idea that set theory was a methodological element in the New Math, rather than a specific content area like arithmetic and geometry. However, the items with concepts and notation from the New Math disappeared from the national exams by the mid-1970s; for grade 3 it was in 1974/75, for grade 6 it was in 1975/76 and for grade 9 it was in 1976/77 (RA 2).

7.2 Implemented Curriculum

The analysis has been divided in two parts: one about textbooks; the other about teachers. Regarding the textbooks and the period of 1969–1980, Diagram 3 shows there was a lot of textbook series and it has not been possible to analyse all books for all nine grades. I have studied textbooks intended for grades 1, 4 and 7 from the period 1969–1979. More precisely, 43 textbooks from 43 series have been analysed; 13 books for grade 1, 11 books for grade 4 and 19 books for grade 7. Each book represents a textbook series for grades 1–3, 4–6 or 7–9. In total, 63 series were published in the period 1969–1979. In order to find out to what extent the authors have applied ideas of the New Math, the analysis focused on methodological ideas regarding set theory. The basic question was, did the authors apply concepts from set theory, and if so, to what extent and did they do so regularly throughout the book?

The analysis posits that the period of 1969–1976 was a success in the sense that all investigated textbooks in that period (33 in total), except one, included set theory. The exception was published in 1975. In 30 of these textbooks, set theory was used in several chapters, not only in the first chapter, but on the other hand, not in every chapter. There were also differences between the textbooks; some comprised more set theory than others. Mainly, the textbooks produced in the early 1970s contained more, but there were no clear trends in this respect. For further details about the textbooks and the results of the analysis, see *Sources*, *Analysed textbooks*.

If the same measure of success and failure is applied to the period 1976–1979, it appears to be more of a failure. All textbooks that did not include set theory (nine in total) were published in the period 1975–1979. Here we can observe that the books that did not include set theory were published after 1974 when the textbook review became voluntary. But it was not a complete failure since six textbooks that included set theory were published in the same period. Note that five of the textbooks that did not include set theory were intended for grade 7. Moreover, none of the investigated textbooks for grade 7 published after 1974 included set theory.

The fact that some publishing companies choose to produce textbooks that did not include set theory also says something about the teachers. The publications of the new books indicate that the companies were aware of dissatisfaction among the teachers regarding the New Math and the early textbooks. By this time, the publishing companies had well-established contacts with the teachers through textbook consultants. As mentioned above, they were trained in marketing, but also in gathering information about how textbooks were used in teaching. Here I assume the production of brand new books means extra costs and the companies would not invest if they did not know about the demand and the probable profit. Therefore, there is reason to believe that the dissatisfaction was not insignificant, especially in grades 7–9.

There were also critical public voices from teachers. In 1973, the National Association of Teachers of Mathematics, Physics and Chemistry required that the New Math should have a less prominent position in the syllabus. They wanted a syllabus similar to the old one (DN 1973-11-17). Before the New Math syllabus took effect, there were critiques against the reform from textbook authors and people involved in the reform (Prytz and Karlberg 2016, pp. 74–76; Kilborn 1977, pp. 1, 28–29).

7.3 *Achieved Curriculum*

During the early 1970s, a couple of investigations (Holmgren 1974; Kristiansson 1979) were made regarding the achievements of students that followed the syllabi of 1962 and 1969. In one of the investigations (Holmberg 1974), students from grades 3, 6 and 9 were tested. The results were not distinct. In number sense, the students in grade 6 (following the syllabus of 1969) achieved significantly better; between the other groups there were no significant differences. In the area of mathematics, which included calculation skills, the students in grades 6 and 9 (taking the basic course) that followed the syllabus of 1962 achieved significantly better. Between the other groups there were no significant differences. In attitudes towards mathematics, there were no significant differences between any of the groups (Holmberg 1974, p. 33). In the other investigation, which only concerned grades 3 and 6, the students in grade 6 following the syllabus of 1962 achieved much better. But, in grade 3 there were only small differences (Kristiansson 1979, pp. 106–120).

An important detail is that there were no differences in grade 3 in none of the two investigations. The students in grade 3 that followed the syllabus of 1969 were special since they were the only group that had been taught exclusively according to the syllabus of 1969. All other students who belonged to the 1969 group had started school when the syllabus of 1962 was in effect.

Apart from these two investigations, there are two investigations of the results of the national exams (standardprov) during the 1970s. The first national exam according to the syllabus of 1969 was organised in 1973. One investigation concerns grades 3, 6 and 9 (Murray 1980) and the other just grade 9 (Westin 1999). It should be observed that it is not possible to compare the total results of the tests since the exams were changed (Westin 1999, p. 7). However, each year there were a number of so-called anchor items which were each identical during the period 1973–1979. It is these items that both investigations concern. Observe that, for many items, the students just had to give an answer without a solution, but it was not multiple-choice exams. It was voluntary for the teachers to give the test, but in principle, all teachers in the 1970s did so (Ljung 2000, p. 20).

The conclusion of the investigation regarding anchor items in grades 3, 6 and 9 (Murray 1980) was that the results did not change much during the period of 1973–1979. The best results were achieved in grade 3 in all parts (geometry, natural numbers and diagrams) with average solutions frequencies about 70% or more each year. In grade 6, the students achieved well in arithmetic, with average solution frequencies of roughly 70% each year, but worse in practical arithmetic and estimations, with average solutions frequencies of roughly 50% each year. The worst area was geometry, where average solutions frequencies were roughly 40% each year. In grade 9, only the advanced course was studied, the results were worse; there is no separation in areas in the report, but the average solutions frequencies varied between 30 and 40% (Murray 1980, p. 9).

In the investigation, which only concerned grade 9 (Westin 1999), different conclusions were obtained. The report concluded that the average score on all items, not just the anchor items, increased only minimally in the period 1973–1979. This is in line with the results of the investigation by Murray (1980). But, when the eleven anchor items on the exams for the advanced course were divided into three groups—three easy, five medium and three difficult—another pattern emerged. The average scores of the easy and medium items improved from 1973 to 1979. The greatest improvement took place between 1973 and 1976. Westin (1999) also investigated another set of anchor items. This was the 15 items used in each exam during the period 1973–1976 in both the basic and advanced courses. These items were also divided into groups: five easy, five medium and five difficult. In this case, the average score improved for all types of items (Westin 1999, pp. 103–106). Consequently, Westin's (1999) results indicate better development in grade 9 in the period 1973–1979 than Murray's results do.

Another source of information about the students' achievements is the first two IEA international mathematics surveys (FIMS and SIMS), which were made in 1964 and 1980. Between these two surveys, the Swedish results in grade 7 improved only minimally (Murray and Liljefors 1983, p. 44). This seems to be in line with the total results of the national exams in the 1970s reported by Westin (1999). Here we should note that the Swedish results were poor in relation to other countries in both 1964 and 1980; in both cases, Sweden tied for last place with the USA. However, the fact that Swedish students had started school one year later is a possible cause of that position.

Apart from the small differences between the total scores of FIMS and SIMS, there were also small, but interesting differences on item level. In 1964, the students were better in arithmetic, but in 1980, the students were better in algebra and geometry. The results of FIMS and SIMS was also analysed with respect to which competencies the items required. Also in that case, there were differences, although quite small. In 1964, the students were better at calculation and applications, while in 1980, the students were better at understanding and analysis (Murray and Liljefors 1983, p. 44).

If we consider all these results regarding the students' achievements, it is possible to discern both the successes and failures of the New Math from the standpoint of achievement. The changes in the results between FIMS and SIMS, reported by Murray and Liljefors (1983), are an indication of the success of the New Math. Being better at the more advanced topics of geometry and algebra, which included functions, but also being better at competencies, understanding and analysis, seems to be in line with the ideas of the New Math—emphasis was put on understanding and structures rather than computational techniques. The improvement of the anchor items in the national exams reported by Westin (1999) can also be considered a success for the New Math, at least if we assume that the changes on the intended and implemented level that can be described as being unfavourable to the New Math did not have an immediate effect. On the other hand, the changes between FIMS and SIMS were small and the anchor items were few, which makes it impossible to speak about a clear success of the New Math.

But there are also indications of failure at the achieved curriculum level. Some of the results on the anchor items in the national exams in the 1970s, reported by Murray (1980), were low, especially in grade 9, but also in grade 6. Moreover, the total scores were stable, which indicates that the good results in grade 3 did not yield good results in higher grade levels. The comparisons between students who followed the syllabi of 1962 and 1969, reported by Holmberg (1974) and Kristiansson (1979), indicate failure more than success. However, it is important to note that there are no indications of a massive failure with really low results at all grades levels and all topics.

8 An Explanation of the Decline of the New Math in Sweden

The explanation starts with an analysis of the relationship between the three curriculum levels discussed in the previous section. Due to the fact that the preparations of the New Math reform concerned the development of new textbooks and in the implementation phase it was pivotal to get teachers to change to this type of textbook, the explanation begins with the implemented level and textbooks.

In relation to the ambition to get the teachers to use New Math textbooks, the reform was initially a success. All textbooks produced between 1970 and 1974 were more or less in line with the New Math syllabus. However, from 1975 onwards, a number of new textbooks deviating from basic ideas of the New Math appeared on the market. This can be considered a failure of the reform plans. However, new New Math textbooks were also published from 1975 onwards. Thus, it is far from a complete failure.

The fact that textbooks deviating from the New Math were published indicates dissatisfaction with the New Math among teachers. Especially in grades 7–9; most of the deviating textbooks were published for those grades. It is not likely that profit-seeking publishing companies with good contacts with teachers would have published such textbooks without a real demand. Still, we cannot consider the dissatisfaction among teachers as very wide spread since new New Math textbooks were also published.

Bad results of teaching can explain why some teachers did not like the New Math textbooks. Indeed poor results on various exams, including national exams, shown in the previous section, indicate that mathematics education in Sweden as a whole did not function in an optimal way, especially in grades 7–9. On the other hand, not all results of the exams were bad, which indicate that parts of the system did indeed work.

As I see it, the diversified production of new textbooks from 1975 onwards was a reflection of the exam results. Some teachers succeeded with the New Math syllabus and the New Math textbooks; they continued to ask for such textbooks. Some teachers did not succeed and asked for something different, especially in grades 7–9.

Another aspect of this is that the central school authorities from 1974 onwards allowed the publishing companies to publish and market textbooks not in line with the New Math. Most likely, making the textbook review voluntary in 1974 did not deter the publishing companies in this respect. And technically, it probably made it more difficult for the central school authorities to force textbook production in any direction. However, even a couple of years earlier, the central school authorities refrained from underscoring the importance of the New Math. This was about the same time the reform affected all students; the reform took effect successively between 1970 and 1972.

So, what caused the central school authorities to adopt a passive stance regarding the New Math at an early stage of the implementation process? On the basis of what has been presented above, it was not likely due to a massive dissatisfaction among teachers in general or really bad student results. A possible explanation concerns ideas and a loss of faith in the New Math.

Phillips (2015) connects the decline of the New Math in the USA with a general decreasing faith in science and rationality; the New Math was supposed to foster a rational (as in scientific) mode of thinking, but it should also enhance scientific as well as industrial progress. Hultén (2008) observes a similar loss of faith in Sweden. I cannot exclude that as a cause, but ideas regarding fostering rational thinking was heavily criticised, due to lack of scientific evidence, in an official investigation published already in 1960. This critique seems to have had an impact on the syllabus of 1962. Thus, the loss of faith about fostering rational thinking through mathematics teaching seems to have occurred too early to explain the decline of the New Math in Sweden in the 1970s.

However, a closely related change concerned governance. In the 1970s, centralised governance of the Swedish school system was being questioned. This type of governance was based on high confidence in science and rationalism. As I see it, this ideological shift was probably of greater importance for the decline of the Swedish New Math reform since it was a centralistic reform that was, to a significant degree, designed according to scientific methods and theories.

Apart from ideological shifts, we can also observe sociological shifts in the reform process. New types of people were involved in the New Math reform—experts on psychology and education—especially during the preparations. Gispert and Schubring (2011) observe a similar phenomenon in Germany and France. Recall that development of new textbooks was essential in the preparation of the Swedish New Math reform. In that context, textbook development was a matter of science, or at least something similar to science.

In the Swedish case, this change was perhaps more radical than in other countries since it included a break with established textbook authors. The people engaged as textbooks authors were, however, not researchers in psychology and education; they were mathematicians and teachers. But, they were not experienced textbooks authors. None of the authors, except one, had authored textbooks before the New Math project began.

This sociological change can of course have aroused critique of the New Math. Established authors were probably critical about something that removed them from their established position. And indeed, critique of the New Math came from old established textbook authors outside the New Math project (Prytz and Karlberg 2016, pp. 74–76).

But, why did the central school authorities not withstand initial setbacks like poor results on national exams, but also outside pressure such as critique? Of course, there can be ideological causes of that; the task to execute a radical reform was put in the hands of people who appear not to have been firm enough in their beliefs in the New Math or centralistic governance.

However, there are other causes than beliefs and ideas to consider. Notice that active centralistic governance that aimed for radical changes in school mathematics was new in Sweden; it had not been tried before. The people in charge of mathematics at the central school administration and their superiors were in a new situation without experience of active centralistic governance. Notice here the distinction between, on one hand, having theoretical knowledge of or believing in centralistic governance and, on the other hand, of having experience of practicing it in an active way. This can explain why the central school authorities already gave up important elements of the New Math reform already at an early stage when difficulties occurred; they had no experience of how to advance in an efficient manner. This explanation is in line with Oftedal Tellhuag et al. (2006, p. 256) characterization of Swedish school governance: at this point in time, centralistic governance was a matter of input rather than output management. The input in this case was the New Math syllabus and new types of textbooks, prepared for years; the output was the not-so-good student achievements.

However, there are other aspects to consider. Early on, the possibility of controlling the textbooks disappeared as the textbook review became voluntary. Additionally, the influx of textbooks deviating from the reform of the New Math began after the textbook review became voluntary. However, the review becoming voluntary is not enough to explain why textbooks deviating from the New Math were produced. One factor is knowledge about demand; it is not likely that the profit-seeking companies would produce new textbooks without knowledge about a demand for new types of textbooks. And we do know that they had good contacts with the teachers. Nevertheless, neither knowledge about a demand is enough to cause an increase in the production of new textbooks; resources or good chances of profit are needed as well. It appears as if the publishing companies had that resources or good chances of profit since textbook production had been a lucrative enterprise for some years. To further illustrate my point, assume the publishing companies as well as the state were poor. I believe that would have dampened the publishing houses' willingness to produce new textbooks. This belief is supported by statistics regarding the influx of new textbooks and the schools spending on textbooks. From 1950 to 1970, school spending on textbooks increased ten times; in the same period, the influx of new textbooks increased significantly.

To a great degree, the conclusions and explanations presented here concern the New Math in Sweden. However, they do contribute to research on the history of the New Math in a wider sense. The comparisons with the studies of Phillips (2015) and Gispert and Schubring (2011) show that more than ideological and sociological aspects need to be considered in order to understand the fate of the New Math. This paper highlights the interplay between factors related to textbooks, governance and economics. It is relevant to consider these factors in other countries as well since textbook development was a part of New Math reforms in many countries (cf. Kilpatrick 2012, pp. 563–564).

Sources

Riksarkivet (RA, The Swedish National Archives) 1
Archive: Nordiska kommittén för modernisering av matematikundervisningen (NK)
F1:11–16 Lärares rapporter om försöksverksamheten till kommittén, 1964–1967
RA 2
Archive: Skolöverstyrelsen (SÖ) 1919–1991
54 Avdelningen för lärarutbildning … (L-avd) 1964–1984
04 Sektionen för prov och utvärdering
F I bie Standardprov i matematik
Volume: 3, 5 and 8

Analysed textbooks, grade 1 (The textbooks which do not contain set theory are marked by *)

Title	Authors (family names)	Year of publication
Ettans nya matematik	Hultman Kristiansson Ljung	1970
Hej Matematik	Håstad Svensson Öreberg	1970
Mängder och Tal	Eicholz Forssblad Malm	1970
Matematikserien	Billing Blomdahl Bodén	1970
Ny matematik	Valinder Kos Holmström	1971
Matematik A	Rosenberg Lindström Natt-o-Dag Högberg	1971
Ettans matematik	Hultman Kristiansson Ljung	1972
Små steg	Carlsson Jönsson	1972
Modern matematik	Malmer	1973
*Matematik	Unenge Alm	1975
*Min matematik	Kilborn Johansson Dahlström	1977
Matematik för oss	Rosenberg	1979
*Räkneboken 1	Gustafsson Persson Ronnheden	1979

Analysed textbooks, grade 4 (The textbooks which do not contain set theory are marked by *)

Title	Authors (family names)	Year of publication
Hej Matematik	Håstad Svensson Öreberg	1970
Matematik 1M	Nilsson Johansson Lindström	1971
Matematikserien	Hellström Paulsson Åberg	1973
Ny matematik 1M	Janson Gerefalk	1973
Mängder och tal	Eicholz O'Daffer Bergström Dannberg	1973
Fyrans matematik	Hultman Kristiansson Ljung	1975
Matte för alla	Amundsson Holmström Nordling Rosén	1975
Matte för dig	Amundsson Holmström Nordling Rosén	1976
Mia 4	Lundgren Paulsson	1976
*Min matematik	Kilborn Johansson Lundin	1977
Hej Matematik (new version)	Håstad Svensson Öreberg	1977

Analysed textbooks, grade 7 (The textbooks which do not contain set theory are marked by *)

Title	Authors (family names)	Year of publication
Högstadiets matematik (Lunda-systemet)	Kvist Larsson	1969
Högstadiets matematik	Fredriksson Göransson Thulin	1970
Matematik för högstadiet	Forssblad	1970
Nya högstadiets matematik	Alvin Anderberg	1970
Hej Matematik	Håstad Svensson Öreberg	1971
Matematik 1h	af Ekenstam Wyndham	1971
Matematik 1H	Nyman Emanuelsson Jeanson Rudolfsson	1972

(continued)

(continued)

Title	Authors (family names)	Year of publication
Matematik 7:1	Bentley Johansson Kilborn Åsberg	1972
Högstadiets matematik (Lunda-systemet)	Kvist Larsson	1972
Matematik åk 7 ht sk	Unenge	1973
Nya Hej Matematik	Håstad Svensson Öreberg	1973
Modern matematik	Forssén Frank Olsén	1973
Matematik åk 7s	af Ekenstam Wyndham	1973
Högstadiets matematik	Alvin Anderberg	1973
*Matematik åk 7 (Lunda-systemet)	Kvist Larsson	1977
*Hej Matematik	Håstad Svensson Öreberg	1977
*Min matematik	Kilborn Johansson Åsberg	1978
*Matematikboken sk 7	Undvall Olofsson Forsberg	1979
*Sjuans Matematik 1s	Alvin Anderberg	1979

References

1957 års skolberedning (SB) (1960). *1957 års skolberedning. 3, Kursplaneundersökningar i matematik och modersmålet: empiriska studier över kursinnehållet i den grundläggande skolan.* [School investigation of 1957. 3, studies of syllabi in mathematics and mother tongue: empirical studies of the course content in compulsory schools.] Diss. Stockholm: Högsk.

1957 års skolberedning SB (1961). *1957 års skolberedning. 5, Hjälpmedel i skolarbetet: synpunkter och förslag.* [School investigation of 1957. 5, Teaching material in school: comments and suggestions.] Stockholm.

Bergqvist-Nordfelt (1910). *Handbok ang. läroverksförfattningar.* [Handbook concerning secondary school statutes] Stockholm: P. A Norstedt & söners förlag.

Bergqvist, Bengt J:son, & Wallin, Harald (Ed.) (1928). *Nya läroverksstadgan jämte undervisningsplaner: med flera nya författningar rörande allmänna läroverken.* [The new secondary school statute together with teaching plans: with several new ordinances regarding public secondary schools.] Stockholm: Norstedt.

Bjarnadóttir, Kristín (2014). History of teaching arithmetic. In Karp, Alexander, & Schubring, Gert (eds.) *Handbook on the History of Mathematics Education* (pp. 431–458). New York: Springer.

Dagens Nyheter (DN) 1973-11-17, Matteundervisningen måste bli bättre. [Math teaching needs to be better.]

Damerow, Peter, & Westbury, Ian (1984). Conclusions drawn from the experiences of the new mathematics movement. In P. Damerow (Ed.), *Mathematics for all: Problems of cultural selectivity and unequal distribution of mathematical education and future perspectives on mathematics teaching for the majority* (Science and Technology Education Document Series No. 20, pp. 22–25). Paris: UNESCO.

Gispert, Hélène, & Schubring, Gert (2011). Societal, Structural, and Conceptual Changes in Mathematics Teaching: Reform Processes in France and Germany over the Twentieth Century and the International Dynamics. *Science in Context, 24*, 73–106.

Holmberg, Ingrid (1974). *Effekter av ny läroplan i matematik (Lgr 1969) på vissa aspekter av talbegreppets utveckling, kunskaper i matematik och attityder mot ämnet. [Effects of the new syllabus in mathematics (Lgr 1969) on certain aspects of number sense development, knowledge in mathematics and attitudes towards the subject.].* Malmö: Lärarhögskolan.

Hultén, Magnus (2008). *Naturens kanon: formering och förändring av innehållet i folkskolans och grundskolans naturvetenskap 1842–2007.* [The canon of nature: the formation and transformation of the contents of science in Folkskolan and Grundskolan 1842–2007.]. Dissertation. Stockholm: Stockholms universitet.

Johnsson Harrie, Anna (2009). *Staten och läromedlen: en studie av den svenska statliga förhandsgranskningen av läromedel 1938–1991.* [The state and the teaching materials: a study of the Swedish state's review of teaching material 1938–1991.] Diss. Linköping: Linköpings universitet, 2009. Linköping.

Kilborn, Wiggo (1977). *Hej Läroplan!: hur man bestämmer vad våra barn ska lära sig i matematik.* [Hello Curriculum!: how to decide what our children shall learn in mathematics.] Mölndal: University of Gothenburg.

Kilpatrick, Jeremy (2012). The new math as an international phenomenon. *ZDM: the international journal on mathematics education, 44*(4), 563–571.

Kline, Morris (1973). *Why Johnny can't add: the failure of the new math.* New York: St. Martin's Press.

Kommittén för den fortsatta verksamheten med radio och television inom utbildningsväsendet (KU) (1975). *Distansundervisning: lägesbeskrivning samt organisatoriska alternativ för högskolan: betänkande.* [Distance education: description of the situation and organisational alternatives for the universities: a report]. Stockholm: LiberFörlag/Allmänna förl.

Kristiansson, Margareta (1979). *Matematikkunskaper Lgr 62, Lgr 69.* [Knowledge of mathematics curriculum 62, curriculum 69.] Diss. Göteborg: Univ. Göteborg.

Kungl. Skolöverstyrelsen (SÖ) (1935). *Metodiska anvisningar till undervisningsplanen för rikets allmänna läroverk.* [Methodological instructions to the teaching plan for the national public secondary schools.] Stockholm: Statens reproduktionsanstalt.

Larsson, Esbjörn, & Westberg, Johannes (eds.) (2011). *Utbildningshistoria: en introduktion.* [History of education: an introduction]. Lund: Studentlitteratur.

Lindensjö, Bo & Lundgren, Ulf P. (2000). *Utbildningsreformer och politisk styrning.* [Educational reforms and political governance.] Stockholm: Liber.

Ljung, Bengt-Olov (2000). *Standardproven – 53 år i skolans tjänst. Stockholm.* [National exams – 53 years in the service of schools.] Stockholm: Lärarhögskolan.

Läromedelsutredningen (LU) (1971). *Samhällsinsatser på läromedelsområdet: betänkande.* [Societal efforts in the field of teaching materials: a report.] Stockholm: Esselte tryck.

Marklund, Sixten (1983). *Skolsverige 1950–1975. D. 3, Från Visbykompromissen till SIA.* [School Sweden 1950–1975. V. 3, From the Visby compromise to SIA.] Stockholm: Liber/Utbildningsförl.

Murray, Åsa (1980). *Standardized achievement tests and comparisons of school achievement.* Stockholm: Skolöverstyr.

Murray, Åsa, & Liljefors, Robert (1983). *Matematik i svensk skola.* [Mathematics in the Swedish schools.] Stockholm: LiberUtbildning.

Nordiska kommittén för modernisering av matematikundervisningen (NK) (1967). *Nordisk skolmatematik.* [Nordic school mathematics.] Stockholm: Nordiska rådet.

Normalplan för undervisningen i folkskolor och småskolor (NP) [Normal plan for the teaching in Folkskolan and Småskolan.] (1900). Stockholm: Norstedt.

Nämnaren: publikation om skolmatematik (1983). No 1, "Vad HÄNDE på SÖ?". [What HAPPENED at National board of education?]. Mölndal: Fortbildningsavdelningen, Lärarhögskolan i Mölndal.

Oftedal Telhaug, Alfred, Mediås, Odd Asbjørn, & Aasen, Petter (2006). The Nordic Model in Education: Education as part of the political system in the last 50 years. *Scandinavian Journal of Educational Research, 50*(3), 245–283.
Phillips, Christopher J. (2015). *The New Math: a political history*. Chicago: University of Chicago Press.
Prytz, Johan (2007). *Speaking of Geometry – a study of geometry textbooks and literature on geometry instruction for elementary and lower secondary levels in Sweden, 1905–1962, with a special focus on professional debates*. Diss. Uppsala: Uppsala universitet.
Prytz, Johan (2015). Swedish mathematics curricula. An overview. In K. Bjarnadóttir, F. Furinghetti, J. Prytz, & G. Schubring (Eds.), *"Dig where you stand" 3: Proceedings of the Third International Conference on the History of Mathematics Education* (pp. 309–325). Uppsala: Uppsala universitet.
Prytz, Johan (2016). *The construction of a database regarding Swedish historical textbooks in mathematics (grades 1–9), 1900–2015: A technical description*. Uppsala: Uppsala universitet.
Prytz, Johan (2017, in press). The production of textbooks in mathematics in Sweden, 1930–1980. In K. Bjarnadóttir, F. Furinghetti, J. Prytz, M. Menghini, & G. Schubring (Eds.), *"Dig where you stand" 4: Proceedings of the Fourth International Conference on the History of Mathematics Education*. Roma: Edizioni Nuova Cultura.
Prytz, Johan, & Karlberg, Martin (2016). Nordic School Mathematics revisited. On the introduction and functionality of New Math, *Nordic Studies in Mathematics Education, 21* (1), 71–94.
Statistiska centralbyrån (SCB) (1973). *Statistisk årsbok för Sverige 1973*. [Statistical Abstract of Sweden 1973] Stockholm: SCB
SCB (1974). *Elever i obligatoriska skolor 1847–1962*. [Students in mandatory schools 1847–1962]. Stockholm: SCB.
SCB (1977). *Elever i icke-obligatoriska skolor 1864–1970* [Students in non-mandatory schools 1864–1970]. Stockholm: Statistiska centralbyrån.
SCB (1980). *Statistisk årsbok för Sverige 1980*. [Statistical Abstract of Sweden 1980] Stockholm: SCB.
Skolöverstyrelsen (SÖ) (1955). *Undervisningsplan för rikets folkskolor*. [Teaching plan for Folkskolan.] Stockholm: Norstedt.
SÖ (1962). *Läroplan för grundskolan* [Curriculum for Grundskolan]. Stockholm: Kungl. Skolöverstyr.
SÖ (1969a). *Läroplan för grundskolan. 1, Allmän del*. [Curriculum for Grundskolan. 1, General part]. Stockholm: Utbildningsförl.
SÖ (1969b). *Läroplan för grundskolan. 2, Supplement: kompletterande anvisningar och kommentarer. Matematik*. [Curriculum for Grundskolan. 2, Supplement: additional instructions and comments. Mathematics.] Stockholm: Utbildningsförl.
SÖ (1973). *Basfärdigheter i matematik*.[Basic skills in mathematics.] Stockholm: Utbildningsförlaget.
SÖ (1980). *Läroplan för grundskolan. Allmän del: mål och riktlinjer, kursplaner, timplaner*. [Curriculum for Grundskolan. General part: goals and guidelines, syllabi, timetables]. Stockholm: LiberLäromedel/Utbildningsförl.
SÖ (1982). *Läroplan för grundskolan. Kommentarmaterial. Att räkna: en grundläggande färdighet*. [Curriculum for Grundskolan. Commentary material. To count: a basic skill]. Stockholm: LiberLäromedel/Utbildningsförl.
Undervisningsplan för rikets folkskolor (UP). [Teaching plan for Folkskolan] (1920). Stockholm: Norstedt.
Unenge, Jan (1978). Mathematics in an undifferentiated school system. In F. Swetz (Ed.), *Socialist mathematics education*. Southampton: Burgundy Press.

Unenge, Jan (1999). *Skolmatematiken i går, i dag och i morgon: med mina ögon sett.* [School mathematics yesterday, today and tomorrow: in my view.] Stockholm: Natur och kultur.

Wallin, Harald, & Grimlund, Hugo (Eds.) (1939). *1933 års förnyade läroverksstadga: med förklaringar och hänvisningar: jämte timplaner och undervisningsplan m. m. rörande allmänna läroverken.* [The renewed secondary school statutes of 1933: with explanations and references etc. regarding the public secondary schools.] Stockholm: Sv. bokförl.

Westin, Hans (1999). *Farväl standardprov: standardproven i matematik 1973–1997 för åk 9: jämförelse av resultat på uppgifter som återkommit genom åren.* [Farewell national exams: national exams in mathematics 1973–1997 for grade 9: a comparison of results on items which have recurred over the years.] Stockholm: Lärarhögskolan, PRIM-gruppen.

Author Biography

Johan Prytz is Associate Professor (Docent) at the Department of Education, Uppsala University, Sweden. He has a Ph.D. in mathematics with specialization in the history and pedagogy of mathematics. His main research interests are mainly in the intersection between mathematics education and history of education. In 2011, he received a four year research grant from the Swedish research council for the project The development of School mathematics and reforms of the Swedish school system in the 20th century. A comparative and historical study of changes of contents, methods and institutional conditions. His chapter in this book is a result of this project.

History of Mathematics Instruction in Colonial and Early Post-colonial Cambodia

Sethykar SamAn

Abstract This paper is devoted to the history of mathematics education in Cambodia, which reflected the history of the country and is difficult to explore due to the destruction of the available sources during the Khmer Rouge regime (1975–1979). In 1863, Cambodia became a protectorate (a colony) of France, and gained independence in 1953. Prior to the colonial period, Cambodian culture shared a number of features with other states of Southeast Asia whose religion, science, technology and educational systems, historically, were influenced by respective Indian traditions; unfortunately, the documents related to mathematics instruction during this period of history are either extremely rare or lost. During the colonial period, an attempt was made to introduce to Cambodia a Western (more specifically, French) system of education. The paper will focus on mathematics education during the last years of Cambodian colonial era and first years of independent Cambodia based mainly upon a series of interviews conducted with individuals who were schoolchildren during the period from the late 1940s to 1960s.

Keywords History of mathematics instruction in Cambodia
French colonial education · Colonial education in Cambodia
Early post-colonial mathematics education in Cambodia

1 Introduction: Mathematics Education in Colonial Cambodia

Education in Cambodia during the colonial period was inscribed in the larger educational project encompassing the entire French Indochina, that is, Vietnam (subdivided into Tonkin, i.e., Northern Vietnam, Annam, corresponding to central part of Vietnam remaining under nominal control of Vietnamese authorities, and Cochinchina, the Southern part of the country), Laos, and Cambodia. The educa-

S. SamAn (✉)
Royal University of Phnom Penh, Phnom Penh, Cambodia
e-mail: sethykar@gmail.com

F. Furinghetti and A. Karp (eds.), *Researching the History of Mathematics Education*, ICME-13 Monographs,
https://doi.org/10.1007/978-3-319-68294-5_11

tional activities of French colonial administration in Vietnam are relatively well documented (see, for example, Trinh 1995), while the efforts to build a system of instruction in two other countries are considerably less studied. Due to the massive destruction of library funds during the Khmers Rouges regime (1975–1979), the textbooks used during the colonial era and other relevant documents can hardly be found.

The secondary sources relevant to the projected study of mathematics education during the colonial era are equally scarce; one of them is a description of the French colonial system of education produced soon after its end found in the report of Hollister (1958). According to this author, only 13 secondary schools were operational in Cambodia, and the number of students who received *Baccalauréat* (French style 6-year secondary school program) in 1956 amounted to 54 only, while the number of students of the first year of secondary school amounted to 2037 (p. 210). Hollister claimed that the main goal of school education was training of future civil servants destined to be employed in governmental offices. Hollister also reports that at the moment when he was writing his article, 13 more “model schools” were planned to be open with the US aid (p. 210); this project thus featured a large-scale educational reform, including a new mathematical curriculum. However, the currently available materials do not allow to verify whether this project was successfully completed. Besides, Hollister’s report is dealing mainly with the transitional period *after* the end of the French colonial rule, and does not provide sufficient information about the mathematics instruction of the colonial period. Moreover, the author does not discuss the textbooks, topics studied, and the process of instruction during the transitional period. Another study that could have been relevant to the present investigation is the monograph of Trinh (1995), but despite its title, the author focused mainly on the history of colonial school education in Vietnam. Works of other authors (e.g., Clayton 1995; Dy 2004; Osborne 1969; Rany et al. 2012) contain only very brief mentions of the elements of colonial instruction I am interested in.

The most complete description of the educational system established by the French colonial administration is found in Delvert (1956). As the author himself says, his report is mainly based on the memories of those who went to school during the colonial time, since “the archives are rare, dispersed, and hardly accessible” (p. 309). It is somewhat surprising to know that the documentation related to education was already not available as early as 1956 in Cambodia (it remains unclear if the author had access to the colonial archives in Cambodia or in France).

In his study, Delvert does not specify how exactly he obtained information from the locals. Most likely, his interviews were conducted as conversations lacking any particular structure. Moreover, he does not provide any information about the interviewees, their age, gender, social position, circumstances of their schooling, etc. This makes it difficult to evaluate, on the basis of his description, how much his data reflected the actual situation in Cambodian education of the past. Below I summarize the parts of his paper relevant to the topic of my study, namely, the history of mathematics education in colonial Cambodia.

Colonial primary schools started appearing in 1911, and in 1912 their program of study was established. It included "elements of computations [and] of metric system". At the same time, the authorities ordered the textbooks to be compiled, printed, and distributed (p. 310). Twenty years later, the number of primary schools reached 74, and the number of students equaled to 3507. The instruction was essentially bilingual (in Khmer and French), and trilingual (in Khmer, French, and Vietnamese or Lao) in provinces adjacent to Vietnam or Laos, respectively (p. 311). At this early stage, the key role was played by Henri Russier (1878–1918), Director of the "Service de l'Enseignement" [Education Service], who himself authored or edited a number of textbooks, including textbooks on arithmetic in French and Khmer (pp. 311–312).[1] In 1924 the colonial government ordered the rural Khmer Buddhist "schools of pagodas" to add "calcul élémentaire" [basic arithmetic] to their curriculum (pp. 313–314).

In 1946, the functions and duties of the office of the French colonial government responsible for education, the "Direction locale de l'Enseignement au Cambodge" [that is, "Local direction of Cambodian education"] were transferred to Cambodian government; the responsibilities of the French authorities consisted in providing instructors and textbooks (p. 315). "Elementary mathematics" was introduced in the curriculum of the newly established colleges [lycées] in 1953 (p. 316).

2 The Methodology

Given the almost complete lack of documents related to the history of mathematics education in colonial and early post-colonial period, I suggest that the study of this topic can be conducted on the basis of the four following groups of sources: (a) the French mathematics textbooks published during the period under investigation that arguably were used by compilers of Khmer textbooks or even used in those classrooms where the students were sufficiently fluent in French; (b) the extant mathematics textbooks translated into Vietnamese and published in Vietnam by the colonial authorities of Indochina, since their contents, hypothetically, were close enough to those studied in Cambodian colonial schools; (c) information about mathematics curriculum officially published by colonial authorities (such as the elements of curriculum published in Khmer language in the *Journal for Lecturer*), and (d) interviews with those Cambodian citizens who went to school prior to the end of colonial regime, that is, before 1953, or soon after it (even in the 1960s the school still had some features of the colonial time and therefore, it makes sense for the purpose of the study to explore the school instruction even after the formal end of the colonial time). In this paper I will explore only the sources of the last group,

[1]These textbooks included Dachary (1915), Pujol and Russier (1912) and Russier and Négrignat (1912–1913).

and will provide a detailed report based upon my interviews with Cambodian citizens who were schoolchildren between the early 1950s and late 1960s.

3 Summary of the Interviews

In this section I will summarize the interviews with six individuals conducted in 2015–2016. The interviews were based on a questionnaire developed in advance; each interview included questions related to the following topics:

1. Personal and educational background of the interviewee (name, age, gender, educational background);
2. Information about primary and secondary schools where the person studied;
3. Information about mathematics instructors in primary and secondary schools (in particular, their age, gender, and educational background, whenever available);
4. Information about topics studied in mathematics classroom;
5. Information about the textbooks used for mathematics instruction;
6. Information about the mathematical terms used in the textbooks and in the classroom;
7. Information about counting instruments and other tools used in teaching;
8. Information about instruction procedure;
9. Information about homework.

The interviews were conducted in Khmer language; they were audiotaped and later transcribed.

3.1 Interview 1

The first interviewee (Mrs. PVT)[2] was a Khmer female born in 1939; she started school in 1948, and completed only three years of Tram Thna primary school located in Ta Kao province. The number of schoolchildren in her class equaled to approximately 40–50; most of them were boys (ca. 80%). Despite the elementary (primary school) curriculum, all these students were between 11 and 17 years old. According to Mrs. PVT, after having finished the primary school, the students would have to go to Chom Bork secondary school located a bit far away from primary school (yet in the same province, Ta Kao). In secondary school, there were only two grades, and after having finished secondary school, students would continue their studies in Phnom Penh or worked as teachers of primary schools in the province.

Mrs. PVT's teachers were around 40–50 years old. Her teachers were all men from first grade to third grade, and school director was a man as well; female

[2]Here and below the actual names of the interviewees are concealed for privacy reasons.

teachers were rare. One of her teachers (who probably taught the first grade) was a Khmer-Vietnamese, but he was able to speak both Khmer and French very well. Most of her teachers were Khmer from the same province. For each grade, there was only one teacher teaching several subjects including mathematics. When asked whether it was possible that teachers did not have mathematics background and did not understand some topics in the textbooks that they were supposed to teach, Mrs. PVT did not recollect the background of the teachers, but said that her teachers could teach mathematics very well.

Mrs. PVT studied arithmetic (simple calculations in the first grade which became harder in the third grade); according to her memories, the subject was similar to algebra, but at the primary level the schoolchildren did not use unknowns x, y, z when solving problems; sometimes they used a box sign instead of symbol of unknown. They also learned measurement (for example, mass, length, …) and basic concepts of geometry. In the third grade, they still learned the same topics but started to learn more geometrical topics, e.g. types of triangles and quadrilaterals. When asked whether the topics were studied separately (for example, one class was devoted to algebra, one class to geometry), or they were mixed up together, she stated that in her school the children studied the topics separately (one class for each topic), but she was not completely certain. When asked which topics were usually considered "difficult" by teachers and students, she was not certain; she claimed that equation problems were easy for her, but not for some students.

There was a new mathematics textbook every year, according to the grade. But only teachers had the textbooks. Mrs. PVT was not sure whether there was a special textbook for each topic or not, but she recollected that her teacher used only one mathematics textbook which looked like teacher notes. She was not sure either about the language used in the textbook; she said that it probably contained texts in both languages, Khmer and French (mostly Khmer), because in mathematic class teacher used mostly Khmer and some technical terms in French, and some Khmer terms were most likely translated from French.

The terms used in mathematic class at primary level were easy to understand and sounded very Khmer to the interviewee (i.e., Khmer words were used), for example 'add បូក (*buk*)', 'divide ចែក (*cheak*)', and 'multiply គុណ (*kun*)'. However, some French technical terms were used (she did not specify what were these terms) because there were no suitable Khmer terms with the same meaning, and the use of French terms apparently helped avoiding wrong usage of the words for primary classes, mathematical terms—which were mostly in Khmer (90%) with addition of some French terms—were easy to understand for her because the difficult terms were explained by her teacher, and students also learned French in school. She was not sure if the textbooks explained the technical words or not. When asked whether teachers gave explanation of the meaning of new mathematical terms, in particular, when students asked for explanations, the interviewee was not certain; she just said that students rarely asked teacher any questions, and when students asked some questions, some teachers would have answered properly, but some teachers wouldn't, that is, it depended on each teacher.

The students and teachers used rulers, sticks, and sometimes their own fingers as manipulatives. These tools were easy to find and were used in the local school for a long time. But some instruments used for measurements (e.g., rulers) were probably introduced by the French (she was not completely sure) because there were French measure units on them (e.g., *cm*, *dm*…). The tools were introduced by teachers who explained how to use them. For example, when solving mathematics problems like adding numbers, the students used sticks to perform their calculations, and if a student could solve the problems without using the tools, the solution was also accepted by the teacher. Older students, like those of the 3rd grade, were not allowed to use these tools anymore. Mrs. PVT was not sure whether the textbooks explained the operations with counting tools or not, but she said that it was probably explained in lecturer's notes; the latter then instructed the students orally.

When asked whether the teachers just gave orders ("do so and so", "do what I do") without any explanations, the interviewee answered that it depended on each teacher, but most of teachers explained lessons. She mentioned that one of her teachers (who was a Khmer-Vietnamese) was a bit cruel, maybe he had some mental problems, and whenever students asked questions, he did not respond and may have become aggressive towards students and even punish them. Later this teacher was dismissed and other teachers were polite and tried to answer students' questions properly.

Her teacher required memorization of multiplication table, and asked students to do exercises on the blackboard or recite the multiplication table without any help. Students may have had written tests more frequently then oral tests.

The interviewee did not remember whether the students had exams; she did not remember either how many hours per week the students studied mathematical subjects.

She was not certain about homework either, just said that all homeworks included memorization and repetition. For example, students would have had to write the same formula to fill two pages of their notebooks, or to do some exercises. In class, teacher could ask student to recite the lesson learned last time orally, or by writing on the blackboard. Then, teacher would correct the answer, or gave the right answer without explanation.

Teachers systematically checked each student's homework and gave the correct answers on the blackboard or asked outstanding students to present their solutions.

3.2 Interview 2

The second interviewee (Mr. A) was a 71-years-old man (born in 1945). He moved from his hometown (Kandal Province) to Phnom Penh in 1950. Then he went to primary school in 1953. For two years, he studied in first and second grades at "Preah Chey Chesda Primary school" (he said that in colonial era it was named "François Baudoin School", but he was not sure). In 1955, he moved to "Ta Prom Meanchey Primary School". At this school, he studied from third grade until sixth grade. He said that students were mostly male and there were only about ten female students among approximately 40–50 students. After having finished primary school, he continued studies in secondary school called "Preah Yukunthor High School" in 1960, and stayed there for four years until 1964 or 1965. The number of students was close to that in the primary school. According to him, after having finished 4-year secondary school, students could continue studies for Bachelor Diploma for two years (BAC I & BAC II), but he only obtained a BAC I diploma.

His teachers in primary and secondary schools were around 30–40 years old, they were all Khmer and mostly male. He said that usually only one teacher was responsible for all subjects in a class. Most of the teachers graduated from High School and held at least BAC II Diploma. Some teachers graduated from universities and majored in education. According to him, all teachers were selected through public examination designed by the ministry of education; the most of his teachers were well prepared and knew mathematics very well.

As far as the curriculum is concerned, the interviewee remembered that from the 1st to 3rd grade, he studied only arithmetic; the topics included simple counting, four arithmetical operations (addition, subtraction, multiplication, and division). In grade 4th, he started learning elementary geometry; in the 5th–6th grades, the geometry curriculum included knowledge of properties of such geometrical figures as circles, squares, triangles, etc. The assignments included word problems and measurements of figures. His memories about it were not very clear; he only remembered that in secondary school mathematic exercises were more difficult and involved more formulas; as for geometry, there were more problems to be solved than simple exercises on drawing figures and recognizing shapes. He remembered that arithmetic and geometry were taught separately, and that the exercises related to geometry were more difficult to understand than those related to arithmetic.

In primary school, only teachers had textbooks and these textbooks were compiled by lecturers themselves on the basis of various sources (mainly from France). However, these textbooks corresponded to the requirements of the mathematics curriculum introduced by the Ministry of Education. The guidelines related to mathematics teaching were sent to the principals of all schools. In secondary school, students were required to have textbooks; if they were not able to buy textbooks, it was possible to rent them from school. In secondary school, for each subject there was a special textbook (for example, one for arithmetic, one for geometry…); the interviewee was not so sure about the primary school.

In secondary school, most of mathematics textbooks were in French; the interviewee did not remember if there were any textbooks in French in primary school, but the instruction language was Khmer. In primary school, starting from the 1st or 2nd grade, teachers explained the mathematical terms and the symbols in Khmer. Students were required to memorize all the mathematic symbols; this, according to the interviewee, was not too difficult. Starting from grade 3, French language was introduced to students, and they had to learn mathematics in both Khmer and French. Some terms were easy to understand because teachers firstly explained them in Khmer before using them in French. But some terms did not have Khmer equivalents, and only their French versions were used; they were difficult to understand.

When asked about the counting instruments, the interviewee replied that counting rods were the most common tool in the 1st and 2nd grades. Additionally, each student was required to have a small blackboard to write with chalk, so he or she used it for mathematics exercises; notebooks were also used. He said that students rarely used tools (even to perform measurement). However, such instruments as rulers, compasses etc. were present in the classroom; they were probably introduced during French colonial era, and most likely not designed by the Khmers. Teachers did explain how to use these tools, and then students practiced using the tools by themselves.

According to Mr. A., teachers provided good explanations before letting students practice and do exercises by themselves. For example, a teacher firstly showed to students how to solve problems before giving them further exercises to practice. In the case when a student requested additional explanation, most of the time teachers would offer it and provide more details. However, it depended on the personality of teacher; some teachers wouldn't answer students' questions, and even punished them for asking.

Every student was required to memorize multiplication table and some formulas. In order to check students' understanding, teachers conducted oral quizzes individually or in groups every time, and sometimes wrote some mathematical exercises on the class blackboard and asked the students to do these exercises on their small boards, or in notebooks; usually these exercises were related to arithmetic.

As far as the tests and exams are concerned, Mr. A. did not remember well how often they were conducted; he was only certain that there were exams every semester as well as finals. The test exercises were written on the blackboard; the students were required to write the exercises and design solutions of problems in their notebooks, or on sheets of paper later submitted for evaluation.

The homework mostly contained arithmetic exercises such as calculations and solving text problems; the interviewee rarely mentioned geometry. The homework was done mainly in written form. To check the homework, the teacher would collect students' notebooks and check them. The students who had high scores would write the answers and solutions on the blackboard.

3.3 *Interview 3*

The third interviewee, a 65-year old female (born in 1952, below referred to as Mrs. PB), studied in primary school from 1958 to 1963. She mentioned that during her last year at primary school (grade 7), the new educational program emphasizing the importance of Khmer language was established. However, at that time the use of French language, especial in Mathematics class, was still in practice.

She studied in Botom Reachea primary school (from 1958 to 1962), located in Phnom Penh. There were around 30–40 students in her class, most of them (ca.

65%) were boys. At that time, in her primary school, students were around 6–14 years old. After that she went to Sonthormok secondary school (in Phnom Penh) for just one year; she was not sure about the ratio of male and female students, but mentioned that the number of male students was higher than female ones.

According to her, almost all her teachers in primary and secondary school were well educated Khmer males around 30–40 years old; her teachers were all males, and most of them were Khmers. She was not sure about her teachers' background, but claimed that her teachers were knowledgeable and well-prepared.

Mrs. PB said that from the first grade until third grade students did not study much about arithmetic, they just studied about counting numbers and four basic operations. They also learned Khmer and French terms for measure units, and how to perform measurements, e.g., how to measure weights, lengths, and volumes. As for geometry, the students learned about different types of shapes such as triangles, rectangles, squares, etc. Arithmetic and geometry were studied separately.

From Grade 4th to upper class, mathematics was harder, she said. Students mainly focused on problem solving such as calculating profits, spendings, lengths, weights, etc. Also, they learned about fractions, fraction, and problems that involved many formulas.

Two subjects of mathematics were studied separately, algebra (including arithmetic), and geometry. These subjects were studied separately and equally in term of schedule; however, she mentioned that geometry was given more attention than algebra.

She said that those two main topics were considered to be equally difficult. But when those exercises (algebra and geometry) involved logical problems and the solutions did not follow directly from the formulas provided, they would be considered the hardest mathematic exercises, because students needed not only to remember formulas and know how to use them, but they also to be smart enough to read the problem and find solutions.

She claimed that there were no specific mathematic textbooks for each grade, but there were lecture notes and course outlines for every class and every year. Those course outlines were prepared and distributed by the ministry of education. Those course outlines told the teachers how many lessons to teach each semester, and what were these lessons had to be focused on. Moreover, a teacher might have prepared his or her own lecture notes. Some lecture notes were approved by the ministry of education and available on the market. Those lecture notes included teaching methodology and additional exercises. Those teachers who wanted to prepare their own lecture notes, needed to find relevant materials which were mostly taken from French mathematics textbooks related to the lessons described in the course outlines provided officially. The materials provided by the teachers in classrooms were supposed to be used to prepare the students for their final exams; the tests for the final examinations were prepared by the ministry, and not by the teachers.

The students mainly followed and noted down the instructions in their notebooks later used as their textbooks. Furthermore, they could find more textbook materials in school library or in bookstores. Those mathematics textbooks were available in both Khmer and French languages (but these two languages were never used simultaneously in the same textbook), but most students used French books since in class they practiced using French in solving mathematic problems. Most of the mathematics materials used by the students were books for mathematics practice with answer keys.

Some mathematical terms studied from the first to third grade were studied both in Khmer and French, and those used in Khmer (for example, such terms as "to multiply", "to divide", "parallel", "perpendicular", and so on) were easy to understand since they were used commonly in daily life; however, the terms studied at upper grades were difficult to use if they were in Khmer because those terms were advanced and not commonly used. Therefore, some terms such as "Logic", "Vector", … were used mainly in French, which would be more natural. Moreover, the students studied French since the first grade and some terms used in French were shorter and easier to understand than Khmer terms. Some advanced terms translated from French were difficult to understand since they sounded unnatural and contained many Khmer consonants; for instance, "logic" was rendered as តក្កវិជ្ជា (*tak kak vichea*).

When new mathematical terms were introduced, teachers always wrote down their definitions (in Khmer) on the blackboard and explained them to students, as far as their correct pronunciation and usages were concerned. Some of those key terms were also explained in textbooks, but teachers' explanations played more important role. When introducing some new difficult terms, teachers firstly told students to use them in French and only after the students understood the concepts of those terms, teachers translated them in Khmer and explained more. On the other hand, some teachers may have chosen the vice versa way by translating the terms in Khmer first and explaining the usages before students became familiar with those terms.

Interestingly, the interviewee claimed that her last grade (1963–64) at primary school was the year of campaign for enhancing the use of Khmer language which meant that all subjects studied in each grade must have been instructed in Khmer; however, in mathematics class, some French terms were still used, but when students wrote down their solutions, they used mainly Khmer, not like before when they mainly used French with only little Khmer. After that new program was established, the interviewee found some terms difficult to understand and remember; it was like she learned new terms in Khmer that she had never known before.

As far as the counting instruments are concerned, in the first grade, some tools (which were commonly used and easy to find, such as chopsticks, shells, stones, candies, pencils, etc.) were employed as counting instruments. These tools, used to help students learn counting the numbers superior to 20, were introduced in a relatively short period (within only two or three months). After students became familiar with those numbers, they were trained to count numbers without those tools, mainly using their memory. Moreover, those tools also assisted students to solve elementary problems such as addition and subtraction. However, those tools were used only for relatively short time since students were required to solve these problems without using tools.

Furthermore, from the first grade, when students studied about measurement, some other tools were introduced such as rulers (different types of rulers for different purposes), scales, tools to measure volumes and weights of liquids, etc. The interviewee claimed that some types of rulers used were Khmer instruments (because they had Khmer measurement units written on them), and the other had French measurement units written on them (they still look familiar since they were commonly used every day). For scales, there were many types of scales, among those, Khmer scales also introduced to study. There were many other Khmer tools or measurement units to measure things (not only their weights, but also the quantity of those things), such as កន្ទាំង (*kan'teng*), កញ្ជើរ (*kenh chur*), ល្អី (*l'ei*) (types of Khmer baskets), etc. There were also Khmer measurement units or tools originally based on the use of hand as tool or unit, such as មួយជីប (*mouy chib*), មួយថ្នាំង (*mouy tneang*), មួយត្របក (*mouy tror bor*). The interviewee claimed that those Khmer tools were very commonly used in measuring things (weights and amounts) in everyday life. They were even more popular than using scales (foreign scales were rarely used). Even most of Khmer measurement units and tools focused on measures or amounts, in class, students also learned how to convert those Khmer measurement units into French measurement units such as *kilograms*, etc.

The interviewee also mentioned Chinese counting (or calculating) instruments, such as Chinese beads abacus. She said that those tools were introduced to students just to let them know those tools' names and functions, but the instructors did not let the students use those tools in class.

When introducing instruments, teacher played a key role in giving explanation. Often teachers would invite two students to do measurement by using those tools, and let the others do it after them or observe the progress. The interviewee did not mention the instruction related to the use of measurements in the textbooks, she only claimed that her teacher was good at explaining how to use those things.

As far as the instruction procedure is concerned, before giving instruction, the teacher firstly wrote down the definition of each key term. If students still did not understand them, the teacher might have explained again. The interviewee claimed clearly that teachers at that time were strict, but they gave clear instructions rather than just giving orders. Sometimes, the situation was different, when students asked the same questions several times; in such a case, the teacher may have blamed them or thought that those students did not pay enough attention. As a result, teachers may have ignored their questions or gave a strict order to them to stop and pay more attention next time.

The interviewee's teacher required students to remember all formulas written on the board. Moreover, some teachers also required students to memorize everything (tables, formulas, some key terms, and measurement instruments and units) that they have studied so far. They may have checked the students' notebooks during the oral quiz.

Students were not only required to memorize the lessons, but also to solve the problems correctly in a short time. For example, after giving problems and explaining their solution to students, a teacher would have changed some concept or main idea of the problem; however, this new problem also used the same formula. In other words, the teacher just formulated the provided problem to measure students' understanding of the way the provided formula was supposed to be used. After creating a new problem on the basis of the original one, the teacher told students to do it and they needed to finish that problem as soon as possible. Those students who could solve the problem correctly and submitted to teacher earlier, would obtain higher grades.

Moreover, teacher may have called out some student's name and ask him or her to do exercises on the blackboard without the help from any means. The student just had to use his or her memory to do it. There were also some oral quizzes conducted individually: a student was called to come to teacher's table and teacher asked him or her some questions (about some arithmetical tables, conversion of measurement units, etc.). Quizzes of this type would happen during every mathematics class, usually before starting a new lesson. Usually students were asked to solve some problem related to the formulas they had studied. This kind of quiz appeared in both arithmetic and geometry classes.

Students had monthly tests, semester exams, and final exams. The monthly tests were designed by teachers, semester exams were designed by the head of school who combined all requested exam papers from teachers, and the final exam was prepared by the ministry. The tests most of the time were written. Teacher wrote the test problems on the board then students wrote down their solution and submitted their papers.

The interviewee did not mention clearly what were the subjects that appeared on the tests. But she said that both arithmetic and geometry appeared equally, the main focus was on problem solving, fractions, probability, long problems that involved measuring lengths and perimeters. She said that since mathematics was considered the most important subject, the students had to study mathematic more than other subject according to the schedule. She was not sure, but said that mathematics classes were taught 6–8 h per week.

As far as the homework was concerned, the interviewee claimed that it did not have a clearly established pattern. Solution of long problems related to geometry and involving difficult geometric formulas would be more common than arithmetical problems. Teacher might have written the exercises on the board, and the students wrote the exercises in their notebooks or on pieces of paper. Sometimes, homework was about memorization of formulas and theorems. For the interviewee, long problems were difficult, because sometimes they involved not only formulas, but also logic.

Teachers would check the homework in each student's notebook that they collected and put on her/his table. If there were many exercises involving the same formula, the teacher would present only one solution on the blackboard, or even sometimes the teacher only gave the answer without solution.

3.4 Interview 4

A lot of information was obtained from the interviewee, Mr. SPS. He was born in 1945 in Takeo Province. He attended Wat Phnom Chor Chok primary school, located in the Northern part of Chiso Mountain range in Takeo province, in 1954. At the age of 16, being unable to pass high school entrance exam, he decided to become a monk. He was a monk for 14 years, during which he was able to learn more about Khmer Literature and other subjects at Puthikak Vichealei Prey Sarameret High School in Phnom Penh where students mainly focused on Buddhist texts written in old Khmer and Pali languages. Students (the young monks) had to enter another program called Adult Education at the same High School to study other subjects (similar to those studied in lay public high schools), namely Biology, Mathematics, Geography, Physics, etc. After having finished the Adult Education program, the monks could sit the national teacher entrance examination, and become teachers. After the fall of the Khmer Rouge Regime, Mr. SPS was promoted to be a Village governor at his hometown. In the 1980s, he became the Vice-Director and teacher of Khmer language in class for foreigners in Royal University in Phnom Penh. In 1987, he was sent to Germany to teach Khmer language for six years. At the same time, he also completed his Ph.D. dissertation in the field of Khmer Literature which he defended in Germany. From 1993 on, he taught Graduate programs of Pali, Sanskrit and Khmer Literature to Cambodian youth, and resumed his teaching of Khmer to foreigners as well. He is now a member of the Khmer Literature Studies Division of the Royal Academy of Cambodia.

When Mr. SPS went to primary school, there were around 25 students (among them 5–8 girls) in each class. In secondary school, there were around 35–40 students (about 10 girls) in each class. He remembers that there was only a small number of students who finished secondary schools due to the strictness of the examinations, and their age was quite advanced (most of them were old enough, they were in their 20s, and many of them had to work to support their families).

As far as the information about mathematics instructors in primary and secondary schools is concerned, Mr. SPS remembers that the teachers in primary school were all Khmer males who were around 40–50 years old. Due to the lack of teachers in his province, there were some monks who taught all the subjects in the first and second grades, while the classes in the upper grades would be taught by more professional teachers who also taught every subject. Moreover, there were some schools that had French teachers who taught in the upper classes in primary schools, and Khmer teachers who taught only in the first and second grades, since their teaching abilities were limited. Each teacher in primary school was supposed to finish at least Primary school and to have passed the national entrance exam for secondary school. The interviewee added that the certificate of Primary Education was so valuable at that time, that whoever finished primary schools could work as teacher, or work for tax authorities, or have other public positions. To continue studies in high school, students were required to pass another national exam.

Mr. SPS also provided some general information about public high schools even though he studied in a Buddhist high school. Each teacher was supposed to teach only one subject, but it was possible that he could teach that subject in several classes of different levels. The teachers were all male, and there were also French teachers teaching some subjects. There were two national examinations, one in the grade 9 (that is, the last year of secondary school) and another in the 13th (in the last year of high school). Unfortunately, he could not give more specific information about the educational background of the teachers.

He added that each teacher was well-prepared for classes and gave clear explanations even if he usually taught many subjects at the same time in primary schools. Noticeably, all teachers were around 40–50 years old, because they had finished their graduation late, and it was also because of the lack of teachers during that time.

As far as the studied topics are concerned, in primary school Mr. SPS studied arithmetical operations (addition, subtraction, multiplication, and division) with integer and fractional numbers. He also studied geometric formulas and did exercises on finding areas/volumes and perimeters/surface areas of squares, rectangles, triangles, cubes, etc. For example, students were asked to solve the problem involving the calculation of the volume of a pond, of the area of a field, dimensions of a house, etc. Moreover, he studied various aspects of measurements, such as measurement units of capacity, length, and weight, as well as methods for calculation of those measurements. All subjects in mathematics were studied separately, but the schedule of studies varied every week depending on the teacher. He claimed that the classes were more focused on arithmetic (but he was not very sure).

In secondary and high school, he did not study much about mathematics since he was a monk, the studies of Khmer and Pali Languages were the most important. However, when he joined the Adult Education Program, he studied operations with fractions, solution of equations using modern algebraic symbols, etc. During the interview he rarely mentioned geometry. Each mathematical subject was studied separately and students were required to have different books for those subjects. He did not make clear comments concerning the difficulty of each subject.

As far as the textbooks are concerned, in his first year at primary school, the number of textbooks was not sufficient for both teachers and students. Therefore, teachers of the same school had to share one main textbook. Sometimes, the teacher compiled his own mathematic textbooks from different sources (especially, French books) based on the established curriculum. As for students, they had to note down everything written on the blackboard. Fortunately, the next year, the school had enough textbooks for teachers and students that could be borrowed from the library. Even when each teacher had a textbook, they still had to have their own lesson plans and compiled additional study materials. Interestingly, each year, there would be an expected visit of the inspectors (Khmer and French) from the Ministry of Education. The inspectors came to the class and observed the instruction, as well as checked the lesson plan books of the teacher. If a teacher did not have a lesson plan or did not prepare it well, his teaching evaluation would be considered failed. Additionally, the lesson plan was also used to evaluate the effectiveness of teaching methods by comparing students' understanding and the contents in the lesson plan. When there were some errors, the teacher would be suggested to change the contents, or he would do it by himself when he noticed the weakness, but the change was to be made the next academic year.

There were no special textbooks for each mathematical subject; instead, the mathematics textbook was divided into two parts, one devoted to arithmetic and one, to geometry. The mathematics textbooks were mainly in Khmer for the second and third grades, but for upper classes, French was used. Those Khmer textbooks were translated and/or compiled by the Ministry of Education.

As for secondary school, Mr. SPS provided similar information as the previous interviewees. However, he did say something interesting about the preparation of the lesson plan for mathematics classes. As each teacher in secondary school was responsible for only one subject, mathematics teacher had more time to focus on preparation of his lesson plan and the study materials. Therefore, when his own textbook was well-compiled and written, he would sell the right of publishing his mathematics textbook to a publishing company. His books would be useful for students who may have used it for additional practice or for remembering difficult formulas. The main textbooks were all in French, or students even used French textbooks, because they wanted to follow the French standards since at that time there were many students who applied for scholarship to go to study in France, and in order to fit the French criteria, they had to use French textbooks.

Concerning the mathematical terms, Mr. SPS provided the same information as the previous interviewees. The explanation of the mathematical key terms were mainly offered by teachers, and not taken from textbooks. The interesting point that he mentioned was about "Khmerization" in the late 1950s. He said that when he studied in the Buddhist high school, Khmer language was used mainly even in mathematics class, but in case when there was no suitable translation, the technical terms were kept the same as in French textbooks (there were also some words written in Khmer that were phonetic transcriptions of French pronunciations of the terms). When the authorities noticed that the monks who studied the mathematics in Khmer had equal or even better understanding than the students who studied at normal public schools, they suggested to implement the Khmerization in primary schools for the first step. He said that the monks who always studied all subjects in Khmer language did not find it too difficult to use Khmer in mathematics (except, probably, for the technical terms that could not be translated directly to Khmer). However, for the students who were accustomed to French, the change to Khmer was very difficult, since they had to use a large number of Khmer words instead of familiar French technical words.

When talking about counting instruments and other tools used for instruction, Mr. SPS provided information similar to that of the previous interviewees. But he also mentioned the method of counting that his fellow villagers used: they made small boards (of wood or stone) to write with chalk (also made from mud from pond). He added that functions and usages of other instruments used in classes were explained by teachers. However, students had little chance to operate with those instruments (especially for measurements) since they were not available to all students; only the teacher could use them in class, and sometimes even the teacher did not use those instruments to explain the lessons. Another interesting point about this calculating tool is that it was not used in the public primary or secondary schools, but was used in the Buddhist schools for monks to teach math. The instrument was similar to Chinese traditional calculating tool (beads abacus). The interviewee claimed that since the public schools were influenced by French style of learning, they did not use that tool, but the monks still used it as the traditional method of calculations. Unfortunately, he did not say at which grade they stopped using the tools.

As far as the instruction procedure is concerned, Mr. SPS was not sure about the schedule and time allocated for each subject of mathematics, he only said that in the last classes of the primary school mathematical subjects were studied 4–5 h per week (five days a week, no classes on Thursday and Sunday). In general, teachers provided good explanations, but because of the traditional strict discipline, students often did not dare to ask teacher any questions; instead they would ask their classmates. It was also because they were afraid that the teacher would blame them when they asked questions by saying something like "Did not you listen to my explanations?"! So, every time when the teacher asked "Do you understand?" the students would say "Yes". Anyways, some teachers, when asked, would try to explain again and again. Additionally, the interviewee said that some of his classmates could not understand the exercises and their mathematics was always poor, and the teacher could not help them improve their math. The interviewee thought that it was so because of the limited teaching techniques possessed by the teachers.

Students were required to memorize everything in the lesson. Before continuing a new lesson teacher would ask any student to recite what had been learned during the previous session. The teachers usually gave an oral quiz during the class because it helped to check students' understanding, but the oral quiz was not formal and weighted less than written quiz or test. The bad part was that students sometimes did not understand the lesson clearly, and they just tried their best to memorize everything to avoid punishments which may have followed if they failed the test. Most of the time, teacher himself presented the solution if the exercise was difficult and many students did it wrong.

The students had two big tests per year and many small quizzes. The test or quiz were about reciting a table or geometric formulas, or doing small exercises on calculation, or solving a text problem. In the big test, exercises of all types would appear in written form. The interviewee did not talk about exercises of this kind in secondary school.

When asked about homework, Mr. SPS gave information very similar to that of other interviewees; he also mentioned that homework was always about solving problems and memorization.

3.5 Interview 5

The name of the fifth interviewee will be abbreviated hereafter as LSC. She was born in 1950 in Kandal Province. In the late 1969, after finished High School, she continued her study (Bachelor of English) at Royal University of Phnom Penh for four years and, at the same time, while studying English in the second year, she also studied teaching skills (for two years) at National Institute of Education (teacher college). During her 4th year at university, she worked as an English teacher at Beoung Trorbeak High School (the high school she used to study at). In the early 1975, she got scholarship to study Teaching English as Second Language at Curtin University (Australia) for three years, and continued to study library management there until 1990s. In 1990, she went to Cambodia to participate in a workshop and to train librarians at National Library. In 2003, she moved to Cambodia and has been working at Royal University of Law and Economics as a library manager until now.

From the first to the third grades (1956–59), LSC studied at Chak Ang Re Kroam primary school in Kandal Province (now in Phnom Penh district). When asked, she was not sure about the number of students in class; she only said that the numbers of male and female students were almost equal. In 1960, she moved to study at Khmer English High School (it was named as High School, but it provided primary education starting from the fourth grade, as well as secondary and high school education) from the 4th grade until her graduation from high school.

This school was created in 1960, and had only two classes (4th and 5th grades) with 45 students in each class (the numbers of female and male students were equal). It had no permanent location since it was the first institution for teaching General education and English with a relatively small number of students. Therefore, for the first year, it shared the building with Decho Dom Den primary school (Phnom Penh). For the second year (1961), due to the increase of students (90 students registered annually), the school moved to a bigger place and shared the premises with Sisowath High School. In 1965, the school moved to another place and shared the building with Beoung Trorbeak High School; during that year, a big building was built and the school was located there permanently.

As for the mathematics instructors in primary and secondary schools, the interviewee said that during her primary education (in both schools), her teachers were women (they had less male and more female teachers, ethnic Khmers, aged around 30–40). She said that there were more female teachers at primary school because teaching was considered a very good job for females, and they tended to have a better relationship with young students. Most of the teachers had finished Primary schools, trained in teaching skills at the National Institute of Education for two years, and underwent a national examination to become teachers at primary schools. She added that it was rare for teachers in 1st–3rd grades to use French or English.

For secondary school teachers, there were both male and female teachers responsible for one subject in several classes (usually they were Khmers whose age was around 30–40), but the teachers who taught English were all males. In the high school, teachers were mainly males. In high school, teachers were mostly males who graduated from overseas universities. Some of them were trained as mathematicians (not mathematic teachers!), and could use both English and French, but they did not have teaching skills.

As for the topics studied, the interviewee provided the same information as the previous ones; however, she did not give clear information about the difficulties she experienced with any subjects, she just said that if one put enough effort to study, one could catch up with what the teacher explained. She claimed that her teachers, especially in primary school, were all well prepared. During vacation, teachers would prepare their lesson plans based on the established curriculum sent from the Ministry of education; in this way, even when one teacher had to teach all subjects, he or she still managed it well.

As far as the textbooks are concerned, LSC said that during her years in primary schools (both schools), there were not enough textbooks provided to students. But teachers had a textbook for every subject. When in class, students had to note down everything written on the black board. She claimed that it was not common for students to buy textbooks from bookstore, because they just wanted to learn what they were instructed in the class. With some uncertainty she said that the textbooks used by teachers were all in Khmer, because the instruction was conducted mostly in Khmer, except the language classes.

In secondary school, students still did not have mathematics textbooks in English; instead, they followed teacher's instructions and lessons, or they could buy French mathematics textbooks to practice at home since the school also used the same textbooks.

The Khmer English High School system had the same curriculum and textbooks, except that the teaching of all subjects was conducted in English (for high school education only). Since there were no mathematics textbooks in English, the teachers had to use French textbooks and translated the lessons from them into English for teaching in class. The interviewee stated that there were no institutions that could provide translations of mathematics textbooks into English for the school. She found it difficult to find mathematic books to practice, any time when she wanted to have additional materials, she always bought French mathematics textbooks.

When asked about mathematical terms, the interviewee said that during her years in primary school, since there were no textbooks for students, she was not sure about the mathematical terms. She claimed that the used terms were all in Khmer. However, when mentioned about the term ប្រភាគ (*bror pheak*) (which in English means "fractions"), she said that there was no such Khmer term used in her classroom, but she did not remember what term was used instead. She continued that some terms were kept in French since there were no suitable Khmer translations for them. From the 4th grade onwards, the students started to have general English class, but other subjects were in Khmer. Starting from High School (the four last years of high school), they studied every subject in English. During mathematics class, the teacher explained terms (which were not translated), and students noted down the explanation because they did not have textbooks. When asked about English-Khmer or English-French dictionary of mathematical terms, the interviewee said that there was no such a dictionary, and even if it existed, she did not think it could have been useful for her, because she preferred the explanations of her teacher and thought that studying on the basis of the lessons taught in the class would have been enough for her. In other words, she rarely bought textbooks (which were in French and used in the standard curriculum) , or any additional mathematics books (in French as well), because learning the terms and doing the exercises taught in class were clear and satisfactory for her.

When asked about the counting instruments and other tools used in teaching, she gave almost exactly the same answer as the previous interviewees. She added that it was rare to use instruments in class in Primary, Secondary, or High School, and that the teachers chose to give more exercises to practice computations mentally and without instruments, and did not give enough explanations concerning the instruments or exercises involving operations with them.

As far as the instruction procedure is concerned, the interviewee said that the teachers of primary school tended to give orders without offering precise demonstrations. Teacher rarely repeated their instructions, and the students needed to pay attention in class. Teachers were very strict and demanded good discipline; therefore, students did not dare to ask any question, they always answered "yes" whenever the teacher asked "Do you understand?".

In secondary and high school, teachers were not so strict, they gave explanations as many times as the students asked questions. Students were encouraged to ask questions, and they even dared to challenge teachers whenever they made mistakes in teaching. Usually, teacher used a blackboard to write the lessons and explain what he had written, then students recorded everything. At the end of the class she might have asked students to do some exercises, or drills, or small written quizzes; the students were doing them on the small boards, in notebooks, or on classroom blackboard.

In primary school, they did not have tests each month and at the end of semester; instead their ability was evaluated by the ongoing assessment scores, but they had one national exam to enter secondary school. The tests were in written form (the interviewee did not remember the contents). In secondary and high school, they had exams every month, every semester, and every year, which were all in written form. In high school, they had exams in English on all subjects, but the contents of the test were exactly the same as in other schools where the exams were in French.

The interviewee did not remember the timetable of subjects she studied; she just said that in primary school, the number of hours allotted for each subject was the same. In secondary school, there were more lessons on mathematics than on other subjects, and the teachers also needed to make more effort and time to explain mathematics to students. She provided the same information about homework as the previous interviewees.

3.6 Interview 6

The sixth interviewee will be referred hereafter as SG; she was born in 1947, in Kandal Province. She finished her study at secondary school in 1964, and continued to study teaching skills for two years at National Institute of Education, Kandal Province. From 1968 to 1975, she taught at Chak Ang Rae Krom primary school. After Pol Pot regime, she started to teach in 1980 at Chak Ang Rae Ler primary school until 2007. Now, she is retired.

She started her schooling in 1953 at Chak Ang Rae primary school, Kandal Province. It was the first year after that school was established, and there were around 30–40 students (mostly male) whose age varied (the youngest was six, the oldest were 15–16). There was only one classroom without chairs or tables, but there was one big blackboard for teacher. The school was governed by only one person (a male in his 30s), who was a teacher, an administrator, and a principal at the same time. After a few years, two more teachers were hired apparently because to the increase of students.

Later the interviewee studied at Preah Norodom High School in Phnom Penh (the school's purpose was to encourage young females to play role in education as teachers and students, therefore, there were more female students and teachers than male ones). She claimed that most of her teachers were females, both Khmer and French, except for one male mathematics teacher who taught her in the first year of secondary school.

When asked about her mathematics instructors in primary and secondary schools, she answered that in primary school, her teachers were all Khmer nationals in their 30s. She was not sure about their educational background, but added that to become a primary teacher, one was supposed to pass national exam of primary education, then pass an entrance exam to study teaching skills, and go through a teacher training program that lasted for at least two years. After that they had to have another national exam that put their ability in range, and those who were the best had the privilege to choose where to teach. Those who did not pass this exam well enough had no right to choose the place and were not provided with the "basic salary" (the teacher's salary was a combination of the basic salary and monthly wages). During their studies at the Teacher College, the future teachers studied psychology, teaching techniques, curriculum design, and other topics. If a primary school teacher wanted to become a secondary school teacher, she or he had to teach in primary school for at least five years and then pass a special examination (for specific subject).

Starting from 1960, there was no national examination for primary education, there was only entrance exam for secondary school. In other words, students who finished primary school were not allowed to become teachers; they had to pursue their education further.

During the first and second year of her primary school, there was only one teacher, but he needed to manage to teach two classes at the same time. However, she said her teacher was well-prepared and could teach all subjects.

Mrs. SG said that in her secondary and high schools, most of teachers were female; only two teachers were male, they taught mathematics and physics. Most of the teachers were French, except for the Khmer Literature Studies teachers. The interviewee thinks that those French teachers, especially, mathematics teachers, were skillful and obtained good education in France.

When asked about topics, the interviewee said that from the first to third grade, the students spent more time to learn about reading and writing than to learn counting from 1 to 1000 and four arithmetical operations with these numbers. From the 4th grade onwards they started learning about Khmer and western measurements system, geometry, and some algebra. They studied each topic separately, e.g. the class of algebra was on Monday, Geometry on Wednesday, Arithmetic on Friday, etc.; the timetable was designed by each teacher individually. The interviewee said that she did not enjoy mathematics, especially geometry (because of more complex wording and drawing shapes). She also added that Geometry was the hardest topic to study because it was abstract. The students understood what they were doing when drawing a line, circle, or square, but they could not formulate concepts or definitions, e.g., what is geometry? What is a square? What is a triangle?

In secondary and high school, each mathematical topic was also studied separately, but the time table needed to be agreed upon by all teachers in the school. The topics were the same as in primary school, but the classes were more difficult, and longer. Noticeably, the number of hours for mathematics classes increased comparing to primary school. For the interviewee, geometry still was the most difficult.

When asked about textbooks, she said that in the primary school, only teachers had textbooks, yet students could buy textbooks from book stores if they could afford them. She said that she did not buy textbooks, because she had those textbooks from her elder brothers. In other words, one and the same textbook for a subject taught at a certain grade could be used for many years. According to Mrs. SG, one mathematics textbook contained parts devoted to different topics such as arithmetic, algebra, and geometry. Moreover, starting from the 5th grade, they studied mathematics in French, but she was not sure if the textbooks used by teacher were in French or not.

In secondary school, there were different textbooks for different branches of mathematics. Students could borrow textbooks from library or buy them from bookstores where some textbooks in Khmer or French could be found; most of textbooks were in French. Again, the textbook on one subject of a certain level could be used for several years. She added that students rarely bought or borrowed textbooks from library, because they thought that teachers' explanations were sufficient; they just needed to note down what was written on the backboard into their notebooks. Moreover, teachers never suggested students to have their own textbooks for self-study.

When answering the question about mathematical terms, the interviewee started by saying that key terms used in primary classroom were mainly in Khmer and were easy to understand since the learners did not study much about math, and only focused on basic arithmetic. In grades 5 and 6, they started to have mathematics class in French, and the terms used in textbooks were all in French. She claimed that those words were not very difficult to understand, because her teacher always explained those terms to students, and she just needed to note down everything explained by her teacher. Noticeably, she stated that there were no textbooks that combined Khmer and French (that is, each textbook was either in Khmer or in French). Moreover, she was not sure whether the terms were translated from a foreign language or not, because the words were commonly used and written in Khmer ("sounded so Khmer"), and the textbooks or the teachers did not inform the students whether the terms were borrowed from other language.

In secondary school, all terms were in French and explained by teachers, because not many students had textbooks. She said that arithmetic, algebra and geometry textbooks were in French and students studied the exercises in those textbooks for exam preparation. When they did not understand the terms or exercises, they could ask their teacher because there were no explanation or answer keys in those textbooks.

When asked about counting instruments and other tools used in teaching, Mrs. SG provided the same information as the other interviewees. As for the instruction procedure, she stated that in primary school, students were expected to be well-disciplined and were required to listen to teachers' orders. For example, teacher gave the students orders to keep silent and pay attention to his explanation, and then he asked students to repeat after him. Students never asked any questions, they preferred asking their friends or siblings at home. Sometimes, one teacher had to teach two classes at the same time; therefore, teacher asked an outstanding student to be a class monitor and control the class like reading out loud the numbers written on the blackboard for other students to repeat, or showing the solution on the blackboard and check if everyone wrote down lesson in their books. When the teacher asked students "Did you understand?" all students would answer "Yes, Sir" even if they did not understand. Memorization was very important; students had to memorize everything. Sometimes, mathematics teacher called out one student and order her or him to revise orally everything that had been learned and written down in her or his notebook. One teacher told the interviewee that "When you memorize everything, even though now you don't get the meaning, you will understand it later." The students usually had exam every three months and final exam at the end of the year. Mathematics exam topics were related to arithmetic and algebra more than geometry.

In secondary school, students were encouraged to ask questions and were allowed to have their own style of learning. She said it was because there were some French teachers who had different style of teaching from Khmer teachers (who were "stricter"). During the first year of secondary school, she studied with Khmer mathematics teacher who was very strict to students; sometimes, he even taught students as in primary school, e.g., asking them to repeat after him and follow his orders. When her Cambodian teacher explained a lesson, all students had to pay attention and listen carefully, and no one tried to note down the lesson in the notebooks until they were told to do so. However, when she studied with French teachers, students were not under such pressure; they could have their own learning patterns as long as they caught up the lesson. As usual, there were monthly exams, end of term exams, and the final exam. She did not remember the contents of these tests.

When asked about the homework, Mrs. SG said that rarely they had written homework on mathematics in primary school; students only needed to remember everything they had learned. In secondary school, homework depended on each lesson, but it was mainly in written form like solving small arithmetic exercises, or long ones. Teacher often collected students' notebooks to check their works.

4 Summary and Directions for Future Research

The presented sample interviews are a part of a pilot study conducted in order to design a final questionnaire for a series of future interviews. These future interviews will make it possible to explore the mathematics instruction during the late colonial period whose study would have been otherwise difficult or impossible because of almost complete loss of textbooks and other teaching materials that occurred during the Khmer Rouge rule. However, even the first results of the preliminary investigation provide a number of interesting and hitherto unknown features of Cambodian colonial and early postcolonial mathematics education. One can notice, first of all, that the mathematics educators at the early stages of schooling insisted on memorization of basic concepts, while in the secondary school classes, apparently influenced by the French educational system, the learners had more freedom and may have developed their own learning strategies. The role of the foreign (in particular, French) mathematical terms was stressed by all interviewees, in one way or another; it appears promising to devote more study to the technical (mathematical) vocabulary of Khmer mathematics textbooks and compare it with contemporaneous French mathematical terms. The transition from the "French" system to the "Khmer" system that started in the late 1950s apparently involved attempts to build "entirely Khmer" system of mathematics instruction; in these circumstances the information obtained from Interviewee 5 is especially interesting, given that she had to study in an English school where the mathematical subjects were taught in Khmer and in some cases French (and not English or American) textbooks were still of certain use.

It would be interesting to conduct a systematic comparison of the results of the interviews with the mathematics curricula officially published in the late 1940s and 1950s. One of the most promising topics for future research appears to be the bilingual mathematics instruction in Cambodian primary and especially secondary schools; it would be especially interesting to study the Khmer mathematical vocabulary that most probably resulted from a close cooperation between French and indigenous educators. The circumstances of this process, to my knowledge, have never been discussed by the historians of mathematics education.

References

Clayton, Thomas (1995). Restriction or resistance? Educational development in French colonial Cambodia. *Education Policy Analysis Archives, 3(19)*, 1–13.
Clayton, Thomas (1998). Building the New Cambodia: Educational destruction and construction under the Khmer Rouge, 1975–1979. *History of Education Quarterly, 38*, 1–16.
Dachary, François (1915). *Premiers éléments d'arithmétique* [First elements of arithmetic]. Hanoï-Haïphong: Imprimerie d'Extrême-Orient.
Delvert, Jean (1956). L'œuvre française d'enseignement au Cambodge. France-Asie: *Revue de Culture et Synthèse Franco-Asiatique, 13*(125), 309–320.
Dy, Sydeth S. (2004). Strategies and policies for basic education in Cambodia: Historical perspectives. *International Education Journal, 5*(1), 90–97.
Hollister, Frederick. J. (1958). Education in Cambodia. *The Clearing House, 33*(4), 209–212.
Journal for Lecturer (in Khmer language) (1949/50–). Vols. 1, 2, 4–5,
Trinh Van Thao (1995). *L'école française en Indochine*. Paris: Karthala.
Osborne, Milton E. (1969). *The French presence in Cochinchina and Cambodia: Rule and response* (1859–1905). Ithaca, N.Y.: Cornell University Press.
Pujol, Victor, & Russier, Henri (1912). *Enseignement du premier degré. Arithmétique. Premier livret* [Teaching of the first degree. Arithmetic. First booklet. (In Khmer language)]. Saïgon: F.-H. Schneider.
Rany, Sam; Ahmad Nurulazam Md Zain; & Hazri Jamil (2012). Cambodia's higher education development in historical perspectives (1863–2012). *International Journal of Learning & Development,* 2(2), 224-241.
Russier, Henri, & Négrignat, Joseph (1912–1913). *Premières notions d'arithmétique, enseignement simultané du calcul et du langage (livret du maître - livret de l'élève)* [First notions of arithmetic, simultaneous teaching of counting and language (booklet of instructor – booklet of learner)], 2 Vols. Saïgon: F.-H. Schneider.

Author Biography

Sethykar SamAn graduated in 2017 from Royal University of Phnom Penh and CamEd Business School. She is interested in the history of education in colonial and post-colonial Cambodia.

Patterns for Studying History of Mathematics: A Case Study of Germany

Gert Schubring

Abstract The history of mathematics teaching and learning presents an already well-established field of interdisciplinary research. What is still necessary, however, is to deepen and improve the methodological approaches to this research. A major issue is constituted by the polarity between the essentially national or regional restriction of the research agendas and the interpretation of the findings in a broader comparative understanding. The polarity and complementarity between the global and the local is used here to unravel general dimensions for the apparently specific cases of mathematics teaching in the German states. Firstly, however, the case of Italy from 1860 is presented to show that a local case is not necessarily a case of a small region and that specific cultural and epistemological visions in a country can effect long-lasting impacts on teaching and learning. Then, a case from pre-modern times in Germany will be analysed, showing the processes driving modernisation. Eventually, the cultural and social divergences between German states in modern times will be used to reveal the role of mathematics education for preparing cultural and technological changes.

Keywords Research methodology · Global versus local · Italy · Germany Prussia · Schleswig-Holstein

1 Introduction

Research on the history of mathematics teaching and learning is not self-contained nor is it autonomous; rather, it constitutes a profoundly interdisciplinary activity. It has major intersections at the very least with history, the history of education, education, sociology, and the history of mathematics. In the first part of this paper,

G. Schubring (✉)
Universität Bielefeld, Bielefeld, Germany
e-mail: gert.schubring@uni-bielefeld.de

G. Schubring
Universidade Federal do Rio de Janeiro, Rio de Janeiro, Brazil

F. Furinghetti and A. Karp (eds.), *Researching the History of Mathematics Education*, ICME-13 Monographs,
https://doi.org/10.1007/978-3-319-68294-5_12

this interdisciplinary character will be analysed and described. In the next part, the conceptual pair of local versus global, recently introduced as a methodological tool in general historiography, will be applied to research into the history of mathematics teaching and learning (while this research field is traditionally characterised by studies on just one nation or culture, which thus explore apparently only local cases). The challenge for this research field is to reveal what is really just of local character and what provides insight into global patterns enabling generalisable results.

To elucidate this constellation, Italy is first presented as an apparently isolated, local case and is contrasted with Germany. Germany, from pre-modern times, did not constitute a homogeneous case regarding education and in particular mathematics teaching; rather, it consisted of a large number of sovereign parts with different educational conceptions and different socio-political structures. Therefore, this seemingly local case proves to reveal various conflicting patterns for mathematics teaching, which already provide clues for generalisable research strategies.

In general, one can state that the field of the history of mathematics teaching and learning is well consolidated; there is no need for sensationalist announcements like a need for rewriting entire national histories. What is necessary, however, is to deepen the understanding of the changes, which have occurred in the course of this history, and the understanding of conditions enabling change. The paper aims to contribute to this task.

This paper is based on earlier research (Schubring 1984, 1991, 2010, 2012), which is used here for elaborating methodological approaches and patterns and consequences for research strategies.

2 Interdisciplinary Research

2.1 The Disciplinary Net

History as science proves to be most relevant to the history of the mathematics education with regard to the methodology of research; one has to rely on its methods for all issues of searching for sources and interpreting and evaluating them, for determining the historical contexts of given situations, for developing methods to elaborate on biographies, and in particular for applying the more recent method of prosopography, that is, collective biographies of relevant groups of persons, which proved fruitful when studying professional profiles of mathematics teachers (Shapin and Thackray 1974).

The history of education, constituting another related discipline, is of the utmost importance for assessing the general development of the educational system(s) in which mathematics education occurs. Together with education itself, it provides the basic patterns of teaching all school subjects and constitutes therefore the major source for dimensions and factors determining, among others, teacher education, school structures, curricular processes and conditions for change.

Clearly, there is an interrelation between the two areas—history of education and history of mathematics education—since results from research on the history of mathematics education can enrich and concretise the history of education or even indicate lacunae of research in the history of education. In general, the history of education is not specifically focused on the content taught, but a subarea has developed there since the 1980s called the history of school disciplines and is particularly prominent in France. One typical example is Chervel's (1988) paper, which launched this conception. Its basic notion is that of "school culture"; school disciplines are studied as expressions of school culture and exclusively in relation to society. Typical disciplines studied within this conception are humanities, mother language, history, geography, and religion. The focus, thus, is on socializing effects —on enculturation. School disciplines are regarded as creations of society; relations to their scientific (academic) disciplines are not approached.

According to this conception, school disciplines are characterised by autonomies, in particular as autonomous with regard to other school disciplines (Chervel 1988, p. 73). However, the history of mathematics education shows that the status and the concrete forms of teaching mathematics depend decisively upon its position within the complex set of all subjects to be taught within a given type of school.

Given the essentially social character of the evolution of educational systems, it is evident that sociology provides numerous approaches and conceptions for researching their history and constitutes therefore another key related discipline. This applies especially to mathematics, which has always functioned at the crossroads between general education and professional training, thus relating its teaching history even to the labour market and career strategies, and thus widening the sociological dimensions implied.

It is quite natural that most research pursued or going on is concentrated on the history within a given nation or a given culture as the history of mathematics teaching and learning first and foremost constitutes a part of the educational history of that country or culture. But in order not to end with a collection of separate, isolated histories without interconnections, one has to establish relations between the different national and/or cultural histories and reveal what is "general" in them and what constitutes, say, cultural, or social, or political peculiarities. Practically all questions of our historical field deserve comparative studies. In a later section, examples from recent research will be analysed.

Thus, comparative history of education constitutes a likewise important tool for references. Since the quality of research results for a given nation is enhanced when reflecting on the generalisable patterns of such specific cases, comparative education provides essential tools for widening the perspectives to international comparisons. Moreover, such comparative issues from mathematics education might enrich research within comparative education, revealing critical open questions there. One example of a study reflecting these relations for the case of Portugal and Brazil is Valente (2012).

At present, methodology of the history as science emphasises the relation between two poles: the local and the global. Both poles have their importance and value, but a meaningful history will emerge when one conceives of both approaches as complimentary ones.

2.2 *Italy—A Local Case?*

That it is not only desirable but even necessary to traverse the borders of one's own country or culture to achieve revealing analyses is shown by the example of the history of Italy after its unification in 1860. Italian historians used to accept whole-heartedly the organisation of mathematics teaching as instituted from 1867 on. At this time, rigour became the key criterion for mathematics teaching, despite the evident failures of this approach (Schubring 2004). Even in the second half of the 20th century these historians have been brought up in the same spirit as inspired the organisation of the second half of the nineteenth century (Vita 1986). Thus, they did not dispose an awareness of the fallacies of rigour. Felix Klein had noted, being disappointed, in 1913 in his diary, regarding the evolution of the international reform movement for mathematics teaching: "Italien ganz ablehnend, seit dem Tod von Vailati" (Schubring 1987, p. 23), that is: "complete rejection [of the reform], since the death of [Giovanni] Vailati".

In fact, Giovanni Vailati, who died quite young in 1909, was one of the rare exceptions among Italian mathematicians who wanted to promote Klein's reform agenda. The extolling of rigour had been the rationale of the then-leading mathematician Luigi Cremona (1830–1903) who is responsible for the 1867 organisation of mathematics teaching.[1] Table 1 is the timetable for the weekly hours for mathematics teaching in the five-year *ginnasio* and the following 3-year *liceo*, both leading exclusively to university studies. Table 1 presents the situation from 1860 to 1865 which reveals the frequent changes, and thus the doubts expressed within educational policy about the value of mathematics teaching.[2]

Table 2 presents the programme elaborated by Cremona in 1867.

Whereas historiography of Italian mathematics teaching had until recently focused only on the *liceo* and thus not shown and discussed the situation in the *ginnasio*, already the first table impressively proves that mathematics constituted there a marginal subject. In contrast to the *ginnasio*, the status of mathematics at the *liceo* ranged somewhat higher, but showed an unstable situation. But what calls, besides this, major attention is Cremona's restriction of mathematics teaching to only three of the eight years of secondary school.

[1]The other members of the committee elaborating the mathematics curriculum for the *ginnasio* and the *liceo* were research mathematicians, too.

[2]Source: the site of the Associazione Subalpina Mathesis, Torino http://www.mathesistorino.it/wordpress/wp-content/uploads/2014/01/VariazioneOrari-GinnasiLicei.pdf.

Table 1 Number of weekly teaching hours for mathematics in unified Italy

School type/grade	1860	1862	1865
Ginnasio I	1	1	1
Ginnasio II	1	1	1
Ginnasio III	1	1	2
Ginnasio IV	3	3	2
Ginnasio V	3		2
Liceo I	8	6	6
Liceo II	0	2	3
Liceo III	3	3	3

Table 2 Weekly hours for mathematics teaching 1867 (Giacardi and Scoth 2014, p. 211)

Ginnasio I	Ginnasio II	Ginnasio III	Ginnasio IV	Ginnasio V	Liceo I	Liceo II	Liceo III
0	0	0	0	5	6	7.5	0

The question can be posed: Why was no teaching of mathematics offered in the first four grades and then a point of accumulation emerged in the fifth grade? The answer is that according to Cremona's conception in the first four grades, students would be not mature enough to learn deductive mathematics—and for him, only deductive mathematics, in particular Euclid's geometry, was of educational value. Only from the fifth grade on, could they be taught such rigorous mathematics! In fact, it was Euclid's book I, which should be taught in that grade (Decreto 1867). Moreover, following the same spirit, an official school mathematics discipline was invented in Italy, I guess unknown in any other country, called *aritmetica razionale*—rational arithmetic, exposing an axiomatic arithmetic (offered from the same fifth grade on). An additional rationale of Cremona had been to integrate mathematics as a likewise classical teaching subject into the secondary Italian schools dominated by veneration for classical humanities (Decreto 1867). For sure, after a few years and numerous bad experiences in school practice, some of these decisions were revised. Mathematics continued, however, to be a marginal subject in the *ginnasio* and in a doubtful position in the *liceo* (Schubring 2004).

Italy, therefore, seems to constitute an isolated case. It is probably more correct to view it as an element at the extreme end of a somewhat continuous scale characterising the educational value of mathematics—from technological and professional exigencies through classical values of humanities.

3 Germany: A Case for Studying Some Global and Local Elements

3.1 *The German Situation*

Normally, a single country would constitute a local case for studying the history of mathematics education and should need additional input from comparative history of education to reveal global aspects and thus enabling deeper insights. Germany, however, presents an exceptional case where local and global poles can be observed to a certain extent, so that the study of this country is able to contribute to general insights. The reason for this rather exceptional position resides in the political and cultural history of this country.

One has to be aware that Germany did not develop as a country with a unified territorial structure. First it was a part of the Frankish Empire, created by the Emperor Charlemagne in the 8th century—later on, within the Holy Roman Empire of German Nation it became ever more disentangled into a collection of more or less sovereign sub-entities. This process of decentralisation was reinforced with the Protestant Reform by the middle of the 16th century. The map in Fig. 1 shows the major states within the German Empire at the eve of the French Revolution.

Given the enormous number of states within this Empire (about 300 by its dissolution in 1806), one might think at first that these states constituted just a particularly numerous collection of local cases. Having become aware, however, of the cultural, religious and political history of the German Empire since the Protestant Reform, one understands that this diversity reveals a certain global structure: a division into Catholic and Protestant states. In fact, these states not only practiced an impressing richness of cultural differences, the Protestant Reform and the Catholic Counter-Reform effected a split into states of Protestant religion and states of Catholic religion (basically a North–South split)—with an immediate outcome for mathematics teaching.

The Humanism movement since the Renaissance had led to the establishment of mathematics professorships in the arts faculties of the universities, and the Protestant universities continued these disciplinary structures. In the Catholic territories, however, the Jesuit order dissolved the arts faculties—and thus also the mathematics professorships—and instituted in their colleges a teaching of mathematics reduced to only a few months in the last grade. Thus, while in other European countries either one of the two institutionalisations of mathematics, can be observed, in the case of the German states it constitutes a directly observable variable (see Schubring 2014).

The only difference that exists between the German Catholic states and the other European Catholic states is that there did not emerge schools for applications or military schools, which would have initiated an alternative impetus for strengthening mathematics (see Schubring 2002).

Thus, the German diversity does not present just "local" cases, incomparable among them in the vast territory of the Holy Empire, but two opposed systems of

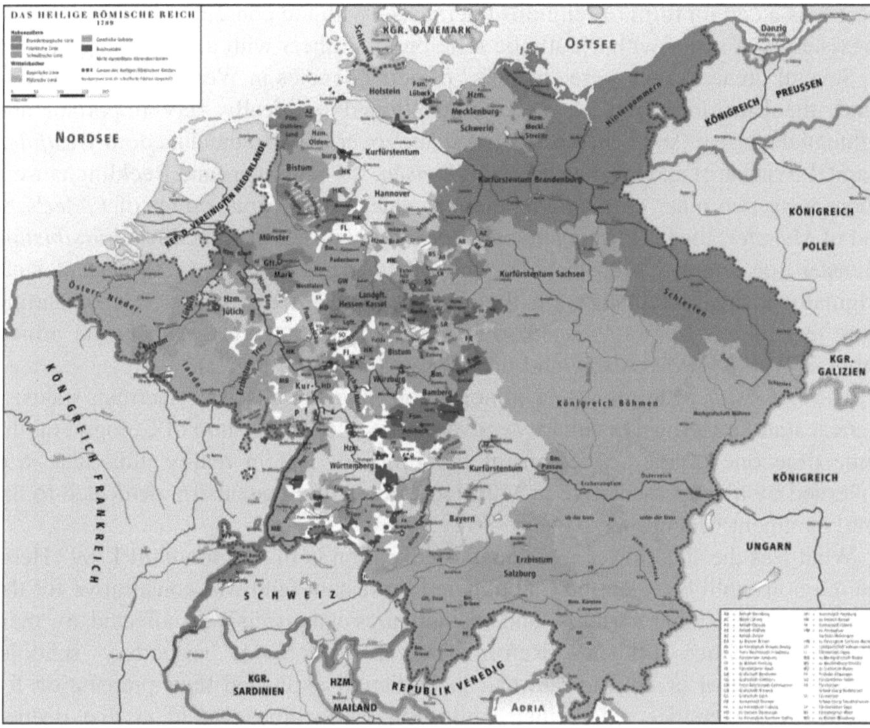

Fig. 1 Map of the territories belonging to the German Empire in 1789 (Kinder 1998a)

Protestant and Catholic political structure and educational conceptions. And for one of these two sides we can ask a research question, which might offer some general insight: what happened after the dissolution of the Jesuit order in 1772: was mathematics teaching strengthened?

3.2 The Case of the Hochstift Münster

There, I should like to evoke an apparently only local case: the *Hochstift* Münster. This *Hochstift* (meaning one of the types of clerical states in Germany, with a bishop or an archbishop exercising the sovereignty of the state), with Münster as its capital, was a part of Westphalia, a region in Northwestern Germany. As a result of the Thirty-Years War (1618–1648), the political landscape of this region of Westphalia represented, as in other regions, a patchwork of numerous sovereign entities—principalities, dukedoms, counties, components of kingdoms, clerical states, independent towns, etc. Without counting even smaller entities, it consisted of 21 "states" within the realm of the Westphalian territory. Besides all its disparities, some structural features can be identified: as another major result of this same war, there was a certain form of religious tolerance—with the consequence of a peaceful coexistence of states with a Catholic religion and others with a Protestant one.[3]

Several greater states were "owners" of dependencies in Westphalia: on the one hand, the Electorate of Cologne, the mentioned politically very important and influential clerical (and hence: catholic) state, governed also the dukedom *Westfalen* (capital Arnsberg) and the *Vest Recklinghausen*, the region around Recklinghausen. There were two other considerable clerical states, the archbishoprics of *Paderborn* and of *Münster*, the latter being named *Hochstift* Münster or sometimes *Fürstbistum* Münster (or simply: Münster). On the other hand, Prussia—a Protestant state originally located exclusively in the east of the German Empire, but expanding more and more—had established some footholds in the West, namely the principality of *Minden-Ravensberg* and the county *Mark* (see Fig. 2).

The *Hochstift* Münster was a minor among the medium size German states; a clerical state (a German peculiarity) governed by the Archbishop of Cologne (at the same time one of the 7 Electors of the Empire). But in reality, this state was governed by his representative in Münster: Franz von Fürstenberg, dedicated to the Enlightenment (Schubring 2010, p. 36).

What was the situation of mathematics education in this region until 1789? Here, the religious split between the Catholic and Protestant faith was constitutive for the divergent educational systems, and mathematics was decisively affected by their differences. Since Luther's Reformation, a system of secondary schools, *Gelehrtenschulen* or *Gymnasien*, had in particular expanded their preparation for studies at universities where their philosophical faculties would at first deepen propaedeutic knowledge, including mathematics, before students went on to professional studies. The focus of teaching at the secondary schools was on classical languages. Mathematics was for a long time, despite Melanchthon's more wide-ranging conception, restricted to reckoning and arithmetic in the lower grades. Only during the eighteenth century did this become gradually supplemented by geometry teaching in the upper grades (see Schubring 2002).

In the Catholic territories, the education system was dominated by the Jesuits and by their Ratio Studiorum, which defined in a uniform manner the structure and the contents of teaching. The teaching of the former faculties of arts had been supplanted by the new colleges, which copied the model of Gymnasien, but radicalised the monopoly of language teaching. Only in the last grades, the philosophical classes—extending, contrary to the preceding grades, over two, sometimes three years—the teaching of philosophy had subdivisions, according to the Jesuit vision of Aristotle's philosophy: in particular ethics and physics. And as a part of physics, a few months of mathematics teaching were prescribed, of a very elementary level, due to the lack of earlier teaching (Schubring 2014, pp. 133–135).

---

[3]The origin of the Thirty-Years War was the intention to establish dominance of Catholic religion in Germany. Its outcome, however, as signed in 1648 by the European powers in the famous *Westfälischer Frieden* (Westphalian Peace), was to establish this coexistence (Kinder 1998a).

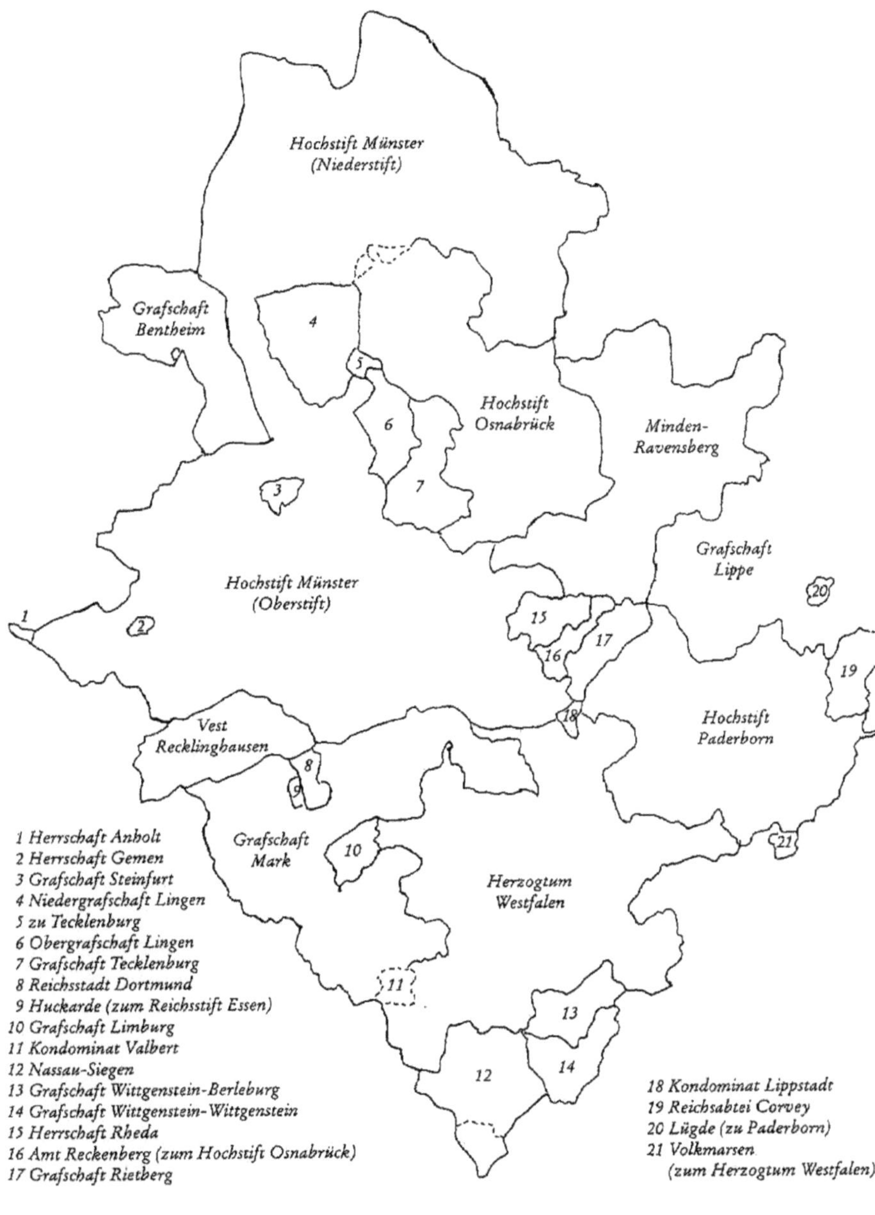

Fig. 2 The political landscape of Westphalia in 1789 (Schubring 2010, p. 72)

The expulsion (from 1759 on) and dissolution (in 1772/73) of the Jesuit order prompted the majority of Catholic states to undertake educational reforms. In general, this was without lasting effects, due to the still unchanged feudal structures of society; there was, however, one state, of medium size, which had already earlier initiated profound reforms independently, following the overall anti-Jesuit policy due to the Enlightenment. This was the clerical archbishopric *Hochstift Münster* where the enlightened governor Franz von Fürstenberg had effected structural and curricular reforms since 1763. Regarding the college/*Gymnasium*, he replaced the existing curriculum, absolutely dominated by the classical languages, with a multiple-subject one, introducing for the first time mathematics as one of the major subjects, to be taught in each grade. Its rationale was not a utilitarian one but rather the ability ascribed to it to promote logical thinking. It was Fürstenberg himself who trained the first mathematics teacher to realise the new curriculum. This teacher, Caspar Zumkley, even became later the director of the highly renowned *Gymnasium* in Münster. And he turned out to be a prolific author of the mathematics textbooks for this new function. The new curriculum also became applied to the other secondary schools in the *Hochstift* and served as a model for the other territories in Westphalia governed by Cologne, but the conditions were lacking there to apply it entirely (Schubring 2010, pp. 36 ff.).

The next research question, already relating our local case more to global issues, is whether the French Revolution, and more concretely the Napoleonic period improved mathematics teaching in Westphalia even more?

Napoleon used the Prussian defeat for a major territorial restructuring. He then created two satellite states in Germany: the *Kingdom of Westphalia* (capital: Kassel) and the *Grand Duchy Berg* (capital: Düsseldorf), both governed by his relatives. Hence, the region of Westphalia was divided among three powers: two direct satellite states of the Napoleonic Empire (*Großherzogtum Berg* and *Königreich Westfalen*) and one genuine German state allied to France—now called *Großherzogtum Hessen* (see the map, Fig. 3). Given the strong position of mathematics in the *lycées* established in France since 1803 (see Schubring 1984), we should therefore expect an immediate and marked reinforcement of mathematics in secondary schooling. In 1810, Napoleon occupied and annexed some parts of the territories governed by his relatives, together with large parts of Northern Germany, directly to the French Empire, so that—for instance—the northern parts of *Minden-Ravensberg* and *Münster* became French. Since a key characteristic of French educational policy in the Napoleonic period had been its strict uniformity for the entire country, and since satellite and allied states copied more or less the French structures, one should hence expect a new status for mathematics teaching in Westphalia.

The surprising result of my research was, however, that this was not the case. The governments in Düsseldorf and in Kassel began only after several years to plan educational reforms, but they missed the necessary conceptions as well as energy to realise any essential changes. One explanation is that these satellite states were victims of the financial burdens of mandatory contributions to finance Napoleon's permanent wars, so that the governments could not spend on educational reforms. Even in the northern parts, which were French from 1810 on, projects for

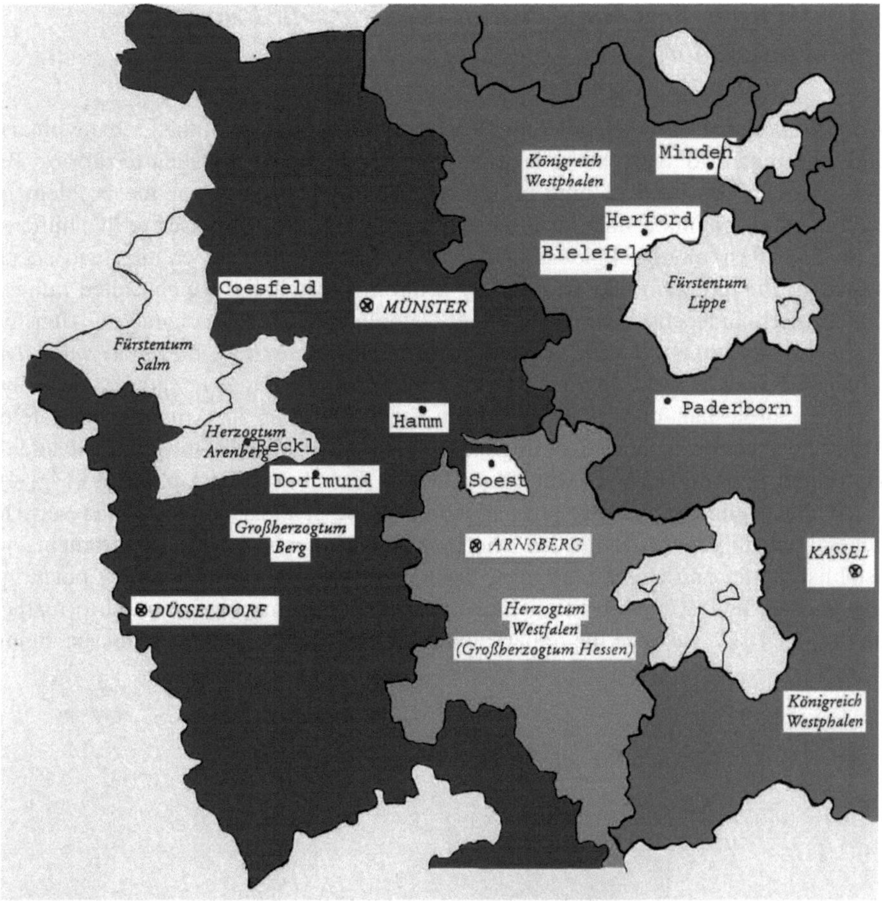

Fig. 3 The political landscape of Westphalia in 1808 (Schubring 2010, title page)

assimilating them to the common structures in the French departments did not become realised until 1813/14 when this entire system collapsed. The only state somewhat reorganising the schools within its new territories was the originally German one: *Hessen*. During its government, mathematics teaching became, in fact, reinforced in Arnsberg, the only Gymnasium in that region (Schubring 2010, p. 45).

We have to state here the revealing paradox that states, which were projected as model states of modern Napoleonic social and cultural structures for Germany as a whole, were not able to reform the educational system accordingly and in particular were unable to promote mathematics teaching, one of the core components of the original French system.

3.3 Münster Becoming Part of the Prussian Province Westphalia

After the collapse of the Napoleonic Empire, Prussia was one of the global winners; and with regard to Westphalia, it now obtained all the Westphalian territories. By now, Prussia consisted of nine provinces and faced in particular the problem of establishing a common state for Protestants and Catholics, who used to live hitherto in more or less mono-religious states with decidedly different cultural and social systems. This problem was particularly acute in Westphalia, which united numerous formerly independent states of either Catholic or Protestant character. After the first reorganisations by the Prussian government, there were eleven *Gymnasien*, which survived or were upgraded from a minor status: five were catholic ones and six were protestant ones. The five were the *Gymnasien* in Arnsberg, Coesfeld, Münster, Paderborn, and Recklinghausen and the six were those in Bielefeld, Dortmund, Hamm, Herford, Minden, and Soest (Schubring 2010, p. 32; see Fig. 4).

At the beginning, in 1815, the situation of the schools in Westphalia was, in general, terribly poor. It needed enormous investments for the governments to establish better and larger buildings, to increase salaries and create new positions for teachers and eventually, to achieve curricular reforms. With regard to mathematics, by 1815, we have the remarkable result that the Protestant Gymnasia, being

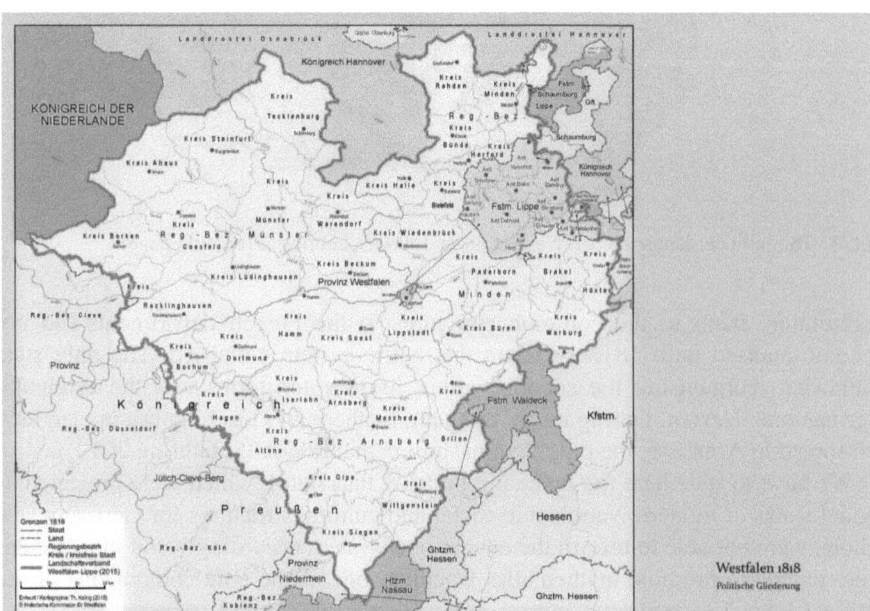

Fig. 4 Westphalia in 1818, now entirely Prussian (internet portal "Westfälische Geschichte" of the Landesverband Westalen-Lippe, http://www.lwl.org/westfaelische-geschichte/portal/Internet/dokumentation/karten/karten_online/haupt.php?urlNeu=Ja)

municipal schools, were badly equipped contrary to the state run Catholic Gymnasia—and, moreover, as a result of the Fürstenberg reforms, mathematics teaching was better established there (Schubring 2010, p. 70). Thus, our primarily very local case, the minor state of Münster, proves to yield a more globally relevant pattern.

3.4 Mathematics Between the States of the Deutscher Bund

From the early nineteenth century, with the establishment of public school systems in the various states, the old polarity between Protestant and Catholic education systems became superseded by a *new polarisation* highly revealing for mathematics teaching. In Prussia, the profound neo-humanist reforms from 1810 on conceived of school knowledge as an interconnected "organism"—with the three major teaching subjects: classical languages, mathematics and the sciences, and history and geography—while the other states maintained the classical dominance of the ancient languages'—with only minor modernisations regarding mathematics (Fig. 5).

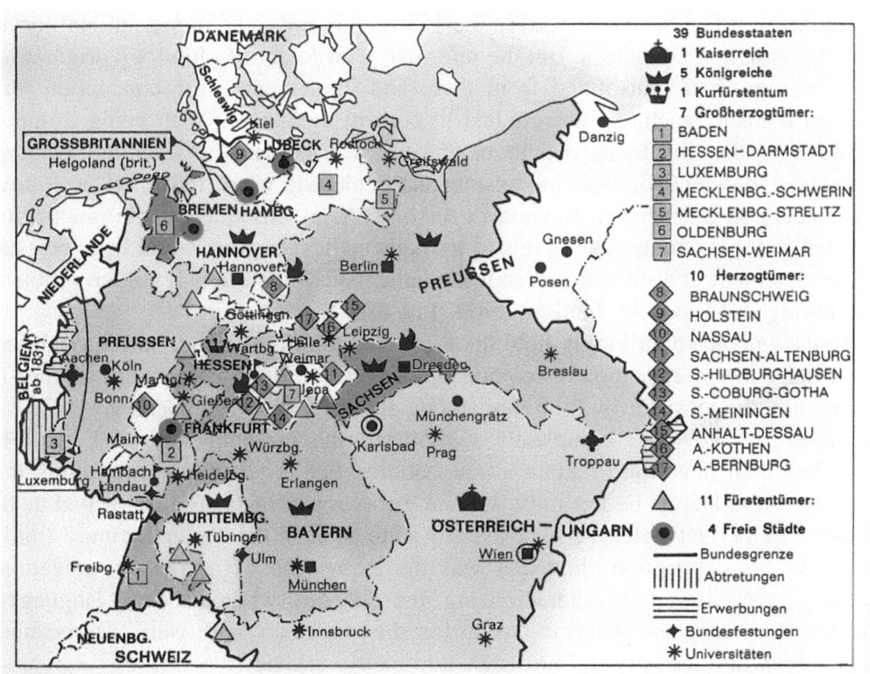

Fig. 5 Map of the *Deutscher Bund*, a loose union of 39 German States, 1815–1866 (Kinder 1998b)

Emblematic of this "moderate humanism"—as opposed to Prussian neo-humanism—became three states: the dukedoms of Schleswig and Holstein in the North, the kingdom of Saxony in the Middle, and the kingdom of Bavaria in the South. Most important aspects like teacher education and the curricula were clearly different. One has to state that the mathematical contents in were quite different all over Germany. Thus we have for the Germany of the 19th century various differing local cases, which in their sum reveal the pertinent global issue: the conflicts about a restructuring of the educational systems—where secondary schools would no longer exclusively lead to university studies, thus no longer preparing students only for the traditional learned professions (in law and medicine) but preparing for a much broader labour market, including carers in engineering and technology (see Schubring 2012).

In fact, with this "moderate humanism", mathematics did not achieve the status of a major teaching discipline. In particular in these three states, the classical languages remained dominant. We have a lively autobiographical report on the educational system in the two dukedoms Schleswig and Holstein (in Northern Germany), by Friedrich Paulsen (1846–1908), a historian of education who moulded the general vision of the history of German education and in particular of its nineteenth century history. From 1863 on, he was a student of the highly traditional Gymnasium in Altona, the *Christianeum*, and graduated there in 1866 with the *Abitur*. As he reported in his autobiography, the secondary schools in the two dukedoms had been reformed in the 1830s, copying some organisational forms from the Prussian Gymnasia. But the reformer, a professor of philology originating from Saxony, had transmitted from there and implemented the conception that modern realistic teaching subjects had to content themselves with being a minor subject and that the focus had to be the classical-humanistic teaching. Paulsen underlined that the Gymnasia in Holstein and Schleswig were "reformed" to follow the conception practiced in Bavaria or Saxony. As a consequence, the practice of the staff in these countries with regard to mathematics teaching could be expressed by the statement: if a student succeeds in mathematics, it is fine—if not, no problem (Schubring 2012, p. 532; Paulsen 1909, 114 ff.).

Paulsen depicted strikingly how his form, and he himself, too, had internalised this conception of education: as complete disregard of their mathematics teacher—calling him the most sorrowful figure among all his teachers (Paulsen 1909, 120 f.). By contrary, they had completely assimilated the pre-eminence of Latin. He described their devotion to attain a true Latinity. For instance, for translating into Latin, the text had to be not only without errors regarding grammar; it had to be written in a perfect style: the major goal was to achieve the "color Latinus" (ibid., 123). As his description shows, it was the expression of a social segregation: Latinity was viewed as characterizing the scholar, while modern languages, mathematics, and the sciences were for the merchant and were depreciated. Technology and industry did not enter into their worldview:

> The precondition for the success of this treatment of Latin evidently consists of the fact that writing Latin is reputed by the students as the first and foremost skill of the scholar. For us,

> it is still completely applied; primarily it was just Dr. Henrichsen who was able to maintain in us this conviction: Latin characterises the scholar – with a good Latin style one will succeed throughout the entire world, he used to say. Nothing occurred, which could have made us doubting in this. We confidently dispraised the modern languages: they are for the merchant, not for the scholar; we disdained mathematics and we knew nothing of the sciences; and of technique and industry nobody in our context ever spoke. [...we regarded] the Antiquity as the courtly world in which to be at home constitutes the privilege of the scholar; the Latin style being the common shibboleth and practically the legitimation of scholarly education (ibid., 123 ff.; transl. G. S.).

In Paulsen's analysis, he established a connection between this type of traditional *Gelehrtenschulen*, preparing exclusively for university studies, and the dominance of theological career orientations of students, resulting from the great number of former pastors working as Gymnasium teachers. Prussia, he said, "had enormously progressed in the de-clericalisation of the rank of Gymnasium teachers" (ibid., 117).

The analysis of the evolution of the status of mathematics teaching in secondary schools clearly shows the utmost importance of investigating mathematics teaching always as an element of the set of all school disciplines. Mathematics has never become established and developed in an autonomous manner; it is the variance in the relations of status within this set, which expresses the dominant values in a given period—thus what one can call, in fact, the "school culture". It constitutes therefore a key methodological requirement to analyse mathematics within the set of all school disciplines (see Schubring 2006).

3.5 Reconciling the Antagonisms Within the New German Reich from 1871

By the second half of the nineteenth century, the antagonism between the Prussian special way of modernising the curricula—in the sense of no longer exclusively preparing for university studies and the traditional "learned professions", but for a broad range of careers—and the other states altered to become less acute. The revolutionary movements of 1848 and the industrial take-off underway undermined the static character of their societies and changed the hitherto basically agrarian character of their economies. As a result, the resistance to a multi-disciplinary secondary school was reduced (Schubring 2012, p. 528).

The decisive impetus for the reconciliation was given by the process of establishing the new German Reich from 1871, under Prussian domination and excluding Austria from this *kleindeutsche*[4] solution of the nation-building process. The educational structures had to become somewhat more homogeneous, to allow

[4]Literally meaning: small German—expressing that Austria—the other major state besides Prussia in the former *Deutscher Bund*, and the leading state during centuries within the Holy German Empire—had been deliberately excluded by Prussia.

flexibility of diplomas and graduations among the states remaining independent within this Confederation. During the former period of the "Deutscher Bund", citizens of one of the 39 states were regarded as "foreigners" within the other 38 states—with the consequence that their diplomas were not acknowledged in the other states and had to be evaluated to check their equivalence to diplomas of the respective state. The point revealing the underlying social structures is that the very first impetus for a reconciliation of the curricula originated from military interests. To understand this point, a particular feature of the Prussian social structures needs to be explained. When a compulsory military service had been introduced there in the wake of the wars against Napoleon, a certain privilege had been granted to young men with a *wissenschaftliche Bildung*—proving a scientific education. The privilege—"Einjährigen Freiwilligen Dienst": voluntary one-year service—meant to serve for only one year, instead of three years. Strangely enough, the quality "wissenschaftlich"—"scientific"—was proved by having passed just six of the nine grades of a Gymnasium—the *Abitur* (the final exam, entitling for university studies) was not necessary! When Prussia defeated Austria in the 1866 war, it had simply annexed a number of states allied to Austria. And for these new Prussian citizens and for more northern German states, allied to Prussia in the new "Norddeutscher Bund"—Northern German Union—, they needed to obtain the same privilege. But given their former differing educational systems and different curricula, a conference of delegates of the states of the *Norddeutscher Bund* took place in 1868 to evaluate the curricula and to define the conditions for equivalence, in order to grant the privilege (BuA I, fol. 2).

The 1868 Conference was presided over by Ludwig (von) Wiese who had been, since the 1850s, the most influential and powerful functionary in the Prussian Ministry of Education and the representative of changed politics about mathematics teaching, which no longer corresponded with its original neo-humanist key role. In fact, Wiese's list of criteria for the *Abitur* diploma was just a rephrasing of the qualifications in the Prussian regulations of 1834 for this exam (Wiese 1867, p. 219), which meant a decisive lowering of the level of mathematics teaching (Schubring 1991, p. 64). These criteria for the *Abitur* exam were not mandatory in the reality of content taught for quite a time, but since the 1860s the authorities insisted ever more on their application. Wiese thus represents the new tendency of the Prussian administration in the second half of the 19th century no longer strictly to adhere to the original neo-humanist conception. In the addendum to the second volume of Wiese's collection of decrees and laws for Prussian secondary schools the results of the conference on the "Militair-Ersatzinstruction" for the *Norddeutscher Bund* of 26.03.1868 were published explaining the conditions for equivalence of diplomas and the applicability of the—relatively low—Prussian criteria published yet in 1867 (Wiese 1868, pp. 389–392). Among the states annexed in 1866 and enjoying now the *Einjährigen-Privileg* extended to them were the two dukedoms Schleswig and Holstein. There, functionaries of the former dukedom government, who continued in the new Prussian provincial government, had obstructed in 1866 to apply the Prussian curriculum for mathematics. It was only in 1870, that this obstruction was overcome and mathematics enacted as a major discipline (Schubring 2012, p. 533).

In 1872, the Prussian Minister of Education addressed the Chancellor of the Empire to organise a new analogous conference for assessing the curricula of all the member states of the Confederation and for defining the conditions for the mutual equivalence of the diplomas (*Abitur* and teacher exams). The conference took place from 15 to 19 October 1872, in Dresden, the capital of Saxony—in order not to appear too much Prussia-centred—, and again was presided over by Wiese. Although the debates showed considerable structural differences between the member states, the decisions essentially extended those taken in 1868 (BuA I, fol. 62–79). This means that the relatively low knowledge required in 1834 for obtaining the *Abitur* became now the general, common denominator for equivalence of diplomas—which was not too difficult to apply even for those states, which had maintained mathematics as a marginal discipline in the first half of the 19th century.

4 Conclusion

It was within this educational system of the new German Reich with its compromise between a humanities oriented curricular conception and a modernising conception, open for technological developments, that in 1902 Felix Klein successfully launched the reform movement for mathematics teaching, aiming to overcome the synthetic view of school mathematics as represented paradigmatically by the Prussian 1834 *Abitur* requirements. Mathematics teachers form the various member states of the German Reich jointly strove for a decisive modernisation of school mathematics. The movement's success in Germany, where some complementarity between global and local aspects had then been achieved, provided the basis for extending this reform movement to an international one, from 1908 on.

Felix Klein was profoundly committed to convey the essence of the modernisation concepts to the other participating countries. The broad spectrum of educational and epistemological visions underlying the structures for mathematics teaching realised in the participating countries determined a new agenda for establishing a new balance between the global and the local. It was in particular the case of Italy, which drastically showed the permanence of local patterns inhibiting a homogeneous global modernisation. Convergent with the Italian dominant culture based on humanities and classical antiquity, Italian research mathematicians continued to extoll rigour as the leading vision for school mathematics.

We have remarked cases, during the Napoleonic period, where local conditions did impede a principally possible globalising generalisation. Likewise, the history of mathematics education reveals the perseverance of cultural and epistemological patterns in certain cultures, which resist tendencies and movements for unifying.

References

Sources

Bundesarchiv, Berlin:
Reichsamt des Innern. Militar Generalia, No. 26, Adhibendum 2, vol. 1 (R 1501, 112690). (BuA I).

Publications

Chervel, André (1988). L'Histoire des Disciplines Scolaires. Réflexions sur un domaine de recherche [The History of School Disciplines. Reflections about a Research Area]. *Histoire de l'Éducation, 38*, 59–119.

Decreto Coppino (1867). Istruzioni e programmi per l'insegnamento della matematica nei ginnasi e nei licei [Instructions and programmes for the teaching of mathematics in the *ginnasi* and *licei*]. *Supplemento alla Gazzetta Ufficiale del Regno d'Italia*, Firenze, 24 ottobre 1867.

Giacardi, Livia, & Scoth, Robert (2014). Secondary School Mathematics Teaching from the Early Nineteenth Century to the Mid-Twentieth Century in Italy. In A. Karp & G. Schubring (Eds.), *Handbook on the history of mathematics education* (pp. 201–228). New York: Springer.

Kinder, Hermann (1998a). *dtv-Atlas zur Weltgeschichte. Band I. Von den Anfängen bis zur Französischen Revolution* [dtv-Atlas on world history. Volume I. From the origins to the French Revolution]. 32. Auflage. München: Deutscher Taschenbuch-Verlag.

Kinder, Hermann (1998b). *dtv-Atlas zur Weltgeschichte. Band II. Von der Französischen Revolution bis zur Gegenwart* [dtv-Atlas on world history. Volume I. From the French Revolution to present times]. 32. Auflage. München: Deutscher Taschenbuch-Verlag.

Paulsen, Friedrich (1909). *Aus meinem Leben. Jugenderinnerungen* [From my youth. Youth memories]. Jena: Diederichs.

Schubring, Gert (1984). Essais sur l'histoire de l'enseignement des mathématiques, particulièrement en France et en Prusse [Attempts to the history of mathematics teaching, in particular in France and in Prussia]. *Recherches en Didactique des Mathématiques, 5*, 343–385.

Schubring, Gert (1987). *The Cross-Cultural 'Transmission' of Concepts - the first international mathematics curricular reform around 1900, with an Appendix on the Biography of F. Klein.* Occasional paper Nr. 92, IDM (Universität Bielefeld).

Schubring, Gert (1991). *Die Entstehung des Mathematiklehrerberufs im 19. Jahrhundert. Studien und Materialien zum Prozeß der Professionalisierung in Preußen (1810–1870)* [The emergence of the profession of mathematics teachers in the 19th century. Studies and documents on the process of professionalisation in Prussia (1810 to 1870)]. Zweite, korrigierte und ergänzte Auflage. Weinheim: Deutscher Studien Verlag.

Schubring, Gert (2002). Aspetti istituzionali della matematica [Institutional aspects of mathematics]. In Sandro Petruccioli (Ed.), *Storia della scienza*, Vol. VI: *L'Etá dei Lumi* (pp. 366–380). Roma: Istituto dell'Enciclopedia Italiana.

Schubring, Gert (2004). Neues über Legendre in Italien [New results about Legendre in Italy]. In W. Hein & P. Ullrich (Eds.), *Mathematik im Fluss der Zeit*. Algorismus, Heft 44 (pp. 256–274). Augsburg: ERV Rauner.

Schubring, Gert (2006). Synthesis. Researching into the history of teaching and learning mathematics: "The State of the Art", *Paedagogica Historica*, Special Issue: History of teaching and learning mathematics, ed. by Gert Schubring, *XLII (IV&V)*, 665–678.

Schubring, Gert (2010). *Die Debatten um einen Mathematiklehrplan in Westfalen 1834. Eine regionale Sozialgeschichte der Einführung von Mathematik als Hauptfach* [The debates about a mathematics syllabus in Westfalia 1834. A regional social history of the introduction of mathematics as a major school discipline]. Münster: WTM Verlag.

Schubring, Gert (2012). Antagonisms between German states regarding the status of mathematics teaching during the 19th century: processes of reconciling them. *ZDM - The International Journal on Mathematics Education, 44*(4), 525–535.

Schubring, Gert (2014). Mathematics Education in Catholic and Protestant Europe. In A. Karp & G. Schubring (Eds.), *Handbook on the history of mathematics education* (pp. 130–143). New York: Springer.

Shapin, Steven, & Thackray Arnold (1974). Prosopography as a research tool in history of science. *History of Science, 22*, 1–28.

Valente, Wagner Rodrigues (2012). Por uma história comparativa da educação matemática [For a comparative history of mathematics education]. *Cadernos de Pesquisa (Fundação Carlos Chagas. Impresso), 42*, 162–179.

Vita, Vincenzo (1986). *I programmi di matematica per le scuole secondarie dall'unità d'Italia al 1986. Rilettura storico-critica* [The programmes for mathematics in the secondary schools since the unity of Italy until 1986. A historico-critical rereading]. Bologna: Pitagora Editrice.

Wiese, Ludwig (1867). *Gesetze und Verordnungen für die höheren Schulen in Preußen. Erste Abtheilung: Die Schule* [Laws and decrees for the secondary schools in Prussia. First Part: The school]. Berlin: Wiegandt und Grieben.

Wiese, Ludwig (1868). *Gesetze und Verordnungen für die höheren Schulen in Preußen. Zweite Abtheilung: Das Lehramt und die Lehrer* [Laws and decrees for the secondary schools in Prussia. Second Part: The teaching post and the teachers]. Berlin: Wiegandt und Grieben.

Author Biography

Gert Schubring is a retired member of the Institut für Didaktik der Mathematik, a research institute at Bielefeld University, and at present is visiting professor at the Universidade Federal do Rio de Janeiro (Brazil). His research interests focus on the history of mathematics and the sciences in the 18th and 19th centuries and their systemic interrelation with social-cultural systems. One of his specializations is history of mathematics education. He has published several books, among which is Conflicts between Generalization, Rigor and Intuition: Number Concepts Underlying the Development of Analysis in 17th–19th century France and Germany (New York, 2005). He was Chief-Editor of the International Journal for the History of Mathematics Education.

A Teacher of Mathematics in Times of Change

Harm Jan Smid

Abstract This paper deals with the life of Michael Isaac Swarts Bevel (1778–1833). Bevel was born in Amsterdam, raised in a Mennonite orphanage and started his career as an assistant teacher at an elementary school. He became headmaster of a school for poor children in Leiden and made himself conversant in modern languages, physics and mathematics. He became teacher and supervisor for mathematics for the Society *Mathesis Scientiarum Genitrix* and the first teacher of mathematics at the Latin school of Leiden. He taught mathematics at two prestigious boarding schools and to the officers of the garrison of Leiden. In 1820 he became the inspector for Leiden and its region for the weights and measures of the new metric system. He wrote several textbooks for arithmetic as well as new editions and translations of well-known works of others. His life and career shows how he took advantage of the social, political and educational changes during his lifetime and how these changes and developments are connected and combined within the life of one single person.

Keywords M.I.S. Bevel · The Netherlands · Mathematics education
Metric system · 19th century

1 Introduction

When Schubring published his *Die Entstehung des Mathematiklehrerberuf* [The emergence of the profession of a teacher of mathematics], he also included some "selected teacher biographies" ; to give, as he wrote "a more clear impression of the course of professional paths of life"[1] (Schubring 1983, p. 199). No doubt, descriptions of the lives and careers of mathematics teachers can be illuminating for

[1]All translations from German or Dutch are by the author.

H.J. Smid (✉)
Delft University of Technology, Delft, The Netherlands
e-mail: harmjansmid45@gmail.com

F. Furinghetti and A. Karp (eds.), *Researching the History of Mathematics Education*, ICME-13 Monographs,
https://doi.org/10.1007/978-3-319-68294-5_13

the history of mathematics education during a certain period, if these careers are described in connection with the circumstances and developments, be it political, social, economic or pedagogical, that were influential on mathematics education in that period.

In many countries, the end of the 18th and the first decades of the 19th century were a period of great changes. Those years brought not only great political, social and economic developments, but it was also the period in which in many countries formal mathematics education at elementary and secondary schools was introduced. As a consequence, being a mathematics teacher became gradually an established and well-respected profession.

The new possibilities for mathematics education in the Netherlands in those years can be demonstrated by the remarkable career of the Dutch teacher Michael Isaäc Swarts Bevel. He was a member of a religious minority and was raised in an orphanage, but he had an impressive career as a teacher of mathematics at various school types, was a successful schoolbook author and ended his career as inspector of weights and measures for the region of Leiden, appointed by the King, to implement the new metric system.

Schubring introduced in his book a classification of mathematics teachers in Prussia in three generations (Schubring 1983, p. 45). We will not discuss here in how far this classification is applicable for the Dutch situation as a whole, but when Schubring characterizes the first generation of mathematics teachers of around 1820 in Prussia as one in which "only a few were qualified as mathematics teacher", this is certainly also the case for the Dutch situation of that time. But as we will see, Bevel was one of the exceptions.

During his lifetime and in the first decades after his death, Bevel was well-known, as is demonstrated by the extensive entry in the *Biographisch Woordenboek der Nederlanden* [Biographical Dictionary of the Netherlands] (Van der Aa 1854, pp. 476–479). But in later years, he passed into oblivion. When Coebergh van den Braak published a detailed history of the Leiden[2] Gymnasium, his only remark about Bevel was: "Still in 1816 a certain Bexel was appointed as mathematician. To him belongs the credit of being the first teacher of mathematics at the Latin School of Leiden" (Coebergh van den Braak 1988, p. 83). Not only did he misspell Bevel's name, but he had clearly no idea who he was.

In the last decades some renewed attention is paid to Bevel. Beckers published a short biographical sketch (Beckers 1998), with a focus on his—of course wrong—attempt to perform the quadrature of the circle. Beckers and Smid published an edition of one of his textbooks on arithmetic (Beckers and Smid 2003), and Smid published an article for a general public about his life and career (Smid 2009). In this article, the focus is on Bevel's position and role in the development of mathematics education in the Netherlands, in the context of the political, social and economical changes that took place during his lifetime. We pay no attention to his career as inspector of weights and measures or to other aspects of his life.

[2]We will use the Dutch name for this city: "Leiden", not the English version: "Leyden".

Bevel's career is also interesting in an international perspective; it illustrates the emergence of the profession of being a teacher of mathematics in the first decades of the 19th century, which occurred in more European countries in that period, in a typical Dutch context.

2 A Time of Great Changes

The end of the 18th and the beginning of the 19th century was a time of changes, revolutions and in the end restorations in many countries. But the return of the old monarchies after the Vienna Congress of 1814–1815 could not prevent that the turbulent years after the French revolution of 1789 would have a deep and lasting effect on most European countries.

That is certainly true for the Netherlands. In 1795 the old Republic, with the Prince of Orange as its Stadtholder, was overthrown by a French revolutionary army. A new Republic was founded, formally independent but largely dominated by France. In 1806 the Republic was replaced by a Kingdom with a younger brother of Napoleon, Louis Bonaparte, as King. In 1810 the Netherlands lost their independence and was annexed by France. Its independence was restored in 1813 and it became a Kingdom again, now under King Willem I of Orange, the son of the last Stadtholder.

In the beginning, the French were greeted with enthusiasm by the so-called *Patriots,* opponents of the conservative Stadtholder and advocates of renewal. But soon the economic misery and the endless wars of Napoleon made an end to all support for the French and the defeat of Napoleon was greeted with relief. The return of the House of Orange did not bring a complete restoration; the new King had relatively modern ideas.

This so-called "French period" brought not only economic misery, but also important changes and modernizations. The old republic, a loose combination of seven largely independent provinces in which the cities did have a great extent of autonomy, was replaced by a centralized State with a more powerful central government and legislation on a national scale. One of the first measures of the new government was the emancipation of the Roman Catholics, Mennonites, Jews and other (religious) minorities. They were tolerated in the old republic, which in itself was already rather exceptional, but they could not hold public offices. Now they were granted full political rights.

Moreover, it became possible for the first time to develop a nationwide policy, for instance on education. In the second half of the 18th century there had been much debate in The Netherlands about the improvement of the school system, especially for primary education, but due to the lack of a central government, nothing was achieved. Now, with a powerful central government, this became possible. The first laws on primary education, those of 1801 and in 1803, were no great success due to local resistance, but the law of 1806 was more generally accepted and remained in force until 1857.

These laws introduced several new elements. In the first place, a corps of inspectors was appointed to control the quality of the schools: the ability of the teachers, the way they taught, but also for instance the hygienic conditions in the school buildings. Secondly, all schoolmasters had to pass an exam to get permission to teach at schools paid by the authorities. In the law of 1806 a system of four degrees was introduced. For schoolmasters of the fourth degree, the lowest, only minimal knowledge of the topics they had to teach, such as reading, writing and arithmetic, was required, but schoolmasters of the first degree, the highest, were really well-educated men, with knowledge for instance of natural history, geography, physics and mathematics. Another important novelty of these laws was that the teaching of arithmetic was made compulsory; until then children were often taught reading and writing only.

The laws on primary education also brought several improvements in teaching methods. At the old schools, all children worked on their own, controlled individually by the schoolmaster and asking for his assistance when needed. In the new system, age groups or grades were formed and teaching was addressed to the respective grades as a whole—although often different grades were housed in the same classroom (Boekholt and de Booy 1987, pp. 95–115).

Moreover, the didactics of teaching arithmetic changed. In the old days, children were trained to follow recipes. There was no need for understanding, as long as they could perform the tricks swiftly and without mistakes. But now, teaching arithmetic was part of a more general purpose: to educate young children to become able and good citizens, who could understand what they were doing. The old recipe books on arithmetic, often still dating from the 17th century, could no longer be used; new ones had to be written (Beckers 2000).

When after the French defeat the new King came into power, he maintained most of the measures taken in the French period, also those on primary education. In the first period of his reign, he strived to further modernization of the country. One of his measures was the introduction of the metric system.

The metric or decimal system was developed in France during the last decades of the 18th century. J.H. van Swinden, the leading Dutch mathematician in those years, was a fervent proponent of the new system, and played a role of some significance at the conference devoted to the system in Paris in 1798/99. He used his considerable political influence to foster the introduction of the metric system in The Netherlands. That seemed successful, for in several laws during the French period the introduction of a uniform system of weights and measures was stipulated, and it seemed highly likely that this would be the metric system.

All kinds of practical and political problems, not in the least the fact that Napoleon himself largely turned back the use of the metric system in France, prevented that the introduction of the system was put into effect. The new King however insisted and he ordered the metric system to be obligatory in 1820; which made the Netherlands the first country in Europe formally adopting the system. Moreover, the King ordered that from 1816 on, as a preparative measure, in all schools where arithmetic was taught the teaching of the metric system should be included.

It would still take a long time before The Netherlands would become really a modern country, but no doubt the years around the turn of the century not only brought misery and poverty, but also offered new possibilities.

3 A Teaching Career

Michael Isaäc Swarts Bevel was born on December the 12th of the year 1778 in Amsterdam, in a family of Mennonites. His father died young, and although his mother was related to a wealthy Amsterdam family of tradesmen, his own family must have been relatively poor, for a few years after the death of his father he and his sister were placed in a Mennonite orphanage.

When he left the orphanage in 1800, he became an assistant to a schoolmaster at a primary school in Amsterdam, the usual way then to become a schoolmaster yourself. He soon became second schoolmaster at a municipal primary school, also in Amsterdam and in December 1802 he was appointed as headmaster at a municipal school for poor children in Leiden.

In the 18th century, the Dutch Reformed Church in Leiden had established three primary schools for poor children, but during the French period these schools were taken over by the municipality, which reduced their number to two (de Gelder 1856). Bevel was appointed at one of these schools, which would not have been possible, since he was a Mennonite, if these schools had still been in the hands of the Dutch Reformed Church.

When Bevel started to work in Leiden, he soon acquired a kind of a network of ministers and professors who helped him with studying a whole range of subjects. He took lessons in Dutch grammar and literature, learned German and French, and studied physics and mathematics. Of special importance for this last subject was his friendship with Simon Speyert van der Eyk, professor in mathematics at the University of Leiden.

In 1804 Bevel had to pass the exam for his teaching degree, which he acquired with ample praise of the examiners. The exam committee not only praised his "exceptional knowledge of mathematics, geography, history and physics", but also his capability to "to express himself on these subjects with great ability and clearness" (*Bijdragen betrekkelijk den staat en de verbetering van het schoolwezen in het Bataafsch Gemeenebest*). Under the new law of 1806 he was awarded the rank of the first degree, without doing any further formal exam. Not only the exam committee was very satisfied with his knowledge, also the local inspector in Leiden praised the way he taught the children, his knowledge, his decent behavior and his diligence (RAL, archief Schoolcommissie).

Bevel was a man with wide interests. He published for instance a translation of a French book by Charles de Villers on the philosophy of Immanuel Kant (Villers and Bevel, 1821), but also a book on the possibilities of gas-illumination (Bevel 1817). But there can be no doubt that mathematics was his favorite subject. In 1807, he published a translation and adaptation of Eulers *Vollständige Anleitung zur Algebra*

[Complete introduction to algebra] (Bevel and Euler 1807). Four years later he published a revised edition of a much used textbook on geometry: Pibo Steenstra's *Grondbeginselen der Meetkunst* [First Principles of Geometry] (Steenstra 1811).

For these and other publications he was in 1812 awarded by the University of Leiden with the degree of *Artes Liberales Magister et Philosophiae Doctor*, a remarkable achievement for someone who had not attended a university or even a Latin school. There is something curious about this award. In those years the University of Leiden had lost its independent status and was made part of the *Université Imperiale*, a centralized institute created by Napoleon. The implementation in practice of these measures was seriously hampered by the political chaos of the day and the impending collapse of the *Empire*. The old administrators of the university were removed from their posts, but not yet replaced by new ones—in fact they would never come. It is not impossible that Speyert van der Eyk took advantage of the chaos to award his friend with this title.

While he was an elementary schoolteacher, he had several part-time jobs as a teacher of mathematics. In 1814 he became teacher of mathematics at a vocational evening school, and in 1816 he became the first teacher of mathematics at the Latin school of Leiden.[3] When he became Inspector for weights and measures, he had to give up all his teaching jobs, but after some years he found time for some new part-time teaching jobs. He became teacher of mathematics to the officers of the garrison of Leiden and he taught mathematics at some prestigious boarding schools in villages in the neighbourhood in Leiden.

Bevel remained, besides his teaching duties, active as a mathematician. In 1820 he published together with Speyert van der Eyk a revised and improved edition of a much used handbook on surveying by J. Morgenster: *Werkdadige Meetkunst.* [Practical Mathematics] (Bevel et al. 1820). In 1828 however he overplayed his hand. He published a solution of the problem of the squaring the circle, which of course was wrong. Speyert van der Eyk could not find the error, but did not trust Bevel's 'solution' and advised him not to publish the result. Bevel ignored the advice, even with mentioning Speyert's name. The latter was not amused and distanced himself publicly from Bevel's last publication (Beckers 1998).

4 Teaching Arithmetic at an Elementary School

When Bevel became headmaster at one of the schools for poor children of Leiden, he became the colleague of Jan Brunt, the headmaster at the other school for poor children. Brunt, born in 1777, so almost of the same age as Bevel, had followed a comparable career in teaching as Bevel. He was also interested in mathematics and like Bevel he had followed lessons of Speyert van der Eyk (Van Geuns 1803). In

[3]The word *gymnasium* came in use only after 1840; in the times of Bevel the term *Latijnse school* [Latin school] was commonly used. We will use consequently the term *Latin school.*

1802/1803 Brunt had started writing a series of arithmetic books: *Eerste beginselen der Rekenkunde, voorgesteld in vragen en antwoorden.* [First principles of arithmetic, proposed in questions and answers] (Brunt 1802).

Jan Brunt died suddenly in the spring of 1803, having finished only four volumes of the intended five. Bevel completed the series and published in 1806 his *Vervolg op de eerste beginselen der Rekenkunde* [Sequel to the first principles of arithmetic] (Bevel 1806).

Beckers has described the pedagogical characteristics of the arithmetic books of the period (Beckers 2000) . Arithmetic books in the vernacular had originated at the end of the 15th century. They contained mainly word problems, arranged in chapters by the professions these problems applied to, so that children could pick only the chapters they needed. Until the end of the 18th century, children were trained only in those types of problems that they would encounter in their future profession.The rules were given without any explanation, the emphasis was not on understanding them, but on speedily and correctly applying the rules. The most famous of these arithmetic books, *De Cijfferinghe* [The Ciphering] by Willem Bartjens, remained in print from 1604 until the end of the 18th century (Beckers and Kool 2004).

Now that the teaching and learning of arithmetic was not intended for a future profession only, but was also an element of a more general education, these old books did no longer satisfy. A new generation of arithmetic books emerged. In these new books, the rules were explained, and the children were supposed to understand what they were doing. Problems, mainly the applications of the rule of three, were no longer arranged by profession, but were in a mixed order. Besides the usual word problems, plain exercises without any context became much more prominent. In the long run, these problems would replace the old word problems almost completely, but in the first half of the 19th century, arithmetic books contained a mixture of both types.

The new books contained also problems on subjects that clearly had no practical significance for a future profession or career at all, such as on history, geography or religion. These problems had not only as goal to offer practice in arithmetic, but also to offer some general education to the children. And in the last case, the authors tried explicitly to encourage their readers to develop good learning habits and to convince them about the beneficial effect of studying mathematics in general. They did this often by writing lengthy introductions full of pedagogical and didactical remarks, to be read by the pupils, and by writing their books in a dialogue form: questions and answers spoken by an ideal teacher and equally ideal pupils (Beckers 2000).

The books of Brunt and Bevel were written in agreement with these modern ideas. Their books were well received and had many reprints. The fourth and last volume written by Brunt, and published after his death in 1803, contained an interesting and new element: an introduction to the metric system. When Brunt wrote his booklets on arithmetic, it seemed likely still that the system would be introduced soon, and he must have thought that it was a good idea to explain the system in his schoolbooks on arithmetic.

In the fourth volume of the series, Brunt starts with a lengthy exposition of almost forty pages on decimal fractions, which so far in arithmetic books were hardly treated at all. The next chapter starts with the following question, posed by the ideal pupil that Brunt depicts in his book: *Do the decimal fractions have any use in daily life?* The likewise ideal teacher had to admit that these fractions were not very useful, *since the weights and measures are divided in different ways in our country.* But, he continues, they can be applied to what he calls *the French weights and measures, since their division is based on decimal fractions.*

In the following 10 pages, the metric system is thoroughly treated. Brunt explains the definition of the *mètre* and how the whole system is built on this standard and he shows how easily calculations within this system can be made.

In the next chapter, devoted to the rule of three with the use of common fractions, the problems he demonstrates and the exercises to be made are however mainly based on the old Dutch complicated measurement systems, which were still in use, after all. Only in the end, he gives some problems in which the metric system and decimal fractions should be used, namely for tradesmen doing business with France.

The fifth volume of the series, written by Bevel after Brunts death, is devoted to more complicated problems using the rule of three. In those years, the political climate in the *Batavian Republic,* as the Netherlands were called then, became more conservative, and drastic steps such as the introduction of the decimal system became less likely (Maenen 2002, p. 68). Bevel might have thought that at this moment there was little reason to pay much attention to the metric system, so he dropped the subject completely.

But as we have seen already, the new King ordered that from 1816 on the metric system should be taught as part of the teaching of arithmetic. One of the consequences was that not only the metric system, but decimal fractions as well became much more important. Schoolbooks had to be adapted to the new situation, or in some cases to be replaced by new ones. Brunts four booklets were adapted by N. Anslijn, a former assistant teacher of Brunt, who for instance transferred the treatment of the decimal fractions from the fourth to the first volume, and the treatment of the decimal system to the second volume (Brunt and Anslijn, 1816–1820). Common fractions were now treated much later, in the fourth volume.

Bevel adapted his own fifth volume. In the preface of the second edition of 1819 he wrote:

> De Uitgevers thans eenen herdruk van dit rekenboekje noodzakelijk vindende, heb ik in hetzelve die veranderingen gemaakt, welke, door het nieuw in te voeren stelsel van munten, maten en gewichten, noodzakelijk zijn geworden. Hierbij echter heb ik het oude stelsel niet geheel kunnen achterwege laten, naardien onderscheidene zaken, naar dit stelsel aangevangen, nog niet geheel zullen afgedaan zijn, wanneer het nieuwe stelsel ingevoerd wordt. [Since the publisher was of the opinion that now a second edition was needed, I made those changes that are necessary due to the introduction of the new system of coins, weights and measures. I could however not put completely aside the old system, since some matters, dealt with under the old system, are not yet completed] (Bevel 1817).

The adaptations might be considered as successful; all five volumes stayed in print until the forties of the century.

5 Teaching Arithmetic at a Vocational School

One can argue that in fact Bevel was too highly educated to be a schoolmaster at a primary school for poor children, but he had other teaching activities which might have been more challenging. One of them was teaching mathematics at what we would now call a vocational school.

The 18th century is in the Netherlands sometimes called the age of the Societies, for in that century many of these were founded in that country. Some of them could be called *Learned* Societies with the purpose of publishing journals, books etc. of some supposed scientific significance, others were *Educational* Societies, with the aim to foster the education of young people. The importance of such Educational Societies for the education of the middle and lower classes can hardly be overestimated. Many generations of young man received during the 19th and first half of the 20th century, after having finished primary school, their further vocational training on (evening) schools founded by such Societies.

In 1785 such an Educational Society was established in Leiden, with the name *Mathesis Scientiarium Genitrix* [Mathematics is the Mother of the Sciences] (Peeperkorn—van Donselaar, 1985). Its aim was to provide young boys, who had finished primary education and had already started working, with a good education for technical professions, such as building contractors and architects. Mathematics played, as the name already makes clear, an important role in this Society.

One of the founders, and no doubt the most important figure in the first decades of the Society, was Pieter van Campen (1750–1820). He was an "admitted surveyor", which means that he had passed an exam for surveyor, but maybe he can be described best as a free-lance mathematician, who tried to earn his living by surveying, teaching mathematics as a private teacher, writing textbooks, etc. One of these textbooks was his *Grondbeginselen der Bouwkundige Reekenkunde* [First Principles of Architectural Arithmetic], published already in 1780, but later on also used at the Society (Van Campen 1780).

When the Society started its teaching it had to use already existing elementary textbooks for arithmetic, but it was soon decided that Van Campen should write such a textbook himself, to be used at the Society. The result was his *Gronden der Rekenkunde* [Principles of Arithmetic] (now lost), published in 1788. Considering their subtitle *for the first beginners*, they were intended for those pupils who had received no or only little instruction in arithmetic.

Bevel became a member of the Society in 1811 and in 1814 he became one of the mathematics teachers of the Society. The lessons were given in the evening, so he could combine it with his job at the primary school. His activities for *Mathesis*, as the Society was usually called, led to some publications. The books by Van Campen needed a thorough revision after the introduction of the metric system. Van Campen had left the Society in 1794 in the midst of a considerable quarrel, so there was little inclination to entrust this revision to him. When he died in 1820, his old books were still in use.

The first book to be revised was the *Grondbeginselen der Bouwkundige Reekenkunde.* It was a book of more than 200 pages, treating not only arithmetic with many exercises on building problems, but it also contained chapters on architecture, on building materials and chapters on specific parts of geometry such as circles, helical lines and solid geometry. Much attention was paid to weights and measures, of course using old Dutch systems (Bevel et al. 1826).

It was that aspect that made the book outdated in the first place; from 1820 onwards such a book should use the new metric system. Bevel did not prepare the new edition on his own: he had a co-author in the person of P.E. Rijk. This was the son of one of the other founders of the Society, and by profession, as the title page says: a 'master bricklayer', in fact the head of an important bricklayers company in Leiden. He had received the theoretical part of his training at the Society, and he evidently liked mathematics and teaching, for in his spare time he taught arithmetic there. No doubt he had specific knowledge on building and architecture that Bevel might have missed, and the result of the cooperation was a greatly enlarged new edition of more than 450 pages, published in 1826. Decimal fractions and the metric system were treated thoroughly, but also other chapters were rewritten and new ones added, for instance on trigonometry (Bevel and Rijk 1826).

The cooperation between Rijk and Bevel must have given satisfaction to them, since immediately afterwards they started another project: producing a new elementary arithmetic book. The board of the Society had considered at first to ask them to compose a revised edition of Van Campens textbook, but in the end it was decided that it was better to write a completely new book. The first three volumes of their *Grondbeginselen der Rekenkunde* [Basic Principles of Arithmetic] were published in 1826, 1828 and 1830. A fourth volume was published in 1840 by Bevel's son, C.S. Bevel (Bevel and Rijk 1826, 1828, 1830).

The series of textbooks for the elementary school by Brunt and Bevel clearly show the characteristics described by Beckers, as we have seen. The textbooks for the Society were written for a different category of pupils than the textbooks for elementary schools. The boys who attended the lessons of the Society had already received some form of elementary education and they were following lessons to make progress in their occupation. One might expect therefore that the books for the Society differ in these aspects from the books for elementary school.

That is certainly the case with the *Grondbeginselen der Bouwkundige Reekenkunde.* The four characteristics which are formulated by Beckers: a mixture of word problems devoted to various occupations, abundant plain exercises, word exercises that have no connection with a practical application whatsoever and pedagogical advices and encouragements, are absent for the most part. The book is clearly more congenial to 18th century textbooks, though not completely. The authors take pains to explain the rules and methods, and they are honest when that would lead too far. In such cases for instance, they refer to the books of Brunt and Bevel, or to the well-known geometry book of Steenstra.

The *Grondbeginselen der Rekenkunde* is more in conformity with the modern arithmetic books for elementary schools of the period. The word problems pertain to a mixed variety of occupations or have no practical application at all, and there are many plain exercises. The explicit pedagogical remarks however are missing, and it is striking that Bevel and Rijk did not use the questions and answers form, that not only Van Campen had used in his version of 1786, but that Bevel had also used himself. One could argue that this form went slowly out of fashion. Nevertheless, books in this form were reprinted until the forties. It will be not coincidental that the style of the book is rather objective, not aiming at young children but at pupils that were older and had already attended some form of elementary education.

In 1820, when Bevel became inspector of weights and measures, he stopped teaching for *Mathesis*. Instead he became a kind of supervisor for all lessons in mathematics and acted as a substitute-teacher in case of problems and, as we have seen, he engaged himself as a textbook author for the Society. He remained active for the Society until his death in 1833.

6 Teaching Mathematics at a Latin School

It was not accidental that the first Dutch laws on educational matters referred to primary education. In The Netherlands, the ideals of the Enlightenment had a strong social component. Education of the lower classes had been for instance the main purpose of an influential organization like the *Maatschappij tot 't Nut van het Algemeen* [The Company for the General Wellbeing]. This Company, which still exists, was founded in 1784 by a Mennonite minister, and had departments in dozens of Dutch cities and thousands of members. Its main aim was to educate the Dutch population as a whole, for which purpose it published modern textbooks, to be used in primary schools. There was also some debate on modernization of the Latin schools, but the resistance of conservative powers to these ideas was considerable.

In the years that The Netherlands were a part of France, plans were developed to integrate the Dutch Latin Schools into the French system of the *Lycées*, but before anything could be put into practice, Napoleon was defeated and nothing came of it. Although the new King, Willem I, had generally speaking modern ideas, his measures concerning the Latin schools and universities were rather disappointing. In 1815 he issued a Royal Decree, with only few and small modernizations of the Universities and the Latin schools. One of his civil servants wrote afterwards: "One has missed a nice opportunity" (Bolkestein 1914, p. 108). For Latin schools, only some extra topics, apart from Latin and Greek, were prescribed, one of them "the principles of mathematics", with no further indication what was to be understood by this rather cryptic description.

The obligation to teach mathematics at the Latin school was met with much resistance. Some schools simply refused to do so, others taught only some arithmetics. There were also some schools that developed ambitious programs, but these were exceptions. Henricus Wijnbeek, inspector of the Latin schools, who tried to visit as many schools as he could to oversee the quality of their teaching, complained in his reports to the central government about the low level or even the absence of the teaching of mathematics in many schools. It was the intention of the government that the lessons in mathematics should be taught by the regular teachers of Latin and Greek, but after some years Wijnbeek became convinced that the appointment of special mathematics teachers was inevitable (Smid 1997).

The obligation to teach mathematics caused many practical problems. Many teachers at the Latin schools, trained as philologists, were simply not capable to teach mathematics, other bluntly refused, considering this below their dignity. The solution chosen by some schoolboards was to hire a teacher from outside to do the job. Sometimes the board hired a mathematician who was working on specialized vocational schools, for instance for the training of sailors, sometimes a schoolmaster from an elementary school with an interest in mathematics. These teachers had often a difficult time at a Latin school. Class consciousness was high in the 19th century, and teachers from elementary or vocational schools usually belonged to lower social classes than the pupils of the Latin school. The topic they taught was met with resistance by both the other teachers and their pupils.

In Leiden the rector and praeceptor used a part of the money they were paid for doing this extra teaching job to hire Michael Bevel. He became the first teacher of mathematics at the Latin school of Leiden. The fact that Bevel was employed by the other teachers and not directly by the schoolboard made his position even more vulnerable, as his successor soon would experience, but it seems that it did not have negative effects for Bevel.

The reports of inspector Wijnbeek on the Latin school of Leiden for those years are lost, and we don't know what mathematics Bevel taught and for how many hours, since also the school records give no information. Although there are no direct sources, we can deduce something from the events that occurred after 1820, when Bevel had resigned from this job. Then the board, on advice of Speyert van der Eyk, asked his colleague Jacob de Gelder, professor of mathematics at the Leiden University and a strong advocate of the teaching of mathematics at Latin schools, to teach mathematics at their school. They must have thought that it was a good idea to employ such a well-known mathematician, but they would soon regret their step.

De Gelder, who formally also was not hired by the board but by the rector F.A. Bosse, had outspoken ideas on teaching mathematics in a Latin school, which were completely unacceptable for Bosse, and the two men got involved in a series of conflicts that ended with the resignation of De Gelder. Details of the conflict are described elsewhere, but the exchange of letters during the conflict shows some glimpses of the way things were going in the years before, when Bevel was teaching there (Smid 1993).

The Latin school counted in those years about fifty pupils, divided in six grades; the three highest grades were taught by the rector, the others by the *praeceptor*. In Bosse's opinion, mathematics should be taught only to the pupils of the three highest grades; but when he proposed that to De Gelder, that sparked to everyone's surprise a conflict with his praeceptor Delfos, who suddenly wanted that his pupils also could follow lessons in mathematics. It seems that this had not been the case before.

As a compromise, De Gelder taught two different groups: one group consisting of the combined three grades taught by the rector, one group made up by the highest grade taught by the praeceptor. De Gelder taught each group two hours a week. According to De Gelder it was impossible to achieve anything with that number of hours available, even more so since the pupils could be promoted to a higher grade every half year, which implicated that the constitution of the mathematics groups also changed every half year. De Gelder wanted to have special mathematics grades, separate from the grades in use for the other teaching, but that was unacceptable to Bosse, who received full support by his board of governors.

The biography of Bosse by Bergman underlines the fact that Bosse wanted the teaching of mathematics at the Latin schools to be as minimal as possible, restricted to the elder pupils, since "the teaching of mathematics to young children is highly damaging" (Bergman 1841, p. 96). It seems highly unlikely that Bosse had, in the early years of teaching of mathematics in Latin schools, allocated to Bevel more hours than he, Bosse, clearly considered as the maximum acceptable. Two hours a week, only to the highest grades, most likely Bevel will not have taught more than that.

Although we do not know what content was taught by Bevel, we know at least one part of it. In 1816, the government wrote a letter to all Latin schools that they were obliged to teach the new metric system that had just been introduced by the King.[4] Some months later, the provincial representative of the King asked the schoolboard if this indeed was the case. The board answered that the system was taught by the "most able teacher Bevel", and that they were completely satisfied with the way he was doing this.[5] One has to add that schoolboards often answered to the central government that all was going well, while that was not really the case: they wanted to keep the government off their back. Ten years later for instance, the board spoke highly of the teaching of De Gelder's successor, Jacobus de Pecker, but the inspector of the Latin Schools, Henricus Wijnbeek, characterized his teaching as "of little significance".

The obligation to teach the metric system was not unproblematic; the board of the Latin school of Dordrecht for instance wrote to the government that "a well suited book for that purpose would highly facilitate the teaching of the system". But the government was of a different opinion since it wrote in the letter of September 11th also to the schools that "The newer books on arithmetic, which are now in use at the better equipped elementary schools, can be used as a guideline, if a special chapter is devoted to decimal fractions". There were at least some Latin schools that indeed used books for elementary schools for that purpose; the Latin school of Hoorn used volume four of the arithmetic book by Brunt (so the old version) to teach the decimal fractions, the school in Gouda used a geometry book for elementary schools, that also treated decimal fractions and the metric system, by

[4]Letter of September 11th, 1816, by Repelaer van Driel, Ministry of the Interior.

[5]RAL, archief curatoren.

P.J. Prinsen for that purpose (Smid 1997, p. 74). It seems not unlikely that Bevel used also the books of Brunt, perhaps in the revised edition by Anslijn, and maybe also his own sequel to these books, since he published a second edition in 1819, adapted to the new metric system.

7 Teaching Mathematics for the Military and on Boarding Schools

Bevel's knowledge of physics and mathematics and the old and new metric systems brought him a surprising new job. In 1820 he was appointed by the King as the official inspector of weights and measures for Leiden and its region. Before, cities had had their own gaugers, making use of different regional systems of weights and measures. Now there was a nationwide uniform standard and its implementation and maintenance was an enormous and important task. Almost all the gaugers who had been in office so far were dismissed and a complete new corps of inspectors was appointed. Bevel was not the only mathematics teacher appointed as inspector. O.S. Bangma in Amsterdam and J.G. Arbon in Rotterdam for instance, who were also teachers of mathematics, were also appointed as inspectors. In the first years these inspectors traveled around in their regions to verify and stamp all the weights and measures to be used.

Bevel had to give up his jobs at the primary school for poor children and at the Latin school and he stopped teaching for *Mathesis Scientiarium Genitrix*, although as we have seen he remained active for the Society. But after the first hectic years as inspector, when he got some more spare time, he started teaching mathematics again. He became teacher of mathematics at two prestigious boarding schools in the neighborhood of Leiden and taught mathematics to the officers of the Leiden garrison. Although these were part-time jobs, presumably only a few hours a week, they are nevertheless meaningful for the history of the teaching of mathematics in the Netherlands, because they are indications of important developments.

After the defeat of Napoleon, the government under King Willem wanted to train the Dutch army according to the French example. Officers were no longer selected and appointed by their noble birth only and should receive a serious education in which mathematics played an important role. To that purpose, the government founded in 1814 a military academy in Delft, which was moved to Breda in 1828. Jacob de Gelder whom we met already as professor of mathematics in Leiden and as teacher at the local Latin school, was teacher of mathematics in the first years of the academy in Delft. Just as in Leiden, his ideas were not acceptable for the commander of the academy, he got involved in a row and in 1819 he was fired. But he clearly had powerful supporters, since after some months he was appointed as professor in Leiden. In both cases moreover, he got his way in the end: after 1828 the teaching of mathematics in Breda was reorganized according to his ideas, and he was himself appointed in 1828 as member of the board of governors of the Latin school in Leiden to reorganize the teaching of mathematics there.

Many officers who were already serving in the army were lacking such a training based on scientific ideas. The government decided therefore in 1826, that if such officers wanted to obtain a higher rank, they should follow a kind of in-service training, including mathematics. Bevel was appointed to teach mathematics to these officers in Leiden. Most likely he taught geometry, since the garrison bought in 1827 a copy of a geometry book written by De Gelder.

The military academy would play an important role in the development of mathematics teaching in the Netherlands. From 1819 on, the academy organized each year an entrance exam, mathematics was a substantial part of these exams. After some years, private schools were founded where pupils received special training in mathematics to become more successful at these exams. In the forties, the teachers of mathematics at the academy started to write special textbooks themselves to prepare pupils for these exams, these textbooks became quite influential and were not only used at schools preparing for these entrance exams, but also in most Latin schools.

Private boarding schools belonged to a category of secondary schools usually called *French schools*. These schools had already originated in the 16th century, offering education for boys who wanted to enter the trading business. The topics they taught were usually French (hence the name), arithmetic and bookkeeping. Since the local, and from 1815 onwards also the national government, took only responsibility for the Latin schools, the French schools were mainly the result of private initiative.[6] During the 18th century they had developed into a wide range of schools of all sorts, including for instance French schools for girls. In the 19th century, some French schools offered a curriculum cognate to the later *Higher Burgher School*, ususally abbreviated as HBS, the Dutch variant of the German *Realschule*. These schools disappeared after the introduction of the HBS in 1863. Many Latin schools attached after 1840 a "second department" to their schools, in fact also a kind of French school. These second departments disappeared also after 1863.

Others French schools even became themselves in fact modern private gymnasia, offering not only modern languages, mathematics and physics, but also Latin and Greek. These schools were usually boarding schools for the élite; their pupils could obtain entrance to the university by passing an entrance exam. After the modernization of the gymnasia in 1876, these schools also went out of fashion.

The most famous of these boarding schools was a school named *Noorthey*, located in Veur, a small village halfway between Leiden and The Hague.[7] Bevel taught at this school mathematics and physics for some years, but in 1827 he had to give up this position. The director of Noorthey wanted to appoint a full time mathematics teacher, which of course was not possible for Bevel.

---

[6]Some cities, including Leiden, founded in the 19th century a municipal French school.

[7]Some decades later, even the eldest son of King Willem III, heir to the throne, attended the school.

The man who was appointed as his successor, J.H. van Sillevoldt, was perhaps the first full time mathematics teacher in the Netherlands; he ended his career as the first director of the HBS in The Hague. Bevel accepted a position as mathematics teacher at a comparable, but smaller boarding school in Noordwijk, another village in the neighborhood of Leiden. In 1833 when his health declined, he had to give up teaching there. Bevel died on December the 26th of the year 1833.

8 An Epilogue

Bevel was not an important mathematician and his reputation on a national scale was limited. He played a role of significance in Leiden, where he, although he did not belong to the real élite, certainly belonged to the upper middle class. That in itself was a remarkable achievement, taking into account that he was raised in an orphanage of a religious minority, but there are other reasons why it makes sense to pay attention to him.

He took advantage of the rapid developments around the turn of the century: political changes that offered new opportunities to religious minorities, the rise of arithmetic and mathematics as school topics in various school types, the new ideas on pedagogic, the introduction of the metric system. By studying his life and career we can envisage these developments not as isolated elements, but we can see how they came together into the life of one person, who had the talent and the perseverance to take full advantage of them. Bevel's career connects figures as Van Campen, a typical 18th century free-lance mathematician, with J.H. van Sillevoldt, a full time mathematics teacher and director of one of the first Dutch *Realschule.* His career equally connects men like Simon Speyert van der Eyk, a professor in fact typical for the 18th century who taught a whole range of topics besides mathematics, with Jacob de Gelder, who concentrated on mathematics only, was the dominant figure in the rise of mathematics education and who transformed the Dutch mathematical society from an organization mainly for amateurs into one for professionals. Bevel taught at various types of schools: elementary schools, vocational schools, Latin schools, and French schools, all of which played a significant role in the development of mathematics education in the Netherlands.

Michael Isaac Swarts Bevel was a teacher of mathematics in times of change. That is the reason why it is worthwhile to pay attention to him. We did not say much about Bevel as a person; it was not our goal to write a personal biography. It seems only fair to mention that to all testimonies of his contemporaries he was an excellent teacher and an amiable man; who "always in good company contributed most to sociability".

References

Aa, Abraham Jacob van der (1854). *Biographisch Woordenboek der Nederlanden, Tweede Deel* [Biographical Dictionary of the Netherlands, Second Volume]. Haarlem: J.J. van Brederode.

Beckers, Daniël Jeroen (1998). *'Uitnemend in beschaafdheid en verstand' Een biografische schets van de onderwijzer en wiskundige M.I.S. Bevel* ['Excellent in refinement and understanding' A biograhical sketch of the teacher and mathematician M.I.S. Bevel]. Nijmegen: KUN.

Beckers, Daniël Jeroen (2000). "My little arithmeticians!" Pedagogic Ideals in Dutch Mathematics Textbooks, 1790–1850. *Paedagogica Historica, 36*(3), 979–1001.

Beckers, Daniël Jeroen, & Smid, Harm Jan (Eds.) (2003). *Grondbeginselen der Rekenkunde* [First principles of arithmetic]. Hilversum: Verloren.

Beckers, Daniël Jeroen, & Kool, Margaretha Johanna Helena (Eds.) (2004). *Willem Bartjens. De Cijfferinghe* [Willem Bartjens. The ciphering]. Hilversum: Verloren.

Bergman, Jean Théodore (1841). *Levensschets van Frans Antoni Bosse* [Biographical sketch of Frans Antoni Bosse]. Haarlem: De erven Vincent Loosjes.

Bevel, Michael Isaac Swarts (1806, 1819). *Vervolg op de eerste beginselen der rekenkunde, voorgesteld in vragen en antwoorden, ten dienste der scholen, door Jan Brunt, in leven stadsschoolhouder te Leyden* [Sequel to the first principles of arithemetic, presented in questions and answers, for the use of schools, by Jan Brunt, late municipal schoolmaster in Leyden]. Leiden: D. du Mortier en Zoon.

Bevel, Michael Isaac Swarts, & Euler, Leonard (1807). *Volledige inleiding tot de algebra, aan de hand gevende, eene gemakkelijke oplossing, van alle soorten van rekeningen; zoo in de wiskunde, koophandel als andere zaken* [Complete introduction to algebra, providing an easy solution to all kinds of calculations, in mathematics as well as in trade and other business]. Dordrecht: A. Blussé en zoon.

Bevel, Michael Isaac Swarts (1817). *Verhandeling over de gasverlichting* [Treatise on gas lighting]. Amsterdam: G. van Dijk.

Bevel, Michael Isaac Swarts; & Rijk, Pieter Engelbrecht (1826, 1828, 1830). *Grondbeginselen der rekenkunde* [First principles of arithemetic]. Leiden: J. Cijfveer.

Bevel, Michael Isaac Swarts; Morgenster, Johannes; Speyert van der Eyck, Simon (1820). *Werkdadige meetkunst, ten dienste van ingenieurs en landmeters* [Practical geometry, for the use of engineers and surveyors]. Amsterdam: Gartman.

Bevel, Michael Isaac Swarts; Campen, Pieter van; & Rijk, Pieter Engelbrecht (1826). *Grondbeginselen der bouwkundige reekenkunde, op de gemakkelijkste wijze voorgesteld Tweede druk* [First principles of architectural arithemetic, presented in the easiest way. 2th. Edit]. Leiden: S. en J. Luchtmans en D. Du Mortier.

Bijdragen betrekkelijk den staat en de verbetering van het schoolwezen in het Bataafsch Gemeenebest [Contributions concerning the state and the improvement of the school-affairs in the Batavian Commonwealth]. (1801–1809). Leiden: D. Du Mortier en Zoon.

Boekholt, Petrus Theodorus Franciscus Maria, & de Booy, Engelina Petronella (1987). *Geschiedenis van de school in Nederland* [History of the school in the Netherlands]. Assen: Van Gorcum.

Bolkestein, Gerrit (1914). *De voorgeschiedenis van het middelbaar onderwijs 1796–1863* [The previous history of the secondary education 1796-1863]. Amersfoort: Vereeniging van Leeraren bij het Middelbaar Onderwijs.

Brunt, Jan (1802–1803). *Eerste beginselen der rekenkunde, voorgesteld in vragen en antwoorden, ten dienste der scholen* [First principles of arithmetic, presented in questions and answers, for the use in schools]. Leiden: D. du Mortier en Zoon.

Brunt, Jan, & Anslijn, Nicolaas (1816–1820). *Eerste beginselen der rekenkunde, voorgesteld in vragen en antwoorden, ten dienste der scholen. Tweede druk* [First principles of arithmetic, presented in questions and answers. 2th. Edit]. Leiden: D. du Mortier en Zoon.

Campen, Pieter van (1780). *Grondbeginselen der bouwkundige reekenkunde, op de gemakkelijkste wijze voorgesteld* [First principles of architectural arithmetic, presented in the easiest way]. Leiden: S. en J. Luchtmans en D. Du Mortier.

Coebergh van den Braak, Antonius Marie (1988). *Meer dan zes eeuwen Leids Gymnasium* [More than six centuries Leiden's gymnasium]. Leiden: s.n.

Gelder, Jan Jacob de (1856). *Wetgeving op het Lager Onderwijs binnen Leyden van 1324 tot 1796* [Legislation concerning elementary education in Leyden 1324–1796]. Leiden: De Breuk en Smits.

Geuns, J. van (1803). *Bericht aangaande het leven van Jan Brunt* [Obituary of Jan Brunt]. Leiden: D du Mortier en Zoon.

Maenen, Johannes Maria Augustinus (2002). *De invoering van het metrieke stelsel in Nederland tussen 1793 en 1880* [The introduction of the metric system in the Netherlands between 1793 and 1880]. Nijmegen: J.M.A. Maenen.

Peeperkorn – van Donselaar, Loes A. (1985). *Twee eeuwen technisch onderwijs, twee eeuwen bij de tijd* [Two centuries technical education, two centuries up to date]. Leiden: Mathesis Scientiarum Genitrix.

RAL: Regionaal Archief Leiden, archief curatoren Latijnse school, inv. nr. 453. [Regional Archives Leiden, archive of the board of governors of the Latin school, inv. nr. 453].

RAL: Regionaal Archief Leiden, archief Schoolcommissie, inv. nr. 7. [Regional Archives Leiden, archive of the School-commission, inv. nr. 7].

Schubring, Gert (1983). *Die Entstehung des Mathematiklehrersberuf im 19. Jahrhundert* [The emergence of the profession of a teacher of mathematics in the 19th century]. Weinheim: Beltz.

Smid, Harm Jan (1993). A conflict at the Leyden Gymnasium. In F. Lalande, F. Jaboeuf & Y. Nouazé (Eds.), *Proceedings of the first European Summer University on 'History and epistemology in mathematics education'* (Montpellier, 1993) (pp. 393–399). Montpellier: IREM de Montpellier.

Smid, Harm Jan (1997). *Een onbekookte nieuwigheid? Invoering, omvang, inhoud en betekenis van het wiskundeonderwijs op de Franse en Latijnse scholen 1815–1863 [An ill-considered novelty? Introduction, extent, content and significance of the teaching of mathematics on French and Latin schools 1815-1863].* Delft: Delft University Press.

Smid, Harm Jan (2009). Michiel Izaäc Swarts Bevel. Een man van gewicht(en) [Michael Izaäc Swarts Bevel. A man of Weight(s)]. *Leids Jaarboekje 2009.* Leiden: Historische Vereniging Oud-Leiden.

Steenstra, Pibo (1811). *Grondbeginselen der meetkunst, 7ᵉ druk, vermeerderd door M.I.S. Bevel* [First principles of geometry, 7th edition, augmented by M.I.S. Bevel]. Leiden: S & J. Luchtmans.

Villers, Charles de, & Bevel, Michael Isaac Swarts (1821). *De wijsbegeerte van Kant, of Grondbeginselen der bespiegelende wijsbegeerte* [The philosophy of Kant, or the first principles of speculative philosophy]. Leiden: A. en J. Honkoop.

Author Biography

Harm Jan Smid obtained a masters degree in mathematics at the Leyden University. He was a mathematics teacher, worked at a College for teacher training and was a lecturer in mathematics and mathematics education at the Delft University of Technology. His Ph.D.-thesis was about Dutch mathematics education in the 19th century. He published many articles on the history of mathematics education, including a history of Dutch mathematics teaching of the last sixty years by means of a series of 10 portraits of important Dutch mathematics teachers, teacher trainers and researchers in mathematics education.

Visual Representations of Arithmetical Operations Performed with Counting Instruments in Chinese Mathematical Treatises

Alexei Volkov

Abstract The chapter is devoted to visual representations of two counting instruments, the counting rods (suanzi 算子, chousuan 籌算) and the abacus (suanpan 算盤), found in Chinese mathematical treatises compiled prior to the early 17th century. The former instrument is generally believed to be replaced by the latter in the mid-second millennium AD. The role of the visual representations of these two instruments in didactical practices has never been given due attention by historians of mathematics education. The author provides a preliminary analysis of the images of the instruments in two mathematical textbooks, the Shendao dabian lizong suanhui 神道大編暦宗算會 compiled in 1558 by Zhou Shuxue 周述學 and Panzhu suanfa 盤珠算法 completed in 1573 by Xu Xinlu 徐心魯 and conjectures that these images were used mainly for didactical purposes.

Keywords History of Chinese mathematics · Chinese counting rods (suanzi 算子, chousuan 籌算) · Chinese abacus (suanpan 算盤) · Zhou Shuxue 周述學 Xu Xinlu 徐心魯

1 Introduction

The extant Chinese mathematical treatises representing indigenous mathematical tradition (that is, the treatises compiled prior to the introduction of Western mathematics in the early 17th century) contain problems required to be solved with counting instruments, namely, with the counting rods (*suanzi* 算子, *chousuan* 籌算) and (Chinese) beads abacus (*suanpan* 算盤).[1] It is generally assumed that the

[1]For brief descriptions of the instruments see, for example, Martzloff (1997), Jami (1998), and Volkov (1998, 2017).

A. Volkov (✉)
National Tsing Hua University, Hsinchu, Taiwan
e-mail: alexei.volkov@gmail.com

F. Furinghetti and A. Karp (eds.), *Researching the History of Mathematics Education*, ICME-13 Monographs,
https://doi.org/10.1007/978-3-319-68294-5_14

counting rods were used in China from the first millennium BC to the mid-second millennium AD, when they were gradually replaced by the beads abacus.

The present chapter is devoted to the descriptions of arithmetical operations supposed to be performed with the counting instruments provided in mathematical treatises presumably used for instruction; a special attention will be paid to the diagrams representing configurations of counters on the counting instruments (rods and beads, respectively) arguably introduced in order to explain to the learners manipulations with the instruments. More specifically, I will discuss two sets of diagrams accompanying descriptions of arithmetical operations performed with two computing instruments, the counting rods and the abacus. These diagrams found in mathematical treatises of the 16th century were arguably provided for educational purposes.

The extant Chinese mathematical treatises compiled prior to the end of the first millennium AD were used in the state educational institutions in China and, some time later, in similar institutions in Korea and Japan; while the audience originally targeted by their compilers remains unknown, their extensive use in educational institutions as mathematical textbooks is well documented.[2] Interestingly enough, the number of references to the counting rods in these texts is relatively small (they are explicitly mentioned, for example, in descriptions of the methods of multiplication and division, of extraction of square and cube roots, and of solution of simultaneous equations in several unknowns), their descriptions in mathematical texts are rare,[3] the methods of representation of numbers with them are mentioned only briefly,[4] and certain elementary operations (for example, addition and subtraction) are not described.

Descriptions of some operations with counting rods can be found in the following mathematical textbooks of the first millennium AD:

- *Sunzi suanjing* 孫子算經 (Computational treatise of Master Sun), an anonymous (despite its title)[5] treatise compiled between the late third and the early fifth century AD; for its history and contents see translations and studies of Berezkina (1963) and Lam and Ang (2004). It contains descriptions of operations of multiplication and division of (integer) numbers (the described procedures could be applied to numbers having a fractional part expressed in the form of decimal fraction).[6]

---

[2]For a description of mathematics education in China, Korea, and Japan prior to 1500 see Volkov (2014).

[3]Only a very short description of the instrument is found in the *Shu shu ji yi* 數術記遺 (Procedures of "numbering" recorded to be preserved [for posterity]) compiled in the 2nd century AD, see Guo and Liu (2001, p. 448); for descriptions of the counting rods found in texts not directly related to mathematics, see Li (1929) and Volkov (2017).

[4]See, for example, a description found in the *Sun zi suan jing* 孫子算經, see Guo and Liu (2001, p. 262); for translations, see Berezkina (1963, p. 23), Lam and Ang (2004, p. 193).

[5]For a hypothesis concerning the author of the treatuse see Volkov (2012, p. 517, note a).

[6]See Guo and Liu (2001, p. 262); for translations and interpretations, see Berezkina (1963, pp. 23–24), Lam and Ang (2004, pp. 193–195).

– *Jiuzhang suanshu* 九章算術 (Computational procedures of Nine Categories, also known in Western translations as "Nine Chapters on Mathematical Art", etc.). The date of compilation of this treatise is unknown; most probably it was compiled no later than the early first century AD on the basis of earlier materials.[7] It contains descriptions of operations performed with the counting rods including extraction of square and cube roots and of solution of linear simultaneous equations.[8]

In the earliest extant printed editions of these treatises dated of the early second millennium AD the descriptions of operations are not accompanied by any diagrams showing configurations of counting rods in the course of computations. It remains unknown if diagrams of this kind were provided in the earliest (handwritten) editions of the treatises, and if they were, why were they omitted by later editors; for instance, were they considered superfluous, or there were some other reasons. It is possible that the diagrams showing these configurations circulated separately from the early handwritten copies of the treatises and were lost by the end of the first millennium AD.

The mathematical treatises of the early second millennium differed from their precursors considerably. The targeted audience of these texts is not well known; theoretically, they also may have been used for instruction, but there is no supporting evidence to this effect. Moreover, some of them featured rather complex mathematical matters (in particular, polynomial algebra) and could have been hardly used as mathematics handbooks in large-scale educational projects, such as, for instance, the program in mathematics and mathematical astronomy implemented at the State Mathematics College of the Song dynasty (960–1279).[9] These texts also contained references to operations performed with counting rods, and in some cases provided diagrams featuring annotated configurations of counting rods. For instance, in the *Shushu jiuzhang* 數書九章 (Nine chapters of writings on computations) compiled in 1247 by Qin Jiushao 秦九韶 (ca. 1202–ca. 1261) one finds numerous diagrams representing configurations of counting rods provided to explain the solutions.[10]

[7]For the history and contents of the treatise see Berezkina (1957), Kawahara (1980), Chemla and Guo (2004), and Dauben et al. (2013).

[8]See Guo and Liu (2001, pp. 121–125, 173–183). For translations, see Chemla and Guo (2004, pp. 361–368, 371–385, 617–659), Dauben et al. (2013, pp. 373–409, 415–455, 905–1033).

[9]On this College see, for example, Volkov (2014, pp. 66–68); on the history of the College and its functioning in the first millennium AD see Volkov (2012, pp. 514–523).

[10]See Qin ([1247] 1993). For a comprehensive study of this treatise see Libbrecht (1973).

2 Procedures of Multiplication Performed with Counting Rods Found in the *Shendao Dabian Lizong Suanhui*

The mathematical treatise *Shendao dabian lizong suanhui* 神道大編曆宗算會 (Grand Compendium of Divine Dao: [Chapters presenting methods] of lineages [of experts] in calendrical computations and of assembly of mathematicians [of the past]) (Zhou [1558] 1995) was compiled by Zhou Shuxue 周述學 (dates unknown; active in the mid-16th century) in 1558; it remains practically unknown to the Western reader, and only a small number of works in Chinese have been devoted to the treatise and its author.[11] At the time when Zhou was writing his book the counting rods were falling into oblivion, and the abacus (*suanpan* 算盤) was gaining popularity among Chinese mathematicians; suffice it to mention the treatise *Suan fa tong zong* 算法統宗 (Unified Origins of the Computational Methods) featuring abacus computations and published by Cheng Dawei 程大位 (1533–1606) in 1592. Zhou's treatise contains three series of diagrams representing operations with counting rods: two of them are provided to explain two different algorithms of multiplication of integer numbers, and the third one, an algorithm of division.

2.1 *The First Procedure of Multiplication Described in Zhou's Treatise*

The described procedure differs from the classical one featured in the *Sun zi suan jing*, since the products of the digits are placed in the line occupied by the upper operand (according to the method described in the *Sun zi suan jing*, the products should be placed in the third line located between the operands); however, despite this difference the two algorithms are virtually identical. To explain the first algorithm, Zhou multiplies 236 by 342; the first diagram he provided shows the initial position of the two operands on the counting surface (Fig. 1). The lower multiplicand (called *fa* 法, lit. "norm, model,"and "divisor" in the context of division) is shifted leftwards for two positions so that its last digit is placed under the first digit of the upper operand (called *shi* 實, lit. "actual [amount]," and "dividend" in the context of division).

The author provides captions explaining the meaning of the digits represented as configurations of counting rods. Each step of the procedure is accompanied by a diagram of the same kind.

The second step is shown in Fig. 2: the first digit of the upper operand, 2, is marked with the character *ding* 頂 (lit. "top of the head"); this digit multiplies (lit. "calls" *hu* 呼) the first digit, 3 of the lower operand marked in the diagram with the

[11]See, for example, Li (1999, pp. 483–485); for a systematic study of the treatise see Yang (2003).

		2	3	6
3	4	2		

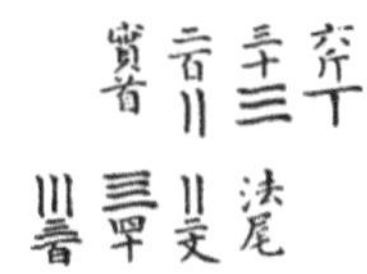

Fig. 1 Step 1: initial positions of the operands (236 and 342) on the counting surface (right) and their transcription with modern (arabic) numerals (left)

6		2	3	6
3	4	2		

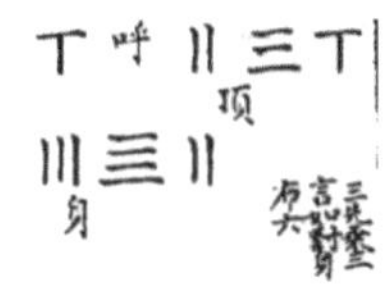

Fig. 2 Step 2: multiplication of 3 of the lower operand by 2 of the upper operand. Note a misprint in the caption: "3 multiplies 3" instead of "2 multiplies 3"

6	8	2	3	6
3	4	2		

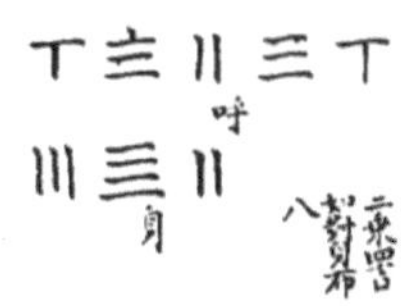

Fig. 3 .

character *shen* 身 "body". The result of this multiplication, 6, is placed above the operand 3.

At step 3 the first digit, 2, of the upper operand multiplies the second digit, 4, of the lower operand; the digit 4 is marked with the character "body". The result, 8, is placed in the upper row above the operand 4 (Fig. 3).

Step 4: the first digit, 2, of the upper operand multiplies the last digit, 2, of the lower operand; the last digit 2 in lower position is marked with the character "body". After that the first digit of the operand, 2, is removed from the counting surface, and the product, 4, is put in this position (Fig. 4).

There are two operations performed at step 5. Firstly, all the counting rods representing the lower operand, 342, are shifted rightwards for one position. When doing so, the person performing the computation has to alternate the orientations of rods in each position, e.g., when two vertical rods placed in the position of hundreds are shifted to the position of dozens, they should be placed horizontally. Secondly, the first (leftmost) remaining digit of the upper operand, 3, now marked with the character "top of the head", multiplies the first digit, 3, of the lower operand. The result of this multiplication, 9, is to be placed right above the lower operand, i.e. in the fourth position occupied by digit 8. Since 8 + 9 = 17, one unit is carried to the fifth position from the right (Fig. 5).

At step 6 the digit 3 of the upper operand placed in the second position multiplies the digit 4 of the operand below; the product, 12, should be placed above this latter digit: 74 + 12 = 86 (Fig. 6).

At step 7 the digit 3 of the upper operand multiplies the digit 2 of the lower operand; the result, 6, is placed in the second position of the upper row, thus replacing the digit 3 which is no longer necessary (Fig. 7).

Again, there are two operations performed at step 8: the lower operand is shifted to the right for one position (the orientation of the rods is changed one more time), and then the only remaining digit of the upper operand, 6, multiplies the first digit, 3, of the lower operand. The result, 18, is to be added to 786 found in the upper line; the sum 804 replaces the latter number. It is interesting that the text contains a written symbol of zero in 4th position from the right; on the counting surface zero in a position does not need any special marker (the position is just left empty) (Fig. 8).

6	8	4	3	6
3	4	2		

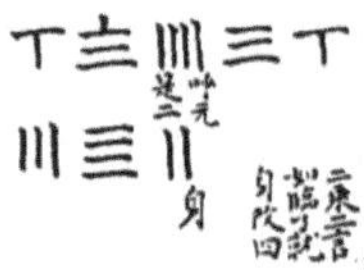

Fig. 4 .

7	7	4	3	6
	3	4	2	

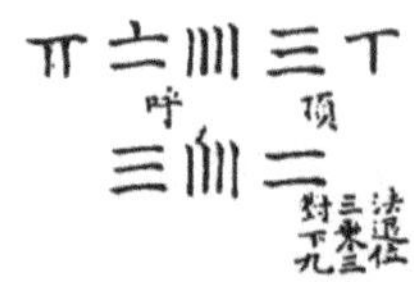

Fig. 5 .

7	8	6	3	6
	3	4	2	

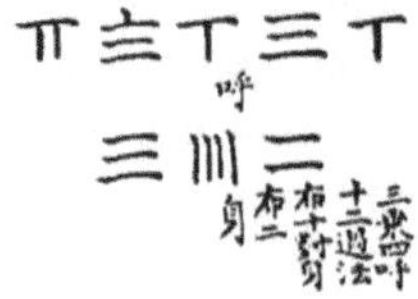

Fig. 6 .

7	8	6	6	6
	3	4	2	

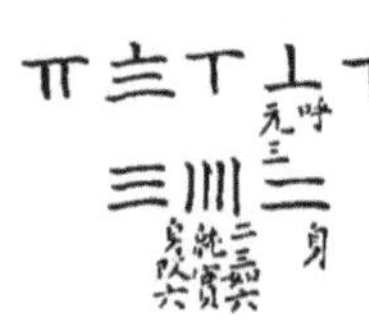

Fig. 7 .

8	0	4	6	6
		3	4	2

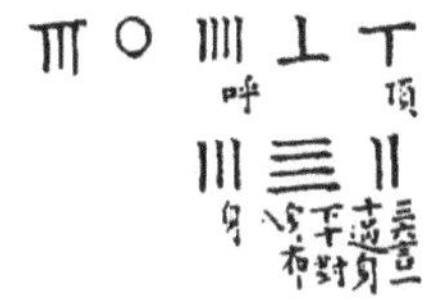

Fig. 8 .

8	0	7	0	6
		3	4	2

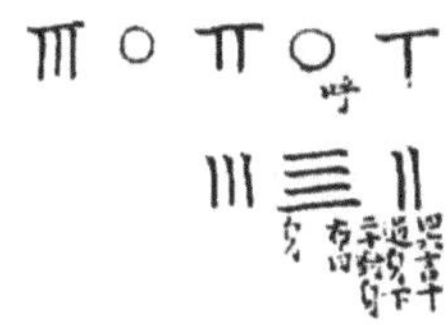

Fig. 9 .

8	0	7	1	2
		3	4	2

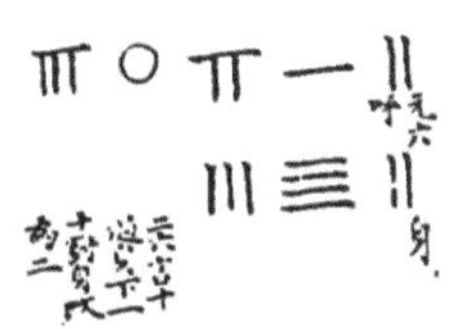

Fig. 10 The final configuration. The product (80712) occupies the upper row. Note the use of the round symbol for an empty position on the counting surface (i.e., zero)

At step 9, the last remaining digit of the upper operand, 6, multiplies digit 4 of the lower row, and the result, 24, has to be added to 46 found in the upper row in the second and third positions from the right. The text contains one more circular sign of zero, this time to mark the empty position of dozens (Fig. 9).

At step 10, the digit 6 in the position of units of the upper row multiplies 2 found below it, and the product, 12, is placed in the first and second position from the right. The final result, 80712, is now found in the upper row. There are no further instructions concerning the operand 342 remaining on the counting surface (for example, there is no instruction prescribing to remove it) (Fig. 10).

2.2 *The Second Procedure of Multiplication Described in Zhou's Treatise*

The second procedure of multiplication is also supposed to be performed with the operands placed on the counting surface one above the other. The procedure is exemplified with the multiplication of the same numbers, 236 and 342.

Step 1. At this step, the operands are placed on the counting surface one above the other. Unlike the "traditional" method described in 2.3.1., the position of units is not determined in advance by the position of any of these two numbers (Fig. 11).

The second step of multiplication procedure comprises a sequence of three operations; only the final disposition of the counting rods supposed to be obtained at the end of this step is shown as a diagram. The detailed instructions are provided under the diagram. The operations to be performed are as follows (below I provide additional diagrams not found in the treatise; they are marked with an asterisk "*"):

Step 2, operation 1: digit 6 from the upper operand multiplies digit 4 of the lower one; the result, 24, is placed in the upper row above the digit 4 (Fig. 12a).

Step 2, operation 2: digit 6 from the upper row multiplies digit 2 of the lower one; the result, 12, is placed above the digit 2 (here 5 results from 4 that was placed on the surface earlier and added to the newly obtained "1" from 12) (Fig. 12b).

Step 2, operation 3: digit 6 of the upper row multiplies digit 3 of the lower one; the result is 18. The digit 6 is removed from the surface, 2 is added to 18, and the sum, 20, is placed above "34":

Step 3. At this step, the first operation is a shift of one of the operands performed to make the disposition on the counting surface look as follows (Fig. 13a).

Then the following operations have to be performed:

Step 3, operation 2: digit 3 of the upper row multiplies digit 4 of the lower one; the result, 12, is added to the "20" of the upper row (Fig. 13b).

Step 3, operation 3: digit 3 of the upper row multiplies digit 2 of the lower one; the result, 6, is added to "5" in the upper row (Fig. 13c).

Step 3, operation 4: digit 3 of the upper row multiples digit 3 of the lower one; after that, this digit, 3, is removed. The result of multiplication, 9, is added to the second "3" in the upper row, and the sum, 12, is placed on the surface above "34" (Fig. 13d).

It is not specified which operand was supposed to be shifted before the beginning of these operations. A comment is due here: the upper operand in the configuration shown above contains the leftmost digit "2" represented with two horizontal rods, but the digit "1" next to it is also represented with a horizontal rod, which violates the basic principle of representation of numbers with counting rods. Therefore here we are dealing with an error of the copyist.

Step 4. At this step, the first operation is to shift the lower operand to the left for one position; the orientation of the rods shown in the diagram again remained unchanged (Fig. 14a).

2	3	6					
		3	4	2			

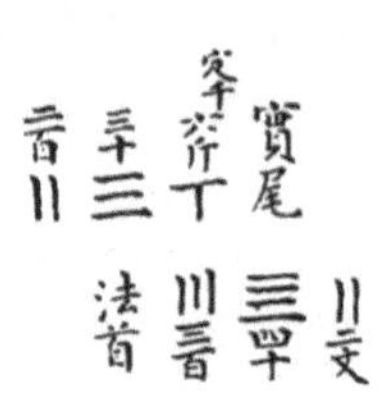

Fig. 11 Second method of multiplication; the original positions of the operands

a*

2	3	6	*2*	*4*			
		3	4	2			

b*

2	3	6	2	*5*	*2*		
		3	4	2			

c

2	3	*2*	*0*	5	2		
		3	4	2			

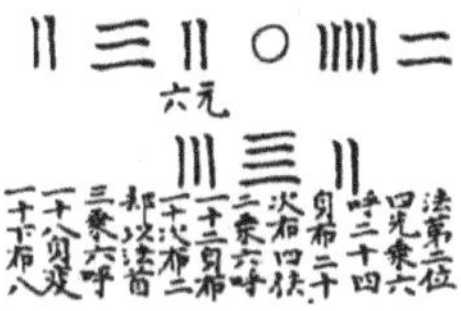

Fig. 12 Tables **a***, **b***, **c** show the configurations on the counting surface after the three operations included in Step 2

a*

	2	3	2	0	5	2	
		3	4	2			

b*

	2	3	*3*	*2*	5	2	
		3	4	2			

c*

	2	3	3	*3*	*1*	2	
		3	4	2			

d

	2	*1*	*2*	3	1	2	
		3	4	2			

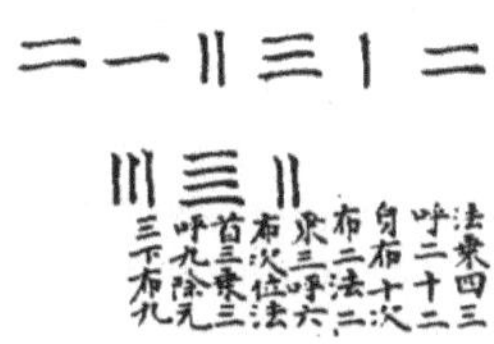

Fig. 13 Tables **a***, **b***, **c***, **d** show the configurations on the counting surface resulting from the operations included in Step 3

The last remaining digit of the upper operand, 2 in the leftmost position, is going to be used to multiply the lower operand.

Step 4, operation 2: the result of multiplication of 2 and 4, 8, is placed in the upper row in the cell above "2", since that cell already contained "2", one unit will be carried to the next position to the left (Fig. 14b).

Step 4, operation 3: the leftmost digit "2" of the upper row multiplies the rightmost "2" of the lower one; the result, 4, is to be added to "3" in the upper row (Fig. 14c).

Step 4, operation 4: the leftmost "2" of the upper row multiplies "3" of the lower row, and has to be removed after that; the result, 6, is added to the "2" in the upper row found above the "4" in the lower one (Fig. 14d).

The result of the multiplication, 80712, is found in the upper row. Note again the zeros found in the transcription of the configuration of the counting rods; the left one is not necessary, and yet it was kept by the author (or added by a later editor or copyist).

a*

	2	1	2	3	1	2	
	3	4	2				

b*

	2	2	0	3	1	2	
	3	4	2				

c*

	2	2	0	7	1	2	
	3	4	2				

d

		0	8	0	7	1	2
		3	4	2			

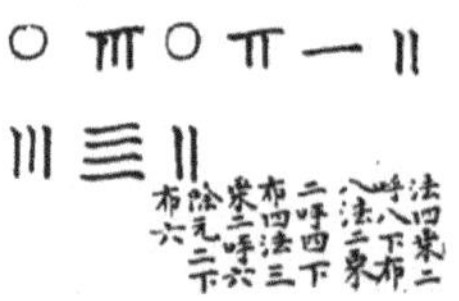

Fig. 14 Tables **a***, **b***, **c***, **d** show the operations leading to the final configuration on the counting surface. Note two round symbols for zero, one of them in the leftmost position

3 Procedure of Division Performed with Counting Rods Found in the *Shendao Dabian Lizong Suanhui*

The procedure of division is identical with that described in the *Sun zi suan jing*,[12] that is, three horizontal rows of the counting surface are used (the divisor is placed in the lower row, the dividend, in the middle one, and the result is placed in the upper row). The procedure is explained with the example of division directly related with the two examples used to explain the procedure of multiplication: the dividend is 80712 and the divisor, 236; it is easy for the reader to guess that the expected result equals to 342.

Step 1. The dividend (80712) is placed in the middle line, the divisor (236) in the lower line. The divisor is shifted to the left for two positions. Accordingly, the first digit of the quotient will be placed in the upper line in the position of hundreds (above digit "7" in the middle row); this position is marked with one counting rod (it is interesting that a rod is used here as a mere marker, and is not supposed to express any particular numerical value) (Fig. 15).

Step 2. The first digit of the quotient is estimated as equal to 3; it remains unknown how this estimation was performed. One may suggest that the procedure used to estimate this digit consisted in two operations: (a) division of first digit of the dividend (8) by the first digit of the divisor (2); the result, 4, was interpreted as suggesting that the first digit of the quotient could not be larger than 4. Apparently,

[12]For translations and discussions of the latter procedure see Berezkina (1963, pp. 23–24), Lam and Ang (2004, pp. 194–195).

		1		
8	0	7	1	2
2	3	6		

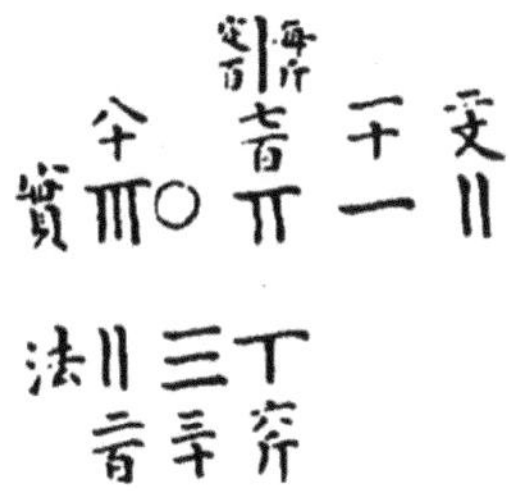

Fig. 15 .

the digit 4 was tried but the product of 4 and 236 happened to be larger than 807 (actually, to discard the option "4" it would suffice to notice that 4 multiplied by 23 is larger than 80); (b) verification of the fact that the product $3 \cdot 236$ is smaller than 807.[13] Once "3" is selected as the first digit of the quotient, the product $3 \cdot 2$ is subtracted from the highest position of the middle line (Fig. 16).

Step 3. At this step the first found digit of the quotient, 3, multiplies the second digit of the divisor, 3, and the product, 9, is subtracted from "20" located in the first two non-empty positions of the middle row (Fig. 17).[14]

Step 4. At this step the first found digit of the quotient, 3, multiplies the third digit of the divisor, 6, and the product, 18, is subtracted from "117" located in the first three non-empty positions of the middle row (Fig. 18).

Step 5. At this step two operations are performed: (1) the divisor is shifted to the right, and (2) the second digit of the quotient is found. As far as (1) is concerned, it is important to note that the orientation of the rods forming the digits of the quotient was changed in order to correspond the traditional convention (units are vertical, dozens horizontal, etc.). Once the divisor is shifted, the disposition on the counting surface would look as follows (no diagram is provided in the treatise) (Fig. 19a).

To identify the second digit of the quotient, it would be necessary to test the largest integer number smaller than 9/2, that is, 4. This number turns out to be good, since $4 \cdot 236$ is less than 991.[15] The product $4 \cdot 2$ is subtracted from 9 in the leftmost position of the number in the middle row, and the resulting configuration would look as shown in Fig. 19b.

Steps 6 and 7. At these two steps, the subsequent digits of 236, namely 3 and 6, are multiplied by the second digit of the quotient, 4, and subtracted from the number in the middle row; the configurations obtained at these two steps would look as in Figs. 20 and 21:

In Fig. 20 "7" is obtained as $19 - 12$ $(= 4 \cdot 3)$ and "47" in Fig. 21 is obtained as $71 - 4 \cdot 6$.

[13]This reconstruction has one obvious weak point: the operations necessary to conduct it appear to be more complex than those used at the following steps for the division itself.

[14]The digit "3" in the upper row was missing in the copy of the treatise used for this translation.

[15]Once again, the actual process used to test the digit 4 is not known.

		3		
2	0	7	1	2
2	3	6		

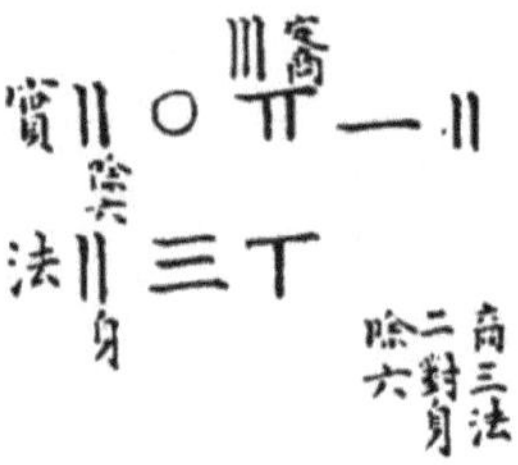

Fig. 16 .

		[3]		
1	1	7	1	2
2	3	6		

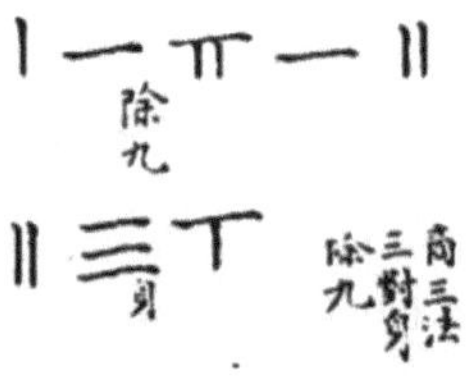

Fig. 17 .

		3		
	9	***9***	1	2
2	3	6		

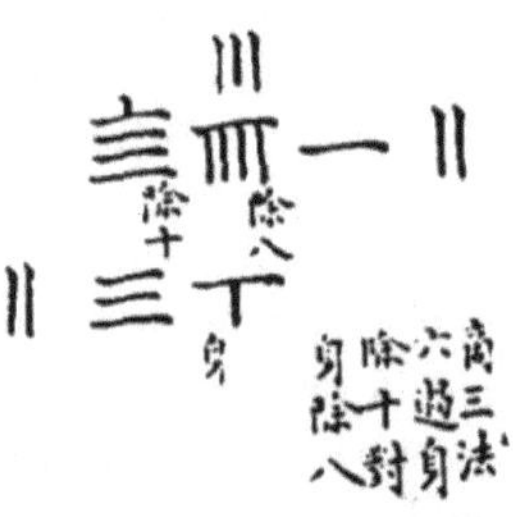

Fig. 18 .

a*

		3		
	9	9	1	2
	2	3	6	

b

		3	4	
	1	9	1	2
	2	3	6	

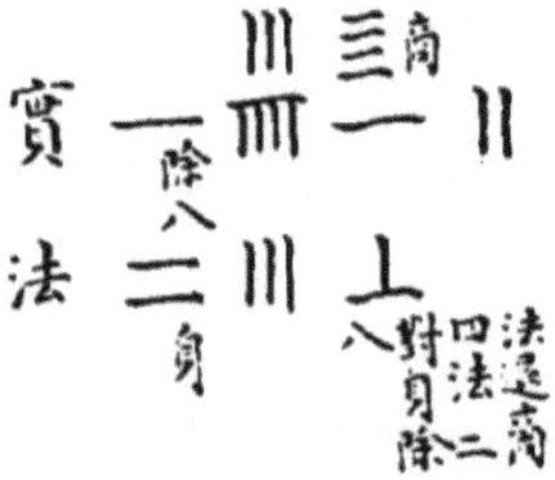

Fig. 19 **a***, **b**

		3	4	
		7	1	2
	2	3	6	

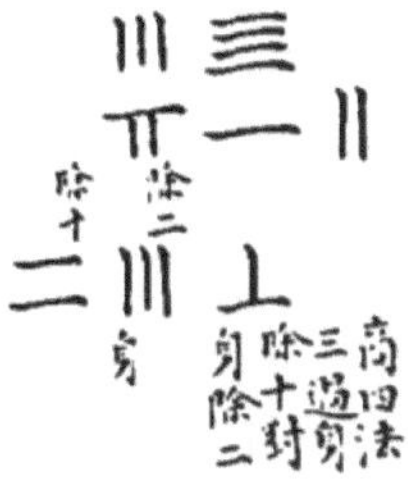

Fig. 20 .

		3	4	
		4	*7*	2
	2	3	6	

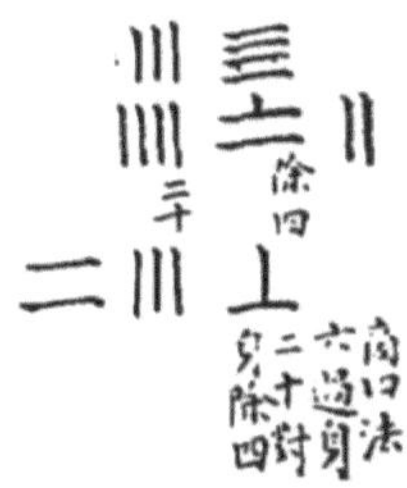

Fig. 21 .

Step 8. At this step, the divisor is again shifted rightwards for one position; it should be noted that the rods changed their orientation once again. No diagram showing the configuration on the counting surface is provided, but this diagram must have looked as in Fig. 22a.

The calculator now has to find the last digit of the quotient. To evaluate this digit, the first digit of the amount remaining in the middle row, 4, can be divided by the first digit of the divisor, 2; since the result equals to 2, the last digit of the quotient can be 2 or 1. The record provided in the treatise shows the computations performed for the digit 2. At step 8, this 2 multiplies the first digit of the divisor, also equal to 2, and the result, 4, is subtracted from the digit found in the position of hundreds of the middle row, which also happened to be equal to 4. All the counting rods therefore have to be removed from this position, as in Fig. 22b.

Step 9. At this step, the last digit of the quotient, 2, multiplies the second digit, 3, of the divisor, and the obtained product, 6, is subtracted from the digit 7 above it (Fig. 23).

Step 10. At this step (that happened to be the last one) the last digit of the quotient, 2, multiplies the last digit of the divisor, 6, and the obtained product, 12, is subtracted from the number 12 located in the middle row. The middle row is empty now; this means that the operation of division is completed, and the result is found in the upper row (Fig. 24).

a*

		3	4	
		4	7	2
		2	3	6

b

		3	4	2
			7	2
		2	3	6

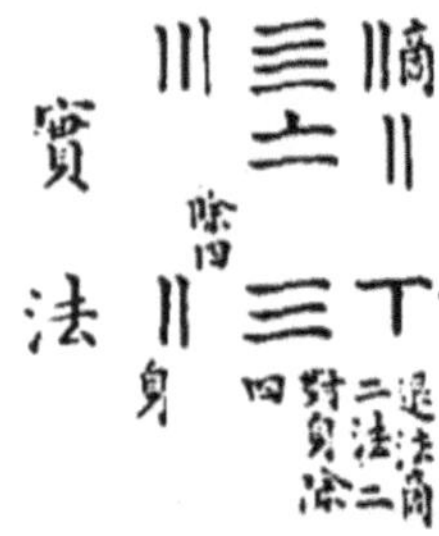

Fig. 22 a*, b

		3	4	2
			1	2
		2	3	6

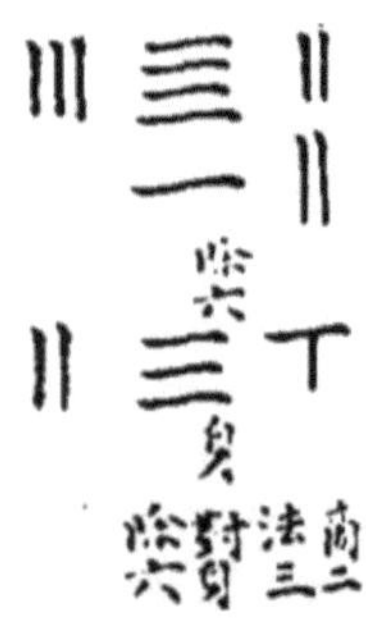

Fig. 23 .

		3	4	2
		2	3	6

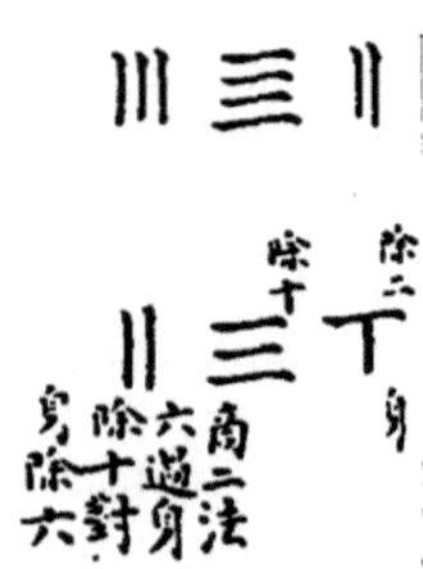

Fig. 24 .

4 Images of Configurations of Beads on an Abacus

4.1 The Earliest Chinese Treatises Containing Visual Representations of Operations with the Abacus

The mathematical texts that started appearing in China since the early second millennium AD contained references to operations supposed to be performed with the abacus (*suanpan* 算盤) , that is, the well-known instrument designed as a rectangular frame with several (vertical) bars and beads sliding on them. The origin of the instrument remains obscure, and will not be discussed here (the interested reader is referred to Li (1931)) . The extant texts contained numerous explanations as to how the basic arithmetical operations were supposed to be performed with the instrument, and these explanations were usually provided in the opening sections of the textbooks. Even though the targeted audience of these treatises and circumstances of their use are unknown, it would be safe to conjecture that at least the opening parts (sometimes rather voluminous) of them were designed as elementary arithmetical primers featuring arithmetical operations supposed to be performed on the abacus. The earliest extant text of this kind containing depictions of the beads abacus is the *Panzhu suanfa* 盤珠算法 (Computational methods for pearls in a tray [=abacus]), 1573, by Xu Xinlu 徐心魯; the other two are the *Shuxue tonggui* 數學通軌 (Comprehensive guide to mathematical studies), 1578, by Ke Shangqian 柯尚遷 and the *Jiulong yijue suanfa* 九龍易訣算法 (Computational methods [using] easy mnemonic formulae of Nine Dragons) (author unknown), compiled most probably in the late 16th century. The explanations of operations with the abacus in these texts are richly illustrated; the drawings feature configurations of beads on the counting device provided, apparently, for didactical purposes. Later similar diagrams systematically appeared in Japanese mathematical treatises. Below I will briefly discuss the diagrams explaining manipulation with the abacus found in the treatise *Panzhu suanfa*.

4.2 Panzhu Suanfa 盤珠算法 (Computational Methods for "Pearls in a Tray" [= Abacus]): A Brief Introduction

The introductory notice written by Han Qi 韓琦 for a recent reprint of this treatise in Guo (1993) informs that the full title of this treatise is *Xinke dingzheng jiachuan mique panzhu suanfa shimin liyong* 新刻訂正家傳秘訣盤珠算法士民利用, that is, *Computational methods for abacus, secretly transmitted within families (or "a family?"), newly blockprinted* [*lit. "carved"*] *with emendations and corrections [to be] beneficially used by noble men and* [*simple*] *people.*[16] The dates of life and

[16]Han (1993, p. 1141).

biographical data of the presumed compiler (or editor?) of the treatise, Xu Xinlu 徐心魯, are unknown. In his introductory note another modern publisher of the treatise, Jing Yushu 靖玉樹, suggests that Xu (who presumably lived in the mid- and late 16th century) was only the editor of the original version of the treatise compiled much earlier (Jing 1994b, p. 2037).

Hua (1987, p. 74), informs about two more treatises with slightly different titles and, apparently, identical or almost identical contents; he suggests that these three versions may have been editions of one and the same original text.[17] Hua also informs that this treatise does not exist any longer in China, and that one copy of it is preserved in Japan, in the Imperial Cabinet Library (Naikaku bunko 內閣文庫).[18] The text reproduced in Kodama (1970) [that is, Xu ([1573a] 1970/1994)] indeed bears a seal of this library on the first page, yet the copy reproduced in Guo (1993) (that is, Xu [1573b] 1993) does not have the seal;[19] therefore, one may conclude that there still exist at least two copies of the treatise.[20]

The treatise consists of two chapters (*juan* 卷). The first one contains explanations of operations with the abacus (Xu [1573b] 1993, pp. 1143–1152), lists of "large" and "small" numbers (pp. 1152–1153), multiplication table (ibid.), and brief information about monetary units, measures of volume, etc. (p. 1153). It also contains seven problems accompanied with diagrams showing positions of beads; they will be discussed below. The second chapter contains a number of problems to be solved with the abacus, a large part of which are calculations of areas of fields. This relatively small treatise is rather copiously illustrated; the first part contains 37 diagrams showing the positions of beads on the abacus; on average, almost each page devoted to operations with the instrument contains at least one diagram. Diagrams showing positions of beads are found also in the second chapter of the treatise (here there are 16 of them); they are apparently introduced to explain operations necessary to be performed for solutions of problems found in this chapter. The abacus featured in this treatise is of the type 1 + 5, that is, it has one bead in the upper section and five beads in the lower one. However, Hua Yinchun 華印椿 quotes the Japanese scholar Totani Shinichi 户谷清一 who on the basis of his analysis of one problem of this treatise concluded that the abacus used for the solution of this problem must have had two beads in the upper section (Hua

[17]Kodama (1970) on p. 23 reproduces two pages of each of these two versions.

[18]Li (1999, p. 486, n.1) mentions this copy of the treatise.

[19]My preliminary investigation of the versions Xu ([1573a] 1970/1994) and Xu ([1573b] 1993) suggests that they were printed with the same printing blocks, that is, that they are textually identical. However, their full titles differ, which means that the title on the wooden block used for printing the first page was re-carved.

[20]The whereabouts of the copy reproduced in Guo (1993) [that is, (Xu [1573b] 1993)] are not specified in the introduction by Han (1993) (even though Han Qi mentions that a copy of it is preserved in the Cabinet Library, the copy Xu ([1573b] 1993) does not have the seal of this library on page 1b, unlike the copy Xu ([1573a] 1970). It is known that one more copy of the treatise is preserved in the Institute for Research for Humanities, Kyoto University (Japan), but this copy remains unavailable to me.

1987, pp. 74–75). Hua thus suggests that the illustrations showing one bead in the upper section were provided only to simplify the diagrams of the author. This opinion was contested by Jing Yushu, who claimed that it would not have been difficult for the publisher to carve the second bead in the upper part if it were necessary (Jing 1994b, p. 2038).

In all the depictions of abacus in this treatise the beads are shown in two different colors, white and black. At the first sight at the very first diagram (Fig. 25) the colors are added to distinguish between "active" (or "activated") beads (that is, beads involved in representation of a number on the instrument and moved to the horizontal bar) and the "passive" (not representing any number and shifted to the frame of the instrument).

However, there is no correspondence between the color of the beads and their "active" or "passive" status, as the second diagram suggests (see Fig. 26): the beads pushed to the horizontal bar (that is, "activated" ones) are colored black.[21]

As it becomes clear from a preliminary analysis of the contents of the chapter, the colors indeed had a special function, but their interpretation should not be based on the distinction between the "active" and "passive" status of the beads; instead, the white color was used to designate the beads that were shifted during the last operation performed on the instrument, while the black color was used for the beads left untouched.

The page layout of the treatise deserves at least a brief mention. Each page is divided into two parts by a horizontal line. The upper sections of the majority of pages of the first chapter contain depictions of the abacus accompanied by brief prescriptions concerning operations (hereafter referred to as "captions"), while the lower sections contains more detailed verbal explanations, and, in some cases, depictions of the abacus. To analyze the interrelationships between the diagrams of the abacus and the contents of the treatise, a description of the operations listed in the treatise appears to be necessary. This description is provided in the following section.

4.3 *Representation of Operations to be Performed with the Abacus Found in the Panzhu Suanfa* 盤珠算法

The upper parts of pages 1b–10a of the first chapter contain diagrams titled "*Di K shang fa* (第K上法)" and "*Di L tui fa* (第L退法)", that is, "Rule for moving

[21]This use of colors is repeated throughout the entire first chapter: in the diagrams placed in the upper sections of pp. 1b, 2b, 3b, 4b, 5b, 6b, 7b, 8b, 12a–b, 13a–b, 14a–b, 15a–b, 16a–b, 17a–b, 18a–b, 19a–b the "active" beads are white, while in the diagrams in the upper sections of pp. 2a, 3a, 4a, 5a, 6a, 7a, 8a, 9a the "active" beads are black.

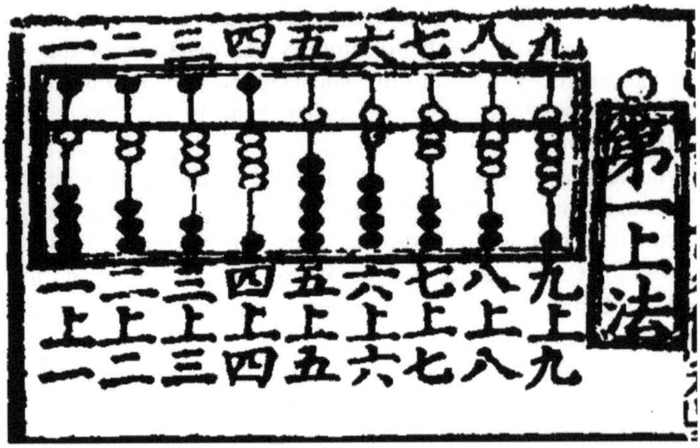

Fig. 25 .

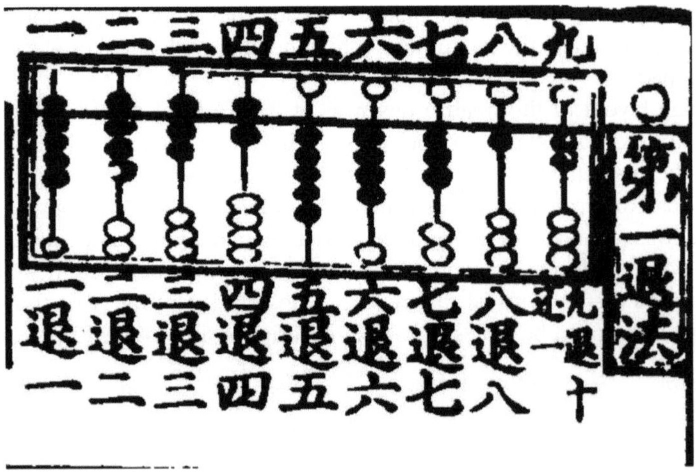

Fig. 26 .

up/raising corresponding to K" for $K = 1, \ldots 9$, and "Rule for retreating corresponding to L") for $L = 1, \ldots 9$, respectively. The upper parts of pages 12a–19b contain diagrams titled "*M yin fa* (M因法)" and "*N gui fa* (*N* 歸法)", that is, "Rule for multiplication [by] *M*" ($M = 1, 2, \ldots 9$) and "Rule for division [by] *N*" ($N = 2, \ldots 9$).

4.3.1 Diagrams Related to the "Rules for Moving up"

The following diagrams are placed in the upper parts of the pages; the beads shown in white are those being moved to perform the operation.[22]

4.3.2 Diagrams Related to the "Rules for Retreating"

The operations mentioned as "Rules for retreating" deal with successive subtractions of 123456789 from 1111111101.[23] The configurations of beads resulting from these operations are shown in the diagrams inserted between those listed in Table 1. The first obtained value thus equals to 1111111101 − 123456789, the second one, to 1111111101 – 2 * 123456789, and so on. Apparently, the numbers obtained at each step should be equal to those obtained via "Rules for moving up"; interestingly enough, in some cases their representations *differ* from those listed in Table 1. The diagrams, once again, are placed in the upper parts of the pages and the beads shown in white are those being moved to perform the operation (Table 2).

4.3.3 Diagrams Accompanying Solutions of Problems

The first chapter of the treatise contains a series of problem; the first of them reads as follows:

> 如有田九百一十四畝八分九厘。每畝收粮二升九合。問該粮若干。荅曰。二十六石五斗三升一合八勺一秒。(Xu [1573b] 1993, p. 1145).
>
> If there is a field [of the area] 914 *mu* 8 *fen* 9 *li*, [and it is known that] from 1 *mu* one harvested 2 *sheng* 9 *he* of grain. Question: what is the amount [of the harvested grain]? Answer: 26 *dan* 5 *dou* 3 *sheng* 1 *he* 8 *shao* 1 *miao*.

This text is followed by a series of computations and a picture of an abacus on which the combination of digits 2653181 (apparently representing the answer) is set. Above the picture of the abacus the area of the field (914 *mu* 8 *fen* 9 *li*) is written. Apparently, the solution consists of the multiplication of the area of the field by the volume of grain collected from one unit of area, that is, in modern terms, the multiplication of 914.89 by 2.9 resulting in 2653.181. It is interesting that the only picture provided for this problem features the configuration of the beads representing the result of the multiplication, and there are no pictures showing the intermediate stages of computations. It turns out that the other diagrams in both chapters of this treatise follow the same principle: they show only the final

[22]In Tables 1, 2, 3 "(10)" means that one upper bead and five lower beads are "active"; "(5)" means that the five is represented with five active beads in the lower section, while "5" means that five is represented as one active bead in the upper section.

[23]The number 1111111101 is not explicitly mentioned in the treatise; it can be restored as the sum of the amount set on the abacus (9876(5)4312) plus 123456789. See also Chen (2013, p. 25).

Table 1 The diagrams for the "rules for moving up"

Diagram no.	Page no.	Title	Represented number	Comments
1	1b	*Di yi shang fa* 第一上法 (rule for moving up corresponding to one)	123456789	
3	2b	*Di er shang fa* 第二上法 (rule for moving up corresponding to two)	246913(5)78 [=123456789*2]	
5	3b	*Di san shang fa* 第三上法 (rule for moving up corresponding to three)	36(10)36(10)367 [that is, 370370367 = 123456789 * 3]	The number is misprinted as 36(10)36(10)***2***67
7	4b	*Di si shang fa* 第四上法 (rule for moving up corresponding to four)	4938271(5)6 [=123456789 * 4]	
9	5b	*Di wu shang fa*第五上法 (rule for moving up corresponding to five)	617283945 [=123456789 * 5]	The number is misprinted as 6***0***7***1***83945
11	6b	*Di liu shang fa* 第六上法 (rule for moving up corresponding to six)	74073(10)734 [that is, 740740734 = 123456789*6]	
13	7b	*Di qi shang fa* 第七上法 (rule for moving up corresponding to seven)	864197(5)23 [=123456789 * 7]	The number is misprinted as 86419***8***(5)23
15	8b	*Di ba shang fa* 第八上法 (rule for moving up corresponding to eight)	9876(5)4312 [=123456789 * 8]	
17	9b	*Di jiu shang fa* 第九上法 (rule for moving up corresponding to nine)	1111111101 [=123456789 * 9]	The number set on the depicted abacus is ***(10)(10)(10)(10)(10) (10)(10)(10)(10)***, that is,11111111***10*** and not 11111111***01***

Table 2 The diagrams for the "rules for retreating"

Diagram no.	Page no.	Title	Represented number	Comments
2	2a	*Di yi tui fa* 第一退法 (rule for retreating corresponding to one)	9876(5)4312 [=1111111101 − 123456789 = (9 − 1) * 123456789]	Same number as in diagram 15; yet the 8th rod contains only 1 + 4 beads
4	3a	*Di er tui fa* 第二退法 (rule for retreating corresponding to two)	864197(5)23 [=987654312 − 123456789 = (9 − 2) * 123456789]	Same number as in diagram 13, misprinted as 864197(5)2***4***
6	4a	*Di san tui fa* 第三退法 (rule for retreating corresponding to three)	740740734 [=864197523 − 123456789 = (9 − 3) * 123456789]	Same number as in diagram 11, but represented *differently*
8	5a	*Di si tui fa* 第四退法 (rule for retreating corresponding to four)	61728394(5) [=740740734 − 123456789 = (9 − 4) * 123456789]	The number is misprinted as 617283*8*4(5); five is represented as five beads below
10	6a	*Di wu tui fa* 第五退法 (rule for retreating corresponding to five)	493827156 [=617283945 − 123456789 = (9 − 5) * 123456789]	Five is represented differently, as one bead above
12	7a	*Di liu tui fa* 第六退法 (rule for retreating corresponding to six)	36(10)370367 [=493827156 − 123456789 = (9 − 6) * 123456789]	Same number as in diagram 5, but represented *differently and without misprint*
14	8a	*Di qi tui fa* 第七退法 (rule for retreating corresponding to seven)	246913578 [=370370367 − 123456789 = (9 − 7) * 123456789]	The number is misprinted as 246*8*13578; five is represented as one bead above
16	9a	*Di ba tui fa* 第八退法 (rule for retreating corresponding to eight)	123456789 [=246913578 − 123456789 = (9 − 8) * 123456789]	
18	10a	*Di jiu tui fa* 第九退法 (rule for retreating corresponding to nine)	0 [=123456789 − 123456789 = (9 − 9) * 123456789]	

Table 3 The problems from the first chapter of the treatise accompanied by diagrams showing configurations of beads

Problem no.	Contents of the problem	A diagram represents the result?	Comments
1	Multiplication of 914.89 by 2.9	Yes	The answer is 2653.181 (correct)
2	Multiplication of 914.89 by 290	No	The answer (26531) is the integer part of the product 26531.81 (it is possible that the text contains a misprint and the answer was supposed to be 26532, that is, the rounded value 26531.81)
3	Division of 26532 by 29	Yes	The stated answer is 914.8969 (the correct answer should be 914.89655…); the picture shows an abacus with the configuration 9148914; apparently this 14 misrepresents 69 (the beads in upper section of these two positions are mistakenly pushed away from the bar)
4	Division of 26532 by 29	No	The stated answer is 914.8919 (this number may have resulted from the partly adjusted diagram for the previous problem, if the digit 6 of 914.8969 was misread as 1)
5	Division of 2653.2 by 515	Yes	The stated answer is 5.151823, while the number set on the abacus is 5.106823; the transformation of 5 into 0 and of 1 into 6 can be easily explained. The correct answer is 5.15184466…[a]
6	Multiplication of 5.151823 (from the previous problem) by 515	Yes	The stated answer is 2653.177, this number is shown correctly on the accompanying picture of abacus. The correct answer, however, should have been 2653.188845
7	Multiplication of 515.1823 by 0.515[b]	No	The stated (and again wrong) answer is 265.3177, that may be represented with the same configuration of beads, which explains why this problem does not have a special diagram

[a]This problem may have some connections with the extraction of square root, since the quotient (5.15…) is approximately equal to the divisor (515) divided by 100

[b]The value 515 is misprinted as 525

configurations of the beads; the second chapter also contains the configurations resulting from performing check-up calculations.

Other problems in the first chapter involve different matters (for example, computation of land tax, purchase of rice, etc.), but the numerical data they involve do not change much, as Table 3.

5 Conclusions

The goal of the present chapter was to discuss two categories of diagrams found in Chinese mathematical treatises: (1) those representing configurations of counting rods in the course of computations, and (2) those featuring positions of beads on an abacus. Diagrams of both types arguably were designed for educational purposes, and, most probably, were supposed to be used by the learners when studying the texts in which they were incorporated. It remains unknown whether the learners had to reenact the operations with the counting instruments described in the texts exactly as they were shown in the diagrams, or simple inspection of the provided diagrams would have sufficed; both options appear possible. However, it can be argued that at least *some* operations with the counting rods described in certain treatises of the second millennium most likely were assumed to be performed only in *written* form, without manipulations with actual counting instrument. If this conjecture is valid, the visual representations of this type should be given a special attention, because in this case we are dealing with a particular category of mathematical notation.

It can be also argued that, despite the conventional view, diagrams showing counting rods can be found in mathematical texts designed as late as the mid-16th century, that is, at the time when, according to a number of authors, this instrument was generally replaced by the *suanpan*. As for the diagrams representing configurations of beads on abacus, it should be noted that they became especially widely used only relatively late, in the second half of the 16th century, while early textbooks produced in the first half of the second millennium AD and featuring abacus computation contained only descriptions of operations without depictions of the configurations of beads on the abacus.[24] This phenomenon, to my knowledge, has not been properly discussed by historians of mathematics and mathematical didactics; it may have reflected a considerable change in the didactical practices related to teaching arithmetic, as well as a change of the audience targeted by the authors of these arithmetical manuals.

[24]See, for example, the opening part of the *Jiuzhang suanfa bilei daquan* 九章算法比類大全 by Wu Jing 吳敬 (fl. 1450) (Wu [1450] 1993, p. 16); for a translation and discussion see Chen (2013, pp. 54–58).

References

A. Primary Sources

Guo Shuchun 郭書春 (Ed.) (1993). *Shuxue juan* 數學卷 (Mathematical section) of Ren Jiyu 任繼愈 (gen. ed.), *Zhongguo kexue zhishu dianji tonghui*中國科學技術典籍通彙 (Comprehensive compendium of classical texts related to science and technology in China). Zhengzhou: Henan Jiaoyu Publishers.

Guo Shuchun 郭書春 and Liu Dun 劉鈍 (Eds.) (2001). *Suanjing shi shu* 算經十書 (Ten mathematical treatises). Taibei: Jiuzhang Publishers.

Jing Yushu 靖玉樹 (Ed.) (1994a). *Zhongguo lidai suanxue jicheng* 中國歷代算學集成 (Collected materials on traditional Chinese mathematics). Jinan: Shandong Renmin Publishers.

Kodama Akihito 兒玉明人 (1970). *Jūroku seiki matsu Min kan no shuzan sho*十六世紀末明刊の珠算書 (The abacus books blockprinted [in China] during the late Ming [dynasty] in the 16th century). Tokyo: Fuji Tanki Daigaku shuppanbu.

Qin Jiushao 秦九韶 ([1247] 1993). *Shu shu jiu zhang* 數書九章 (Nine chapters of writings on computations). In Guo 1993, Vol. 1, pp. 439–724.

Wu Jing 吳敬([1450] 1993). *Jiuzhang suanfa bilei daquan* 九章算法比類大全 (The Great Compendium of the Generic [Methods Related to] the *Computational Methods of Nine Categories*). In Guo 1993, Vol. 2, pp. 5–333.

Xu Xinlu 徐心魯 ([1573a] 1970/1994). *Panzhu suanfa* 盤珠算法 (Computational methods for pearls in a tray). Reproduced in Kodama 1970, pp. 1–23 and in Jing 1994a, Vol. 2, pp. 2039–2060.

Xu Xinlu 徐心魯 ([1573b] 1993). *Panzhu suanfa* 盤珠算法 (Computational methods for pearls in a tray). Reproduced in Guo 1993, Vol. 2, pp. 1143-1164.

Zhou Shuxue周述學 ([1558] 1995). *Shendao dabian lizong suanhui*神道大編曆宗算會(Grand Compendium of Divine Dao: [Chapters presenting methods] of lineages [of experts] in calendrical computations and of assembly of mathematicians [of the past]). *Xuxiu Si ku quan shu* 續修四庫全書, Vol. 1043, pp. 555-832. Shanghai: Shanghai guji chubanse.

B. Translations and Secondary Works

Berezkina, Él'vira I. (1957). 九章算術 Математика в девяти книгах (Matematika v devyati knigakh [Mathematics in nine books]), *Историко-математические исследования* (Istoriko-matematicheskie issledovaniya [Studies in the history of mathematics]), Vol. 10, pp. 427–584.

Berezkina, Él'vira I. (1963). О математическом труде Сунь-цзы. Сунь-цзы: Математический трактат. Примечания к трактату Сунь-цзы (O matematicheskom trude Sun'-tszy. Sun'-tszy: matematicheskii traktat. Primechaniya k traktatu Sun'-tszy [On the mathematical work of Sun zi. Sun zi: Mathematical treatise. Annotations for Sun zi's treatise]). *Из истории науки и техники в странах Востока* (Iz istorii nauki i tekhniki v stranakh Vostoka [Studies and materials from the history of science and technology in the countries of the East]), Vol. 3, pp. 5–70.

Chemla, Karine, & Guo Shuchun (2004). 九章算術。*Les Neuf Chapitres: le classique mathématique de la Chine ancienne et ses commentaires.* Paris: Dunod.

Chen Yifu (2013). L'étude des différents modes de déplacement de boules du boulier et de l'invention de la méthode de multiplication *Kongpan Qianchengfa* et son lien avec le calcul mental. Thèse de doctorat. Paris: Université Diderot (Paris-7).

Dauben, Joseph W. 道本周, Guo Shuchun 郭書春, & Xu Yibao 徐義保 (2013). 九章筭术 *Nine Chapters on the Art of Mathematics*. Shenyang: Liaoning jiaoyu chubanshe.

Han Qi 韓琦 (1993) *Panzhu suanfa* tiyao盤珠算法提要 (Presentation of the *Panzhu suanfa*). In Guo (1993), vol. 2, p. 1141.

Hua Yinchun 華印椿 (1987). *Zhongguo zhusuan shigao* 中國珠算史稿 (Outline of the history of Chinese abacus arithmetic). Beijing: Zhongguo caizheng jingji chubanshe.

Jami, Catherine (1998). Abacus (Eastern). In R. Bud & D. J. Warner (Eds.), *Instruments of Science: an Historical Encyclopedia* (pp. 3–5). London & New York: Garland.

Jing Yushu 靖玉樹 (1994b). *Panzhu suanfa* jianshu (Short description of the *Panzhu suanfa*). In Jing 1994a, Vol. 2, pp. 2037–2038.

Kawahara Hideki 川原秀城 (1980). *Ryū Ki chū Kyū Shō San Jutsu* 劉徽註九章算術 (The *Jiu zhang suan shu* commented by Liu Hui), in: Yabuuti Kiyosi 藪內清 (Ed.) *Kagaku no meicho* 科學の名著 (Classics of Science) (vol. 2, pp. 75–264). Tokyo: Asahi, 1980.

Lam Lay Yong & Ang Tian Se (2004). *Fleeting footsteps. Tracing the conception of arithmetic and algebra in ancient China* (Revised Edition). Singapore: World Scientific.

Li Di 李迪 (volume editor) (1999). *Xi Xia, Jin, Yuan, Ming* 西夏金圓明 ([Mathematics of the state of] Xi Xia and of the Jin, Yuan and Ming [dynasties]). Volume 6 of Wu Wenjun 吳文俊 (general editor). *Zhongguo shuxue daxi* 中國數學史大系 (Encyclopedia of the history of Chinese mathematics). Beijing: Beijing Shifan daxue chubanshe.

Li Yan 李儼 (1929). Chousuan zhidu kao 籌算制度考 (A study of the system of counting rods). *Yanjing xuebao* 燕京學報, no. 6 (December), 1129–1134.

Li Yan 李儼 (1931). Zhusuan zhidu kao 珠算制度考 (A study of the system of [Chinese] abacus). *Yanjing xuebao* 燕京學報, no. 10 (December), 2123–2138.

Libbrecht, Ulrich (1973). *Chinese Mathematics in the Thirteenth Century. The Shu-shu Chiu-chang of Ch'in Chiu-shao*. Cambridge, Mass.: The MIT Press.

Martzloff, Jean-Claude (1997). *A history of Chinese mathematics*. Berlin etc.: Springer.

Volkov, Alexei (1998). Counting rods. In R. Bud & D. J. Warner (Eds.), *Instruments of Science: an Historical Encyclopedia* (pp. 155–156). London & New York: Garland.

Volkov, Alexei (2012). Argumentation for state examinations: demonstration in traditional Chinese and Vietnamese mathematics. In Karine Chemla (Ed.), *The history of mathematical proof in ancient traditions* (pp. 509–551). Cambridge: Cambridge University Press.

Volkov, Alexei (2014). History of mathematics education in Oriental Antiquity and Middle Ages. In A. Karp and G. Schubring (Eds.), *Handbook on the history of mathematics education* (pp. 55–70, 79–82). New York etc: Springer.

Volkov, Alexei (2017). Chinese counting rods: Their history, arithmetic operations, and didactic repercussions. In Volkov and Freiman (2017) (forthcoming).

Volkov, Alexei & Freiman, Viktor (Eds.) (2017). *Computations and Computing Devices in Mathematics Education before the Advent of Electronic Calculators*. New York etc.: Springer (forthcoming).

Yang Qiongru 楊瓊茹 (2003). Ming dai lisuanxuejia Zhou Shuxue ji qi suanxue yanjiu 明代曆算學家周述學及其算學研究 (A study of Zhou Shuxue, an expert in calendrical science of the Ming dynasty, and of his mathematical methods). An unpublished Master thesis. Taipei: National Taiwan Normal University.

Author Biography

Alexei Volkov graduated from the Department of Theoretical Mechanics and Mathematics of Moscow State University (1980) and received his Ph.D. degree in History of Mathematics from the Institute for History of Science and Technology of Academy of Sciences (Moscow 1989). Currently he is a Professor of the National Tsing-Hua University (Taiwan) where he teaches history of science at undergraduate and graduate levels. His publications include journal articles and book chapters focused on the history of mathematics and history of mathematics education in pre-modern China and Vietnam.

Author Index

F. Furinghetti and A. Karp (eds.), *Researching the History of Mathematics Education*, ICME-13 Monographs,
https://doi.org/10.1007/978-3-319-68294-5

Subject Index

F. Furinghetti and A. Karp (eds.), *Researching the History of Mathematics Education*, ICME-13 Monographs,
https://doi.org/10.1007/978-3-319-68294-5